Personalized Nutrition as Medical Therapy for High-Risk Diseases

T0141210

Personalized Nutrition as Medical Therapy for High-Risk Diseases

Edited by
Nilanjana Maulik

CRC Press
Taylor & Francis Group
Boca Raton London New York

CRC Press is an imprint of the
Taylor & Francis Group, an **informa** business

CRC Press
Taylor & Francis Group
6000 Broken Sound Parkway NW, Suite 300
Boca Raton, FL 33487-2742

First issued in paperback 2022

ISBN-13: 978-1-138-08268-7 (hbk)
ISBN-13: 978-1-03-233666-4 (pbk)
DOI: 10.1201/9781315112374

Publisher's Note

The publisher has gone to great lengths to ensure the quality of this reprint but points out that some
imperfections in the original copies may be apparent.

**Visit the Taylor & Francis Web site at
http://www.taylorandfrancis.com**

**and the CRC Press Web site at
http://www.crcpress.com**

This book is dedicated to my late father for his continuous support, love and trust. He taught me to be an independent and determined person. Without his encouragement and dedication, I would never have been able to achieve my objectives and succeed in life.

Contents

Preface

Nutrition has a significant impact on human health. For many years, it has been known that a variety of plants used in traditional medicine and their products have properties that allow them to act as genomic modulators, such as the natural phytochemicals used to reduce toxicity due to radiation or chemotherapy. The Mediterranean Diet has similarly received substantial scientific attention due to its anti-inflammatory effects. Diet influences the expression of the genome, and experts now suggest that personalizing an individual's diet based on his or her biological profile (called personalized nutrition) could improve health and lifestyle.

Evidence suggests that people metabolize nutrients differently based on their genetic, epigenetic and metabolic profiles. For example, nutrition researchers have been long puzzled over the fact that people living in the Arctic, who consume a diet consisting of fatty fish and meat, have a lower incidence of heart attacks. After conducting substantial research, scientists have identified several gene variants in the Inuit population that assist in breaking down fatty acids more easily. Caffeine is another example of an ingredient that frequently makes headlines with news media, either celebrating its health benefits for some or highlighting adverse effects related to heart risk in others.

Since the advent of the human genome project, nutrient-gene connections have become a hot topic of research interest, especially in medical science, and several studies have explored the importance of interactions between genes, diet and lifestyle. Similar interactions have been observed between food and the metabolome, epigenome, exposome (external influences such as the environment and behavior) and even the gut microbiome. The picture that has emerged from all of these studies indicates that not only is it important to eat 'good' food, but it is also essential to eat foods that are 'right' for you. Therefore, medical therapists have started to consider personalized nutrition as a possible therapeutic modality with several preclinical studies demonstrating the potential of this novel approach. However, although interest in the field is growing, it is in its infancy, and more research is necessary to further understand the complex relationships between nutrition, environment, lifestyle and the genome.

Other than those afflicted with certain medical conditions such as gluten intolerance and diabetes, food products currently available in the market are generally based on the specific preferences of consumers rather than genetic and biologic factors. However, as consumers become more aware of the inherent susceptibilities carried in their genes, they have begun to demand products more suitable to their specific circumstances. Nutritional advice by clinicians is similarly generic and guided by a handful of parameters such as lipid profiles, blood glucose levels, blood pressure and BMI. Recent studies focused on understanding personalized dietary advice have found individual-level advice to be more fruitful than population-based nutritional advice, however. Therefore, we believe that a better understanding of the role of personalized nutrition in healthy mindbody and lifestyle modifications is an important challenge because, in the near future, personalized nutrition is likely to change how we determine food choices.

This book addresses the role and importance of personalized nutrition in daily life within 13 relevant and focused contributions from world-renowned scientists in the field. This reference book is likely to be of interest to all healthcare-related professionals, nutritionists, biochemists, biologists, primary care-related physicians and dieticians to heighten their awareness about the importance of this field of study, since personalized nutrition, while not yet ready for practice, may soon become a valuable therapeutic and/or preventative option for individuals at high risk.

Nilanjana Maulik, PhD, FAHA, FACN
Department of Surgery
University of Connecticut School of Medicine
Farmington, CT-06001, USA

Acknowledgments

Developing and editing a book is both harder and more rewarding than I ever could've imagined. This book would not have been possible without the hard work and dedication of my colleague, Mahesh Thirunavukkarasu, PhD, Assistant Professor, Department of Surgery, UConn Health, Farmington, Connecticut. He communicated with the authors on a daily basis and assisted me in editing the volume. His efforts were instrumental in completing the book on time. A very special thanks to Diego Accorsi, MD, the greatest cover page designer I could ever imagine.

I am also eternally grateful to the fellows, residents and colleagues for their help and support. I would also like to extend my heartfelt appreciation to my mother, brothers and my husband who encouraged and helped me to complete the book. Lastly, I would like to express my deepest gratitude to all of the authors for their remarkable contributions. I thank you all.

Editor

Nilanjana Maulik is a well-established and highly reputed cardiovascular scientist. She is an expert in the field of vascularization and cardiac regeneration and develops therapeutic strategies for ischemic heart disease. Her laboratory has identified important pro-angiogenic molecules that play an important role in therapeutic angiogenesis using various preclinical models and disease conditions. Her research has advanced knowledge in the areas of angiogenesis and revascularization of the ischemic myocardium. Dr Maulik received her PhD in Biochemistry from Calcutta University, India. After completion of her PhD, Dr Maulik joined the Department of Surgery at University of Connecticut Health as a research fellow. She has continued her service there as a faculty member, and was promoted to tenured professor. Dr Maulik also serves as a faculty member in the Cell Biology graduate program at the University of Connecticut Health. She has been heavily involved in NIH-funded research all her career. She also serves as an expert (cardiovascular) in the NIH study sections regularly; she frequently gives invited lectures at national and international scientific conferences. She has trained more than 150 scientists/fellows/residents, most of whom are actively engaged in professional careers all over the world. Dr Maulik is a member of several prestigious societies including the Federation of American Societies for Experimental Biology (FASEB), the American Heart Association (AHA), the International Society of Heart Research (ISHR), the American College of Nutrition (ACN) and the International College of Angiology (ICA). Presently, she is on the editorial boards of several major cardiovascular journals and served as Editor-in-Chief of the prestigious journal *Molecular Biology Reports* (Springer Press). She is a Fellow of the International Academy of Cardiovascular Sciences (IACS, Canada), ACN and AHA. She has published more than 210 original peer-reviewed articles and almost 36 book chapters. She has also edited four books on epigenetics, nutrition and cardiovascular diseases for CRC/Springer Press. Lastly, Dr Maulik has organized several international conferences, symposia and delivered more than 125 lectures all around the world.

Nilanjana Maulik, PhD, FAHA, FACN
Professor, Department of Surgery, University of Connecticut Health, Farmington Avenue, Farmington, Connecticut

Contributors

Diego Accorsi
School of Medicine
University of Connecticut
Farmington, Connecticut

Olatunji Anthony Akerele
Department of Biochemistry
Memorial University
St. John's, Newfoundland and Labrador,
    Canada

Michael Arad
Sheba Medical Center
Sackler School of Medicine
Tel Aviv University, Israel

Ebru Arioglu Inan
Ankara University
Ankara, Turkey

Mariann Bombicz
University of Debrecen
Hungary

Sukhinder Kaur Cheema
Department of Biochemistry
Memorial University
St John's, Newfoundland and Labrador,
    Canada

Sean Curley
UCD School of Medicine
University College Dublin
Dublin, Ireland

Andrew S. Day
University of Otago, Christchurch
Christchurch, New Zealand

Dilip Ghosh
Nutriconnect
Western Sydney University,
    Australia

Edith Hochhauser
Felsenstein Medical Research Institute
Sackler School of Medicine
Tel Aviv University, Israel

Jennifer Hubbard
School of Medicine
University of Connecticut
Farmington, Connecticut

Béla Juhász
University of Debrecen
Hungary

Sarina Kajani
UCD School of Medicine
University College Dublin
Dublin, Ireland

Gulsah Kaner Tohtak
İzmir Katip Çelebi University
İzmir, Turkey

Andrea Kurucz
University of Debrecen
Hungary

Rajesh Lakshmanan
School of Medicine
University of Connecticut
Farmington, Connecticut

Sarah Mahayni
UCD School of Public Health
University College Dublin
Dublin, Ireland

Nilanjana Maulik
School of Medicine
University of Connecticut
Farmington, Connecticut

Fiona C. McGillicuddy
UCD School of Medicine
University College Dublin
Dublin, Ireland

Nalini Namasivayam
Annamalai University
Annamalainagar
Tamil Nadu, India

Marcella O'Reilly
UCD School of Public Health
University College Dublin
Dublin, Ireland

Giorgina Barbara Piccoli
Università di Torino
Turin, Italy
Centre Hospitalier Le Mans
Le Mans, France

Helen M. Roche
UCD School of Public Health
University College Dublin
Dublin, Ireland

Seetur R. Pradeep
School of Medicine
University of Connecticut
Farmington, Connecticut

Aktarul Islam Siddique
Annamalai University,
Annamalainagar,
Tamil Nadu, India

Mahesh Thirunavukkarasu
School of Medicine
University of Connecticut
Farmington, Connecticut

Erkan Tuncay
Ankara University
Ankara, Turkey

Belma Turan
Ankara University
Ankara, Turkey

Balazs Varga
University of Debrecen
Hungary

Maayan Waldman
Sheba Medical Center
Sackler School of Medicine
Tel Aviv University, Israel

1 Introduction

Dilip Ghosh

CONTENTS

1.1 INTRODUCTION

Today's nutritional and dietary supplement market is considerably different than it was 10–15 years ago. Consumer demands for healthy foods have been changing considerably every year, particularly in the last decade. More and more, consumers believe that foods with specific functionality contribute directly to their health. Foods today are not only considered as a vehicle to satisfy hunger and to provide necessary nutrients but also to prevent nutrition-related diseases and improve physical and mental well-being. In this scenario, functional food and active ingredients play an outstanding role. From an economic perspective, this increasing demand on such foods/nutrition can be justified by the increasing cost of healthcare, the steady increase in life expectancy, including infants, and the aspiration and desire of older people for improved quality of their later years (Roberfroid 2000).

In the 21st century humankind is facing a global pandemic of diet-related chronic disease and preventable disorders that include cardiovascular disease, obesity and diabetes, cancers, osteoporosis and myriad inflammatory disorders. These are the leading cause of the global healthcare burden. Virtually most of these disorders are diet-related and, not surprisingly, are not responding well to pharmaceutical intervention. The heavily burdened and eroding healthcare system is in need of an etiology-based model that addresses the underlying molecular basis of a patient/consumer's dysfunction and develops therapeutic and preventive strategies that will include the biochemical-molecular individuality of each person. A genetic predisposition model of health and disease is emerging from the Human Genome Project that opens up etiology-based care and will be almost equivalent to the current evidence-based pharmaceutical framework.

The current medical model of genetic determinism is now being challenged by the emerging concept of genetic susceptibility which enables one to change one's health trajectory through the judicious use of diet and lifestyle. In this scenario, innovative, evidence-based food/nutritional and dietary supplements have a significant role in changing our destiny.

1

1.2   WHAT DOES PERSONALIZATION MEAN?

Personalization of nutrition advice is often proposed as one of the most promising approaches. In recent years, most of the health intervention research and methods on the effect of personalization show that advice targeted to an individual's physical parameters, lifestyle and environmental situation is more effective in influencing their health behavior than general information.

It is evident that several dietary components have been recognized to modulate gene and protein expression and thereby metabolic pathways, homeostatic regulation and presumably health and disease. In addition, genes also contribute largely to different responses to diet exposure, including interindividual variations in the occurrence of adverse reactions. Major potential areas where development of personalized foods/nutritionals are realistically possible include type 2 diabetes and obesity, mood foods, Inflammatory Bowel Disease (IBD) and disorders of ageing where diet-gene relationship has been extensively studied.

1.3   WHERE TO GO FROM HERE?

Human desire for individuality is not new. It is embedded in all ancient civilizations and traditional healthcare system such as Traditional Chinese medicine and the Indian Ayurvedic system. All traditional medical systems are descriptive and phenomenological—it typically diagnoses patients using concepts based on the relationship between signs and symptoms. In Western-style modern medicine model the concept of 'one disease—one target—one-size-fits-all' is shifting towards more personalization, including the use of multiple therapeutic agents and the consideration of nutritional, psychological and lifestyle factors when deciding the best course of treatment. This strategic shift in medical practice is being linked with the discipline of systems science and systems biology in the biomedical domain. Systems science aims to understand both the connectivity and interdependency of individual components within a dynamic and non-linear system, as well as the properties that emerge at certain organizational levels. The concepts and practices of systems biology align very closely with those of traditional Asian medicine as well as the very idea of 'health' of the current World Health Organization definition. Now we know individual dietary components can modulate and change gene function. Based on the robust evidence, healthcare professionals are now able to control gene-specific physiological expression with specific dietary intervention. This hypothesis has just become more attainable for more people due to rising prosperity, particularly in emerging markets. Moreover, the tremendous technological advancement reduces the gap between desire and reality by reaching more cost-effective personalized products and services.

1.4   MEDICALIZATION OF FOOD

Since ancient times plants, herbs and other natural products have been used as healing agents. Advances in organic chemistry from the early 19th century have enabled the preparation of numerous synthetic medicines. Yet, the majority of the medicinal substances available today have their origin in natural compounds. The best-known example is aspirin (acetylsalicylic acid), originally derived from the bark of the white willow tree. Other examples include the immunosuppressive cyclosporines, the anthracycline antibiotics and the HMG-CoA reductase inhibitors, commonly known as statins. Traditionally, pharmaceuticals have been used to cure diseases or to alleviate the symptoms of disease. Nutrition, on the other hand, is primarily aimed at preventing diseases by providing the body with the optimal balance of macro- and micronutrients needed for good health.

Due to the emerging knowledge of disease, medicines are now increasingly being used to lower risk factors, and thereby to prevent chronic diseases. Prime examples are blood pressure-lowering and blood lipid-lowering agents which reduce the risk of cardiovascular disease. The appearance of functional foods and dietary supplements on the market has further blurred the distinction between pharma and nutrition. However, it is important to distinguish the target and effective outcome of pharma and nutra intervention.

1.5   THE GENOME-BASED HEALTH CONCEPT

Nutritional genomics is a promising new research and development area and as a young and blue-sky science, it is also associated with intense debate. With high hope to many researcher's nutritional genomics is closely associated with 'personalized nutrition,' in which the diet of an individual is customized based on their own genomic/genetic information, to optimize health and prevent the onset of disease. In this context 'nutritional genomics is largely concerned with elucidating the interactive nature of genomic, dietary and environmental factors and how these interactions impact on health outcomes' (Brown and van der Ouderaa 2007).

Scientists have determined that genetic expression is influenced by 'endogenous and exogenous factors and therefore particularly prone to nutritional imprinting' (Ruemmele and Garnier-Lengline 2012). Moreover, nutrition and genes interact in two different ways. The term 'nutrigenomics,' where the impact of nutritional factors on gene regulation and expression is considered. The other way, 'nutrigenetics,' examines the influence genetic variation has on, or predetermines, nutritional requirements. Both interactions are important considerations for designing a personalized nutrition concept.

Traditionally (and in most of the cases currently), nutrition counseling and recommendations have been offered based on population. An example of such a recommendation is the dietary reference values for calcium: adults between 19 and 60 years of age should consume 1,000 mg per day. This recommendation does not account for individual genetic variations in the ability to absorb and metabolize this mineral. Another example is that of 'dietary interventions as primary prevention to reduce the risk of cardiovascular disease' (de Roos 2013). Without a valid population-based strategy, the population-based intervention does not address the possibility of subgroups' differences in response to the intervention. Personalized nutrition would address such individual variations.

1.6   WAY FORWARD

Currently, personalized medicine and nutrition are not extensively applied on a routine basis at patients' clinics and by their carers: doctors, nurses, dietitians or nutritionists. Recent studies published in last few years not only underline the therapeutic potential of lifestyle interventions but are also generating valuable insights in the complex and dynamic transition from health to disease continuum.

This book discusses the recent developments in the pharmaceutical-nutrition interface and relevant mechanisms, including receptors and other targets. A few clinical practice-based examples are cited in this book.

Several dedicated chapters deal with nutraceutical intervention to manage or treat physiological conditions and diseases such as cardiovascular disease, hypertension, Crohn's disease, chronic kidney disease, hypercholesteremia, maternal and offspring metabolic disorders and psychological disorders. Two more interesting areas covered in this book are the role of caloric restriction in obesity and diabetic heart disease and the effect of high carbohydrate diet-induced metabolic syndrome in the overweight body.

It is evident that pharmaceutical industry will benefit from nutritional genomics knowledge and a physiological approach that puts health above diseases and medical conditions (Ghosh, Skinner et al. 2007). This will help us to get a better understanding about the transition between health, homeostatic resilience and chronic disease, to develop better and more tailored treatment options. The personalized nutrition market is in many ways an unknown space for the 'Big Pharma' players. To overcome this weakness, they are taking on strategic partnerships, collaborations and acquisitions. As a result, we should expect further partnerships between big pharma companies and tech start-ups enabling them to be in this growing trend for personalized products.

Experience shows that commercial providers are keen to proceed to the market with products before the scientific evidence is established. This can only be checked if there are national and international agreed guidelines for using genotype-based advice in personalized nutrition. The Food4Me consortium (Grimaldi, van Ommen et al. 2017) has proposed such guidelines recently, but the research and regulatory communities have to evaluate and agree on the proposed guidelines.

REFERENCES

Brown, L. and F. van der Ouderaa (2007). "Nutritional genomics: food industry applications from farm to fork." *Br J Nutr* 97(6): 1027–1035.
de Roos, B. (2013). "Personalised nutrition: ready for practice?" *Proc Nutr Soc* 72(1): 48–52.
Ghosh, D., M. A. Skinner, et al. (2007). "Pharmacogenomics and nutrigenomics: synergies and differences." *Eur J Clin Nutr* 61(5): 567–574.
Grimaldi, K. A., B. van Ommen, et al. (2017). "Proposed guidelines to evaluate scientific validity and evidence for genotype-based dietary advice." *Genes Nutr* 12: 35.
Roberfroid, M. B. (2000). "Concepts and strategy of functional food science: the European perspective." *Am J Clin Nutr* 71(6 Suppl): 1660S–1664S; discussion 1674S–1665S.
Ruemmele, F. M. and H. Garnier-Lengline (2012). "Why are genetics important for nutrition? Lessons from epigenetic research." *Ann Nutr Metab* 60(3 Suppl): 38–43.

2 Personalized Nutrition
The New Era of Nutrition

*Diego Accorsi, Seetur R. Pradeep, Jennifer
Hubbard, Rajesh Lakshmanan, Nilanjana
Maulik and Mahesh Thirunavukkarasu*

CONTENTS

2.1 INTRODUCTION

Diet and nutrition have historically been regarded as important factors in the promotion and maintenance of health throughout the entire life span and had until very recently played a leading role in the management of disease, according to epidemiological studies (Kussmann and Fay 2008). For example, Hippocrates, heralded as the father of modern medicine, once said: 'Let the food be thy medicine and the medicine be thy food.' Since then our understanding of the specific microbiological mechanisms through which nutrition contributes to overall health and disease continues to grow but, with the advent of pharmacotherapy, nutrition has taken a backseat as a tool to improve health in modern medicine. Nutritional recommendations for the otherwise healthy often follow a generalized, 'one-size-fits-all' approach aimed primarily at weight loss, while formal nutritional guidelines as prescribed by clinicians apply only to subpopulations at risk, such as diabetics, while failing to account for an individual's unique characteristics. Enter the concept of personalized nutrition, which in essence attempts to identify fluctuations from the baseline, and correct these deviations through individualized nutritional recommendations using state-of-the-art technologies to track and process unique and essential biological and environmental variables (Kussmann and Fay 2008). Put more simply; it is not only about eating 'good' food, but also about eating food that is 'right' for you (Rucker 2019). This chapter has two primary objectives: to define personalized nutrition, as well as exploring its relevance and importance in the 21st century by highlighting a few popular topics of discussion.

2.1.1   DEFINING PERSONALIZED NUTRITION

There is no clear-cut definition for personalized nutrition, but in general, it may be thought to include three important tenets. Firstly, it is the provision of nutritional advice adapted to an individual's unique internal and external influences. It is nothing but a specially tailored recommendation of a dietary habit for an individual to maintain good health and lifestyle. Secondly, personalized nutritional interventions can only be effective if they translate into behavioral change towards a healthy lifestyle. And lastly, personalized nutrition should be equally applicable to a healthy individual in order to improve overall public health, as well as patients afflicted with, or susceptible to, specific medical conditions (Ordovas, Ferguson et al. 2018). Because of its origins during the exciting period of genomic sequencing, the term still partially overlaps with some closely related terms such as nutrigenomics, nutrigenetics, gene-food interactions, etc. (Table 2.1), but it is important to note that since then personalized nutrition has become very multifaceted and includes an individual's behavior, dietary habits/cultural influence, food availability, phenotypic and genetic makeup, metabolism and even the microbiome. Some of these terms will be explained in further detail later in this chapter.

2.1.2   ORIGINS OF PERSONALIZED NUTRITION AS A FORMAL SCIENTIFIC IDEOLOGY

American biochemist Roger J. Williams (1893–1988), known for his work on B vitamins, made one of the first mentions of the gene food interaction in an article published in 1950 entitled "Concept of genetotrophic disease." In short, Williams believed that 'a genetotrophic disease is one which occurs if a diet fails to provide sufficient supply of one or more nutrients required at high levels because of the characteristic genetic pattern of the individual concerned' (Kraemer, Cordaro et al. 2016; Williams, 1950). At the time, the influence of genetics was a new medical concept that was thought to be the basis of many conditions, and though this paved the way for the personalization of medical treatment in general, the underlying mechanisms were still unknown, and the tools necessary to put ideas into practice were still undiscovered. Therefore, the concept was to remain in a dormant stage until the advent of human genome sequencing, which heralded a boom in research focusing on gene-phenotype and gene-environment interactions. Soon after, the term 'personalized nutrition' surfaced as an entity unto itself, which extended the use of genetic data from the creation of new therapeutic approaches to the improvement

TABLE 2.1
Descriptors and Definitions

In common with other scientific fields in their early development, multiple concepts and descriptors are
used in personalized nutrition, sometimes without rigorous definition. In addition to the term personalized
nutrition, many other terms are used—for example, precision nutrition, stratified nutrition, tailored
nutrition and individually tailored nutrition. We have attempted to group the descriptors as follows:

- Stratified and tailored nutrition are similar (if not synonymous). These approaches attempt to group
  individuals with shared characteristics and to deliver nutritional intervention/advice that is suited to
  each group.
- Personalized nutrition and individually tailored nutrition mean similar things and go a step further by
  attempting to deliver nutritional intervention/advice suited to each individual.
- Precision nutrition is the most ambitious of the descriptors. It suggests that it is possible to have
  sufficient quantitative understanding about the complex relationships between an individual, his/her
  food consumption and his/her phenotype (including health) to offer nutritional intervention/advice,
  which is known to be individually beneficial. The degree of scientific certainty required for precision
  nutrition is much greater than that required for the other approaches.
- Nutrigenetics is an aspect of personalized nutrition that studies the different phenotypic responses (i.e.
  weight, blood pressure, plasma cholesterol or glucose levels) to a specific diet (i.e. low fat or
  Mediterranean diets), depending on the genotype of the individual.
- Nutrigenomics involves the characterization of all gene products affected by nutrients and their
  metabolic consequences.
- Exposome is the collection of environmental factors, such as stress, physical activity and diet, to
  which an individual is exposed and which may affect health. As one moves from stratified to
  personalized to precision nutrition, it becomes necessary to apply more and more dimensions or
  characteristics to achieve the desired goal. For example, stratification could be undertaken using one,
  or a few, dimensions such as age, gender or health status. In contrast, given the complexity of
  relationships between individual diet and phenotype, deployment of a wide range of dimensions/
  characteristics, perhaps including 'big data' approaches, would be necessary to achieve the goal of
  precision nutrition. An exception to this broad generalization is the management of inborn errors of
  metabolism such as phenylketonuria, where 'precision nutrition' can be achieved using information on
  a single characteristic—that is, genotype.
- Epigenomics is a branch of genomics concerned with the epigenetic changes (methylation, histone
  modification, microRNAs) that modify the expression and function of the genetic material of an
  organism.
- Metabolomics is the scientific study and analysis of the metabolites (usually restricted to small
  molecules, i.e. <900 daltons) produced by a cell, tissue or organism.
- Microbiomics is the study of the microbiome, the totality of microbes in specific environments (i.e.
  the human gut).

Ordovas, Ferguson et al. (2018)

---

of nutritional management. In a report by the Institute for the Future in Palo Alto,
2003, a suggestion was made that within a decade most adults would make at least
part of their nutritional choices based on knowledge of their genetic makeup and
inherent susceptibilities to different foods. While this prediction did not come to
fruition in its entirety by the proposed deadline, that decade did see the beginnings
of formalized, funded personalized nutritional research, including the Food4Me

consortium, which eventually led a five-year project, the largest to date, exploring various elements involved in personalized nutrition using complex multidisciplinary approaches (Kraemer, Cordaro et al. 2016).

2.2 THEORETICAL BASIS FOR PERSONALIZED NUTRITION

In recent years, the emergence of advances in the field of nutrition and health, such as epigenomics, metabolomics, microbiomics, the study of how the environment affects the individual (i.e. the "exposome"), accessible data collection technologies such as mobile apps and fitness trackers, and cutting edge analytical tools, have enabled us to study multiple factors simultaneously to predict an individual's response to dietary interventions (Siroux, Agier et al. 2016). As a result, research has since surfaced to reveal a significant amount of interindividual variability in response to dietary intervention (Mathers 2019). For example, it is widely known that drastic changes in blood glucose concentrations after a meal (glycemia) can be detrimental. Zeevi, Korem et al. (2015) proved in one of their studies that despite the provision of standardized meals, there is significant interindividual variability in blood glucose levels. This is unsurprising as it is well established that glycemic response is determined not only by the type or quantity of food consumed but also by the attributes of the individual consuming the food (Vega-Lopez, Ausman et al. 2007). Selecting glycemia as their parameter/outcome of interest they set out to examine whether or not their predictive tool would be successful at deriving individually tailored nutritional intervention based on individual characteristics such as blood-work, anthropometry and gut microbiome. This predictive tool called the 'Personalized Nutrition Predictor,' was shown to anticipate with fair accuracy the glycemic response of individual subjects to different meals. Very recently, authors Mendes-Soares, Raveh-Sadka et al. (2019) used a similar model in their study which involved 327 individuals. They similarly observed that the inclusion of data gathered from each individual was more likely to accurately predict postprandial blood glucose concentration than using energy and/or carbohydrate content of foods alone (Mendes-Soares, Raveh-Sadka et al. 2019). To date, however, only a handful of high level, human dietary intervention studies have tested the concept of personalized nutrition in a clinically relevant setting.

2.3 CURRENT PERSPECTIVES ON PERSONALIZED NUTRITION

The idea of personalized nutrition is that individual nutritional advice will be more effective at reaching target goals than the more generic approaches which are the basis for most current dietary interventions. This idea may fundamentally change the way consumers make food choices that fit with lifestyle preferences and health goals. Most studies that have contributed to the scientific basis for personalization of nutrition are based on genetics, which, along with the metabolome and microbiome, represent biological or 'intrinsic' factors. This includes an understanding of the biological response to dietary modifications, whether single component (e.g. low fat, low cholesterol, low sodium, etc.) or whole diet interventions, and their effect on indicators of health and disease risk (Biesiekierski, Livingstone et al. 2019). While it is true that in recent years much attention has been paid to deciphering the human genome and its influence on disease risk, genes are only a small part of a very broad picture. There are many other factors, both intrinsic and extrinsic, such as age, stage of life (pregnant, lactating, etc.), sex, race, ethnicity and cultural or religious backgrounds, which can influence the effect of diet on the body (Ordovas, Ferguson et al. 2018). Due to the inherent complexity of analyzing all the factors that make an individual unique, the biggest challenge of personalized nutrition is its study and execution in real-life situations. Therefore, the process will not only require an understanding of an individual's genetic makeup and susceptibility, but also biological substantiation of an individual's responses to food/nutrient consumption manifested as changes in specific measurable parameters (biomarkers, microbiota), analysis of sociobehavioral patterns, food choice and availability, troubleshooting obstacles at all points in the pipeline, and clear demarcation of objectives to inspire and facilitate eating pattern modification (Ordovas, Ferguson et al. 2018).

2.3.1 NUTRIENT-GENE INTERACTIONS

Heralding back to the origins of personalized nutrition, the diet-gene interaction is perhaps the most researched aspect of the field, with several examples of researched gene-food interactions published to date (Table 2.2). Though individuals share the clear majority of their genes in common, the relatively few differences are enough to cause significant alterations in phenotype, including the way individuals respond to their diet. Changes to phenotype may evoke changes in behavior, physiological characteristics or susceptibility to diseases as well. Of these genetic differences held accountable for phenotypic diversity, it is increasingly apparent that single nucleotide polymorphisms (SNPs), the most known and common genetic variations between the human beings, will be a key factor.

The study of diet-gene interactions can be loosely sub-classified into two distinct study areas: nutrigenetics and nutrigenomics. Nutrigenetics deals with how genetic makeup influences the way nutrients are acquired, metabolized and stored, while nutrigenomics is the study of how food components influence gene expression. Nutrigenetics plays more of a preventative role, generally delineating the foods that should be avoided because they could place the individual at risk for disease. On the other hand, nutrigenomics also has the potential to transform diet into a therapeutic tool to treat diseases, similar to pharmaceutical drugs. It is based on identification of genetic risk factors and targeting of key players of gene expression at any given stage, to up- or down-regulate the effects of certain genes.

2.3.1.1 Biomarkers

In order to monitor changes to an individual's physiology and gene expression, it becomes necessary to use biomarkers; these can represent certain physiologic parameters such as blood pressure, components of the metabolome, or may also be one of many molecules involved in the complex cascades responsible for gene expression. The most valuable biomarkers are those that are easy/cheap to obtain and measure. It is also necessary to define a normal range that equates to the "healthy

TABLE 2.2

Summary of Genetic Variants That Modify the Association between Various Dietary Factors and Performance-Related Outcomes

Gene (RS Number)	Function	Dietary Factor	Dietary Sources	Performance-Related Outcome
CYP1A2 (rs762551)	Encodes CYP1A2 liver enzyme: metabolizes caffeine; identifies individuals as fast or slow metabolizers	Caffeine	Coffee, tea, soda, energy drinks, caffeine supplements	Cardiovascular health, endurance (Clenin et al., 2015; Haas & Brownlie, 2001; Palatini et al., 2009; Soares, Schneider, Valle, & Schenkel, 2018)
ADORA2A (rs5751876	Regulates myocardial oxygen demand; increases coronary circulation via vasodilation	Caffeine	Coffee, tea, soda, energy drinks, caffeine supplements	Vigilance when fatigued, sleep quality (Begas, Kouvaras, Tsakalof, Papakosta, & Asprodini, 2007; Ghotbi et al., 2007; Hunter, St Clair Gibson, Collins, Lambert, & Noakes, 2002; Yang, Palmer, & de Wit, 2010)
BCMO1 (rs11645428)	Converts pro-vitamin A carotenoids to Vitamin A	Vitamin A	Bluefin tuna, hard goat cheese, eggs, mackerel, carrots, sweet potato	Visuo-motor skills and immunity (Czarnewski, Das, Parigi, & Villablanca, 2017; Ferrucci et al., 2009; Garvican et al., 2014; Lietz, Lange, & Rimbach, 2010; Lietz, Oxley, Leung, & Hesketh, 2012; Palidis, Wyder-Hodge, Fooken, & Spering, 2017)
MTHFR (rs1801133)	Produces the enzyme methylenetetrahydrofolate reductase, which is involved in the conversion of folic acid and folate into their biologically active form, L-methylfolate	Folate	Edamame, chicken liver, lentils, asparagus, black beans, kale, avocado	Megaloblastic anemia and hyperhomocysteinemia risk (Curro et al., 2016; Dinc, Yucel, Taneli, & Sayin, 2016; Goyette et al., 1994; Guinotte et al., 2003)

(Continued)

TABLE 2.2 (CONTINUED)
Summary of Genetic Variants That Modify the Association between Various Dietary Factors and Performance-Related Outcomes

Gene (RS Number)	Function	Dietary Factor	Dietary Sources	Performance-Related Outcome
HFE (rs1800562 and rs1799945)	Regulates intestinal iron uptake	Iron	Beef, chicken, fish, organ meats (heme iron); almonds, parsley, spinach (non-heme iron)	Hereditary hemochromatosis (Marjot, Collier, & Ryan, 2016; Pantopoulos, Porwal, Tartakoff, & Devireddy, 2012; Recalcati, Minotti, & Cairo, 2010)
TMPRSS6 (rs4820268), TFR2 (rs7385804), TF (rs3811647)	Regulate the peptide hormone, hepcidin, which controls iron absorption	Iron	Beef, chicken, fish, organ meats (heme iron); almonds, parsley, spinach (non-heme iron)	Iron-deficiency anemia risk (Allen et al., 2008; Benyamin et al., 2009; Cannell, Hollis, Sorenson, Taft, & Anderson, 2009; Garcia et al., 2014; Soranzo et al., 2009; Vidoni, Pettee Gabriel, Luo, Simonsick, & Day, 2018)
FUT2 (rs602662)	Involved in vitamin B12 cell transport and absorption	Vitamin B12	Clams, oysters, herring, nutritional yeast, beef, salmon	Megaloblastic anemia and hyperhomocysteinemia (Reardon & Allen, 2009)
GSTT1 (Ins/Del)	Plays a role in vitamin C utilization via glutathione S-transferase enzymes	Vitamin C	Red peppers, strawberries, pineapple, oranges, broccoli	Circulating ascorbic acid levels mitigate exercise-induced ROS production (Braakhuis, 2012; Shaw, Lee-Barthel, Ross, Wang, & Baar, 2017)
GC (rs2282679) and CYP2R1 (rs10741657)	GC encodes vitamin D-binding protein, involved in binding and transporting vitamin D to tissues; CYP2R1 encodes the enzyme vitamin D 25-hydroxylase involved in vitamin D activation	Vitamin D	Salmon, white fish, rainbow trout, halibut, milk	Circulating 25(OH)D levels impacting immunity, bone health, inflammation, strength training and recovery (Barker, Schneider, Dixon, Henriksen, & Weaver, 2013; Larson-Meyer & Willis, 2010; Thomas, Erdman, & Burke, 2016; Wang et al., 2010; Yucha C, 2003)

(Continued)

TABLE 2.2 (CONTINUED)

Summary of Genetic Variants That Modify the Association between Various Dietary Factors and Performance-Related Outcomes

Gene (RS Number)	Function	Dietary Factor	Dietary Sources	Performance-Related Outcome
GC (rs7041 and rs4588)	GC encodes vitamin D-binding protein, involved in binding and transporting vitamin D to tissues; Vitamin D is required for calcium absorption	Calcium	Yogurt, milk, cheese, firm tofu, canned salmon (with bones), edamame	Bone/stress fracture risk Muscle contraction, nerve conduction, blood clotting (Barker et al., 2013; Larson-Meyer & Willis, 2010; Thomas et al., 2016; Wang et al., 2010; Yucha C, 2003)
PEMT (rs12325817)	Involved in endogenous choline synthesis via the hepatic phosphati-dylethanolamine N-methyl-transferase pathway	Choline	Eggs, beef, poultry, fish, shrimp, broccoli, salmon	Muscle or liver damage, reduced neurotransmitters (Elsawy, Abdelrahman, & Hamza, 2014; Kohlmeier, da Costa, Fischer, & Zeisel, 2005; Zeisel, 2006; Zeisel et al., 1991)
MTHFD1 (rs2236225)	Encodes protein involved in trifunctional enzyme activities related to metabolic handling of choline and folate	Folate/ Choline	Folate: Edamame, chicken liver, lentils, asparagus, black beans, kale, avocado Choline: Eggs, beef, poultry, fish, shrimp, broccoli, salmon	Muscle or liver damage, reduced neurotransmitters (Elsawy et al., 2014; Kohlmeier et al., 2005)
FTO (rs1558902/ rs9939609)	Precise function undetermined; plays a role in metabolism and has been consistently linked to weight, BMI and body composition	Protein/ SFA:PUFA	Protein: chicken, beef, tofu, salmon, cottage cheese, lentils, milk, Greek yogurt SFA: cheese, butter, red meat, baked goods PUFA: flaxseed oil, grape seed oil, sunflower oil	Optimizing body composition (Knapik, 2015; Krieger, Sitren, Daniels, & Langkamp-Henken, 2006)

(Continued)

TABLE 2.2 (CONTINUED)

Summary of Genetic Variants That Modify the Association between Various Dietary Factors and Performance-Related Outcomes

Gene (RS Number)	Function	Dietary Factor	Dietary Sources	Performance-Related Outcome
TCF7L2 (rs7903146)	Involved in expression of body fat	Fat	Nuts/seeds, butter, oils, cheese, red meat, high-fat dairy	Optimizing body composition (Goss et al., 2013; Nascimento et al., 2016)
PPARg2 (rs1801282)	Regulates adipocyte differentiation	MUFA	Macadamia nuts, almond butter, peanut butter, olive oil, canola oil, sesame oil	Optimizing body composition (Merritt, Jamnik, & El-Sohemy, 2018)

Guest, Horne et al. (2019)

state" so that interventions can be geared towards normalization of these values. As we continue to unearth complex food-genome/metabolome/microbiome interactions, it is predicted that the number of biomarkers at our disposal, as well as our understanding of specific mechanisms leading to their change, will have to increase. It will become necessary to monitor changes in multiple biomarkers simultaneously as well, as very few processes in the body are a result of simple one-to-one correlations. For example, in a study by Price and colleagues, 107 adults were followed over a period of nine months with laboratory examinations to isolate specific genetic variants, proteins and metabolites, and microbiota readings in hopes that these markers could help guide health recommendations. As more and more biomarker candidates surfaced, dense "data clouds" formed that could then be tested with perturbation experiments (exposing candidates to certain stimuli). These clouds were, therefore, further sub-divided into risk 'communities' or biological networks based on such responses. Using this information and with the help of complex analytical tools, the researchers recognized 'actionable possibilities,' meaning they could identify specific therapeutic actions that had the potential to affect an entire marker community towards normality (Price, Magis et al. 2017).

2.3.2 Altering Human Behavior

As mentioned previously, for any nutritional recommendation to be effective, it must motivate a change in human behavior. The Food4Me study is one of the largest to date to explore whether personalizing nutritional feedback can have a positive effect on human behavior. It is a four-arm, web-based randomized controlled trial conducted across seven European countries and represents one of the most encouraging analyses in favor of the personalization of nutritional advice leading to a change in dietary habit and clinically improved health outcomes, using a self-sustainable and realistic internet-based data-gathering and dissemination system. A total of 1,607 participants were recruited and randomized into a control group that was provided generalized nutritional feedback based on current European guidelines (Level 0), or to one of three personalized nutritional intervention levels (Levels 1–3), which differed in the type of patient data used to generate the dietary plan. The primary endpoint was to determine whether personalized dietary recommendations based on individuals' characteristics were more effective at promoting healthy eating patterns than generic nutritional advice, while secondary endpoints included whether higher-frequency feedback improved adherence to nutritional plans (Celis-Morales, Livingstone et al. 2014). The primary conclusion was that, after six months of dietary interventions, participants randomized to the personalized nutrition groups (Levels 1–3) had improved healthy eating behavior as measured by using the Healthy Eating Index (Guenther, Casavale et al. 2013) when compared to those randomized to the control (Level 0).

Though there is plentiful evidence that genes affect diet and vice versa, whether the inclusion of genomic data in a dietary interventional plan will necessarily influence human behavior is much less understood because very few studies have explored food-gene interactions in realistic clinical settings, and instead tend focus on simple-to-test correlations. Would knowing about genetic susceptibility make an individual more likely to change his or her dietary habits? The Food4Me trial once again exemplifies this. One of the objectives of the Food4Me trial was to investigate which information/data, when divulged to the subject, is more likely to produce a positive behavioral change towards a desired goal. To do this, subjects receiving personalized nutritional therapy were divided into three groups each receiving different types of feedback: feedback based upon diet (Level 1); 2) advice based on diet plus phenotype characteristics (Level 2); or 3) advice based on current diet, phenotype and genotype data (Level 3). Phenotypic feedback was based on anthropometric data (e.g. BMI, waist circumference) as well as blood biomarkers such as glucose and lipid levels. Genotypic data involved feedback based on five major genetic variants selected specifically for the study, all of which are known to be linked with increased susceptibilities to conditions such as obesity and diabetes (MTHFR, FTO, TCF7L2, APOEε4 and FADS1 genes).

The description of the advice and feedback given to participants randomized to all levels of intervention, including level 0, a control, is summarized in Table 2.3. Results showed that, while providing individualized reports to each subject was successful at changing behavioral patterns towards a healthier lifestyle over a period of six months, there was no additive advantage in giving out more information beyond diet-based feedback (Celis-Morales, Livingstone et al. 2016). This is important because the acquisition of phenotypic and especially genotypic data is still relatively expensive. Therefore, some would query the usefulness of including this information in the development of a nutritional plan if its inclusion will not necessarily help surpass arguably the largest hurdle in its implementation, namely

TABLE 2.3
Description of the Advice and Feedback Given to Participants Randomized to levels 0, 1, 2 and 3 of the Intervention

	Level 0	Level 1	Level 2	Level 3
Participant feedback and advice	Non-personalized dietary advice (energy intake, fruits and vegetables, wholegrain products, fish, dairy products, meat, type of fat and salt)	Feedback on how food groups intakes compare with guidelines (to optimize the consumption of fruits and vegetables, whole-grain products, fish, dairy products and meat)	Feedback on how food groups intakes compare with guidelines (to optimize the consumption of fruits and vegetables, whole-grain products, fish, dairy products and meat)	Feedback on how food groups intakes compare with guidelines (to optimize the consumption of fruits and vegetables, whole grain products, fish, dairy products and meat)
	Non-personalized physical activity advice (at least 150 min of moderate intensity per week or 75 min of vigorous intensity per week or an equivalent combination of moderate and vigorous intensity activity) (WHO 2010,)	Participant anthropometric profile (weight, BMI)	Participant anthropometric profile (weight, BMI, WC)	Participant anthropometric profile (weight, BMI, WC)
		Participant physical activity Profile (Baecke Questionnaire and Accelerometry)[a]	Participant physical activity profile (Baecke Questionnaire and Accelerometry)[a]	Participant personalized nutrition profile (Baecke Questionnaire and Accelerometry)[a]
		Participant nutritional profile based on the online-FFQ (protein, carbohydrates, total fat, monounsaturated fat, polyunsaturated fat, saturated fat, salt, omega-3, fiber, calcium, iron, vitamin A, folate, thiamine, riboflavin, vitamin B12, vitamin C)	Participant nutritional profile based on the online-FFQ (protein, carbohydrates, total fat, monounsaturated fat, polyunsaturated fat, saturated fat, salt,omega-3, fiber, calcium, iron, vitamin A, folate, thiamine, riboflavin, vitaminB12, vitamin C)	Participant nutritional profile based on the online-FFQ (protein, carbohydrates, total fat, monounsaturated fat, polyunsaturated fat, saturated fat, salt,omega-3, fiber, calcium, iron, vitamin A, folate, thiamine, riboflavin, vitaminB12, vitamin C)
			Participant blood profile related to nutrition (glucose, total cholesterol, carotenes, n-3 index)[b]	Participant blood profile related to nutrition (glucose, total cholesterol, carotenes, n-3 index)[b]
		Personalized advice for: weight, personalized nutrition and dietary intake	*Personalized advice for:* weight, WC, physical activity, dietary intake and blood markers	Participant genetic profile related to nutrition (MTHFR, FTO, TCF7L2, APOE e4 and FADS1 genes)
				Personalized advice for: weight, WC, physical activity and dietary intake, blood and genetic markers

Celis-Morales, Livingstone et al. (2014)

Feedback provided at month 1 and 2 for L1, L2 and L3 was only for those participants in the high-intensity group.

FFQ = Food Frequency Questionnaire, *BMI* = body mass index, *WC* = waist circumference, *PA* = physical activity.

[a] Feedback on participants PA profile for the 'low-intensity' group was derived from accelerometer. The Baecke Questionnaire (Baecke et al. 1982) was used only when insufficient data were available from the accelerometer. For participants in the 'high-intensity' group, both accelerometry and the Baecke questionnaire were used.

[b] Feedback on blood profile related to nutrition was only available for month 0, 3 and 6 for both low- and high-intensity groups.

compliance. Therefore, while advice based on genetics may be more effective on paper and makes sense based on known gene-food interactions, there needs to be more research on how knowledge of genetic susceptibility influences behavior.

2.3.3 AGE AND STAGES OF LIFE

Nutrition is a critical component of every life stage. However, it is also important to realize that dietary requirements change with age and life stage. Therefore, maintenance of good health requires nutritional approaches that recognize the nuances that affect nutritional requirements with age (Bernstein and McMahon 2017). Early in life, growth and development are the primary biological objectives (Norris, Yin et al. 2007). Macronutrient and micronutrient needs are higher on a per kilogram basis during infancy and childhood when compared to any other life stage. This reflects the rapid cell division characteristic of this stage, which requires high amounts of these nutrients for DNA synthesis and the anabolism of bone, protein and fat. As children grow older, puberty and the transition to reproductive fertility cause a significant change in hormonal status with concomitant physiological and metabolic consequences (Vandeloo, Bruckers et al. 2007). Metabolic rate and activity levels eventually decrease with increasing age which should, in theory, lower nutrient demand. However, it is extremely important to continue to supplement adequate nutrition to those over 65 because this population faces a new set of nutritional challenges, such as inability to chew or digest, lower production of digestive enzymes and lower intestinal absorptive properties. Due to a weakened immunological system as well as general deterioration, these individuals are also usually afflicted with more chronic conditions which predispose them to malnutrition. Polypharmacy, which is common in this age group, may also alter the processing, metabolism and storage of nutrients (Drewnowski and Warren-Mears 2001).

2.3.4 GENDER

Males and females are known to differ in their nutritional requirements; for example, females not only have different basal nutritional requirements such as higher need for iron and other micronutrients (Table 2.4), but they also go through different life stages such as pregnancy, lactation and involution, each with varying nutritional requirements as well (Koletzko, Larque et al. 2007; Olsen, Halldorsson et al. 2007). However, when it comes to personalization of nutritional advice, it is also important to consider behavioral characteristics that, most likely because of societal influence, differentiate men from women. For example, an investigation was conducted by Durga et al., which involved 120 adults (60 male and 60 female) in different professions (20 doctors, 20 lecturers, and 20 software professionals) in studying their attitude, awareness, and interest towards personalized nutrition with the help of a questionnaire developed particularly for this study. The results obtained interestingly showed that awareness and positive attitudes towards personalized nutrition were higher among female participants compared to males (Table 2.5), regardless of profession. Awareness towards personalized nutrition was higher among doctors, followed by lecturers and software professionals in both male and female groups (Durga Priyadarshini 2017).

TABLE 2.4
Different Levels of Recommendation for Women (Not Pregnant or Lactating)

Global Recommendation	Personalized Dietitian Recommendation Based on an Individual's History and Preferences	Personalized Recommendation Based on Individual History, Preferences, and Genetic Information
Zn (8 mg/day): Consume a wide variety of foods containing zinc. Red meat and poultry provide the majority of zinc in the American diet. Other good food sources include beans, nuts, certain types of seafood, whole-grains, fortified breakfast cereals and dairy products	Recommendations vary according to age, sex, pregnancy and lactation (2–13 mg). Personalization will account for these individual characteristics. In addition, consideration should be given to: • People who have had gastrointestinal surgery, such as weight loss surgery, or who have digestive disorders, such as ulcerative colitis or Crohn's disease. Both these conditions can decrease the amount of zinc that the body absorbs and increase the amount lost in the urine • Vegetarians, because they do not eat meat, which is a good source of zinc. Also, the beans and grains they typically eat contain compounds that prevent complete absorption of zinc by the body. For this reason, vegetarians might need to eat as much as 50% more zinc than the recommended amounts • Older infants who are breastfed because breast milk contains insufficient zinc for infants aged > 6 months. Infants taking formula receive sufficient zinc. Older infants who do not take formula should be given foods that contain zinc, such as pureed meats • Alcoholics, because alcoholic beverages decrease the amount of zinc absorbed by the body and increase the amount lost in the urine. Also, many alcoholics eat a limited amount and variety of food, so they may not get enough zinc. • People with sickle cell disease, because they might need more zinc.	SLC30A8: Carriers of the A allele at the rs11558471 SLC30A8 (zinc transporter) variant need supplements containing zinc in addition to a healthy diet to maintain proper glucose homoeostasis. Knowledge of this genetic information will trigger a recommendation for Zn supplementation (Kanoni et al., 2011).

(Continued)

TABLE 2.4 (CONTINUED)
Different Levels of Recommendation for Women (Not Pregnant or Lactating)

Global Recommendation	Personalized Dietitian Recommendation Based on an Individual's History and Preferences	Personalized Recommendation Based on Individual History, Preferences, and Genetic Information
Dietary fat and cholesterol: Choose a diet low in fat, saturated fat, and cholesterol	Use fats and oils sparingly. Use the nutrition facts label to help you choose foods lower in fat, saturated fat, and cholesterol Eat plenty of grain products, vegetables and fruits. Choose low-fat milk products, lean meats, fish, poultry, beans, and peas to get essential nutrients without substantially increasing calories and intake of saturated fat	*TCF7L2:* For carriers of the T allele at the TCF7L2 rs7903146 polymorphism a Mediterranean diet reduces its adverse effect on cardiovascular risk factors and incidence of stroke, but not so a low-fat diet. Therefore, carriers of the T allele will be recommended to: • Eat primarily plant-based foods, such as fruits and vegetables, whole grains, legumes, and nuts • Replace butter with healthy fats such as extra virgin olive oil. • Use herbs and spices instead of salt to flavor foods. • Limit red meat to a few times a month. • Eat fish and poultry at least twice a week. • Drink red wine in moderation. (optional) (Corella et al., 2013).
Vitamin B2 (riboflavin): Consume the appropriate recommended dietary allowance (RDA) from a variety of foods	Recommendations vary according to age, sex, pregnancy and lactation (0.3–1.6 mg/day) Personalization will take account of these individual characteristics. In addition, consideration should be given to: • Vegetarian athletes, as exercise produces stress in the metabolic pathways that use riboflavin People who are vegan or consume little milk, or both, are also at risk of riboflavin inadequacy.	*MTHFR:* Carriers of the TT genotype at the MTHFR C677T polymorphism are at higher risk of hypertension, which may not reach targets (systolic blood pressure <120 mm Hg) with medication. However, they particularly benefit from riboflavin supplementation (~1.6 mg/day) (McAuley, McNulty, Hughes, Strain, & Ward, 2016). *SLC52A3:* Brown-Vialetto-Van Laere syndrome is caused by mutations in the *SLC52A3* gene, which encodes the intestinal riboflavin transporter. As a result, these patients have riboflavin deficiency. Riboflavin supplementation can be lifesaving in this population (Jaeger and Bosch, 2016).

Ordovas, Ferguson et al. (2018)

TABLE 2.5
Difference between Educated Men and Women on Their Attitude, Awareness and Interest Toward Personalized Nutrition

Variable	N		Mean		t-value
	Male	Female	Male	Female	
Attitude	60	60	18.36	19.55	2.35*
Awareness	60	60	23.03	24.48	2.61**
Interest	60	60	20.01	19.9	0.25

Source: Durga Priyadarshini (2017).
* p<.05
** p<.01

2.3.5 Environment: The Exposome

The effects of the environment have been well-documented in the literature (Sabate, Harwatt et al. 2016). The 'environment' includes any type of external influence, and can be either acute or chronic, random/unavoidable (e.g., urban pollution or the effect of solar radiation on vitamin D levels) or volitional (e.g., smoking), generalized to a population (e.g., exposure to fluoridated drinking water) or specific to an individual (e.g., chronic consumption of sweetened beverages) as well as biological or behavioral in nature (e.g., food preferences due to social-religious reasons) (Holick 2006) (Montonen, Jarvinen et al. 2007). Another critical feature affecting an individual's nutritional phenotype is the availability of food resources. Though many consumers have access to balanced and diverse foods throughout the year, others lack affordable diversity based on locality, climate and socioeconomic limitations (Drescher, Thiele et al. 2007). All together, these factors are termed the 'exposome,' and all have the potential to affect an individual's phenotype directly or indirectly.

2.3.5.1 The Microbiome

Recently, one of the environmental factors that has attracted unprecedented attention from the scientific community is the gut microbiome, which has been found to exert significant effects on an individual's nutritional phenotype. The microbiome consists of countless beneficial microorganisms of over 500 species that inhabit the gastrointestinal tract. These organisms play a key role in processes including the digestion and absorption of nutrients, vitamins synthesis, mucosal immunity and breakdown of toxins and carcinogens (Guarner and Malagelada 2003; Sekirov, Russell et al. 2010). The composition and activity of gut bacteria can vary with diet, disease states and pharmacotherapy (e.g., antibiotic use). It also varies greatly with age, from infancy when humans are exposed to important antibodies in breast milk, to certain life events including puberty, the ovarian cycle, pregnancy and menopause (Nicholson, Holmes et al. 2012), and into old age which is associated with drastic shifts in microbial diversity (Conlon and Bird 2014). Therefore, differences in gut microbiome represent a huge source of metabolic variability that should be a target for dietary intervention, with emphasis in obtaining the right balance between 'good' and 'bad' microbes (Kang 2013). Interestingly, stress, another environmental factor, may contribute to conditions like inflammatory bowel syndrome by altering the microbial populations via the central nervous system (CNS). This cross-talk has been termed the gut-brain axis, a bi-directional pathway involving hormonal and neuronal mechanisms (Grenham, Clarke et al. 2011) meaning brain activity can influence the gut microbiota and vice versa (Clarke, Grenham et al. 2013).

2.3.6 PERSONALIZED NUTRITIONAL STRATEGIES FOR ATHLETIC PERFORMANCE

Nutrition has a significant and varied impact on sports and athletic performance. Current dietary advice for the optimization of athletic performance is also generalized, despite it being widely accepted that athletes respond differently to similar foods, nutrients and supplements. This holds true for all ages, ethnicities and level of skill, regardless of ultimate exercise objectives (Guest, Horne et al. 2019). Across many nations, sports represent a progressive, innovative and extremely competitive environment, demanding more and more of young professionals by the day. Providing athletes with tailored dietary recommendation and other performance-related information based on their unique makeup could yield higher competitive success.

2.4 CLOSING REMARKS

Since the advent of genome sequencing, there has been a boom in literature surrounding gene-food interactions. Since then, the floodgates have opened to include research on the interactions between diet and the metabolome, environment and human behavior. The main purpose of this chapter was to highlight selected areas of interest in the growing field of personalized nutritional research; as more tools are created to analyze the complex interactions of an individual with his/her own diet, it is becoming increasingly apparent that tailored dietary advice is now within our grasp, representing the future of the nutritional sciences, and paving the way for more potential therapeutic and preventative interventions aimed at promoting and maintaining social health. However, personalized nutrition, while based on centuries-old beliefs, is still very much in its infancy as an ideology, and there still exist many challenges that must be overcome if it is to have a place in the clinical setting.

ABBREVIATIONS

APOEε4:   Apolipoprotein E epsilon 4 gene
BMI: Body Mass Index
CNS: Central nervous system
FADS1:   Fatty Acid Desaturase 1 gene
FTO: Fat Mass and Obesity Associated gene
MTHFR: Methylenetetrahydrofolate reductase gene
SNPs: Single nucleotide polymorphisms

REFERENCES

Allen, K. J., L. C. Gurrin, et al. (2008). "Iron-overload-related disease in HFE hereditary hemochromatosis." *N Engl J Med* 358(3): 221–230.

Barker, T., E. D. Schneider, et al. (2013). "Supplemental vitamin D enhances the recovery in peak isometric force shortly after intense exercise." *Nutr Metab* (Lond) 10(1): 69.

Begas, E., E. Kouvaras, et al. (2007). "In vivo evaluation of CYP1A2, CYP2A6, NAT-2 and xanthine oxidase activities in a Greek population sample by the RP-HPLC monitoring of caffeine metabolic ratios." *Biomed Chromatogr* 21(2): 190–200.

Benyamin, B., M. A. Ferreira, et al. (2009). "Common variants in TMPRSS6 are associated with iron status and erythrocyte volume." *Nat Genet* 41(11): 1173–1175.

Bernstein, M. and K. McMahon (2017). *Nutrition Across Life Stages*, Jones & Bartlett Learning.

Biesiekierski, J. R., K. M. Livingstone, et al. (2019). "Personalised nutrition: updates, gaps and next steps." *Nutrients* 11(8).

Braakhuis, A. J. (2012). "Effect of vitamin C supplements on physical performance." *Curr Sports Med Rep* 11(4): 180–184.

Cannell, J. J., B. W. Hollis, et al. (2009). "Athletic performance and vitamin D." *Med Sci Sports Exerc* 41(5): 1102–1110.

Celis-Morales, C., K. M. Livingstone, et al. (2014). "Design and baseline characteristics of the Food4Me study: a web-based randomised controlled trial of personalised nutrition in seven European countries." *Genes Nutr* 10(1): 450.

Celis-Morales, C., K. M. Livingstone, et al. (2016). "Effect of personalized nutrition on health-related behaviour change: evidence from the Food4me European randomized controlled trial." *Int J Epidemiol* 46(2): 578–588.

Clarke, G., S. Grenham, et al. (2013). "The microbiome-gut-brain axis during early life regulates the hippocampal serotonergic system in a sex-dependent manner." *Mol Psychiatry* 18(6): 666–673.

Clenin, G., M. Cordes, et al. (2015). "Iron deficiency in sports - definition, influence on performance and therapy." *Swiss Med Wkly* 145: w14196.

Conlon, M. A. and A. R. Bird (2014). "The impact of diet and lifestyle on gut microbiota and human health." *Nutrients* 7(1): 17–44.

Corella, D., P. Carrasco, et al. (2013). "Mediterranean diet reduces the adverse effect of the TCF7L2-rs7903146 polymorphism on cardiovascular risk factors and stroke incidence: a randomized controlled trial in a high-cardiovascular-risk population." *Diabetes Care* 36(11): 3803–3811.

Curro, M., D. Di Mauro, et al. (2016). "Influence of MTHFR polymorphisms on cardiovascular risk markers in elite athletes." *Clin Biochem* 49(1–2): 183–185.

Czarnewski, P., S. Das, et al. (2017). "Retinoic Acid and Its Role in Modulating Intestinal Innate Immunity." *Nutrients* 9(1).

Dinc, N., S. B. Yucel, et al. (2016). "The effect of the MTHFR C677T mutation on athletic performance and the homocysteine level of soccer players and sedentary individuals." *J Hum Kinet 51*: 61–69.

Drescher, L. S., S. Thiele, et al. (2007). "A new index to measure healthy food diversity better reflects a healthy diet than traditional measures." *J Nutr* 137(3): 647–651.

Drewnowski, A. and V. A. Warren-Mears (2001). "Does aging change nutrition requirements?" *J Nutr Health Aging* 5(2): 70–74.

Durga Priyadarshini, R. M. (2017). "Attitude, awareness, and interest towards personalized nutrition among educated men and women in chennai." *J Nutr Hum Health* 1(2): 9–12.

Elsawy, G., O. Abdelrahman, et al. (2014). "Effect of choline supplementation on rapid weight loss and biochemical variables among female taekwondo and judo athletes." *J Hum Kinet* 40: 77–82.

Garcia, M., J. M. Martinez-Moreno, et al. (2014). "Changes in body composition of high competition rugby players during the phases of a regular season; influence of diet and exercise load." *Nutr Hosp* 29(4): 913–921.

Garvican, L. A., P. U. Saunders, et al. (2014). "Intravenous iron supplementation in distance runners with low or suboptimal ferritin." *Med Sci Sports Exerc* 46(2): 376–385.

Ghotbi, R., M. Christensen, et al. (2007). "Comparisons of CYP1A2 genetic polymorphisms, enzyme activity and the genotype-phenotype relationship in Swedes and Koreans." *Eur J Clin Pharmacol* 63(6): 537–546.

Goss, A. M., L. L. Goree, et al. (2013). "Effects of diet macronutrient composition on body composition and fat distribution during weight maintenance and weight loss." *Obesity* (Silver Spring) 21(6): 1139–1142.

Goyette, P., J. S. Sumner, et al. (1994). "Human methylenetetrahydrofolate reductase: isolation of cDNA mapping and mutation identification." *Nat Genet* 7(4): 551.

Grenham, S., G. Clarke, et al. (2011). "Brain-gut-microbe communication in health and disease." *Front Physiol* 2: 94.

Guarner, F. and J. R. Malagelada (2003). "Gut flora in health and disease." *Lancet* 361(9356): 512–519.

Guenther, P. M., K. O. Casavale, et al. (2013). "Update of the healthy eating index: HEI-2010." *J Acad Nutr Diet* 113(4): 569–580.

Guest, N. S., J. Horne, et al. (2019). "Sport nutrigenomics: personalized nutrition for athletic performance." *Front Nutr* 6: 8.

Haas, J. D. and T. t. Brownlie (2001). "Iron deficiency and reduced work capacity: a critical review of the research to determine a causal relationship." *J Nutr* 131(2S-2): 676S–688S; discussion 688S–690S.

Holick, M. F. (2006). "Resurrection of vitamin D deficiency and rickets." *J Clin Invest* 116(8): 2062–2072.

Hunter, A. M., A. St Clair Gibson, et al. (2002). "Caffeine ingestion does not alter performance during a 100-km cycling time-trial performance." *Int J Sport Nutr Exerc Metab* 12(4): 438–452.

Kang, J. X. (2013). "Gut microbiota and personalized nutrition." *J Nutrigenet Nutrigenomics* 6(2): I–II.

Kanoni, S., J. A. Nettleton, et al. (2011). "Total zinc intake may modify the glucose-raising effect of a zinc transporter (SLC30A8) variant: a 14-cohort meta-analysis." *Diabetes* 60(9): 2407–2416.

Knapik, J. J. (2015). "The Importance of Physical Fitness for Injury Prevention: Part 2." *J Spec Oper Med* 15(2): 112–115.

Kohlmeier, M., K. A. da Costa, et al. (2005). "Genetic variation of folate-mediated one-carbon transfer pathway predicts susceptibility to choline deficiency in humans." *Proc Natl Acad Sci USA* 102(44): 16025–16030.

Koletzko, B., E. Larque, et al. (2007). "Placental transfer of long-chain polyunsaturated fatty acids (LC-PUFA)." *J Perinat Med* 35(Suppl 1): S5–S11.

Kraemer, K., J. Cordaro, et al. (2016). *1 Personalized Nutrition: Paving the Way to Better Population Health. Good Nutrition: Perspectives for the 21st Century*, Karger Publishers: 235–248.

Krieger, J. W., H. S. Sitren, et al. (2006). "Effects of variation in protein and carbohydrate intake on body mass and composition during energy restriction: a meta-regression 1." *Am J Clin Nutr* 83(2): 260–274.

Kussmann, M. and L. B. Fay (2008). "Nutrigenomics and personalized nutrition: science and concept." *Per Med* 5(5): 447–455.

Larson-Meyer, D. E. and K. S. Willis (2010). "Vitamin D and athletes." *Curr Sports Med Rep* 9(4): 220–226.

Lietz, G., J. Lange, et al. (2010). "Molecular and dietary regulation of beta,beta-carotene 15,15′-monooxygenase 1 (BCMO1)." *Arch Biochem Biophys* 502(1): 8–16.

Marjot, T., J. Collier, et al. (2016). "What is HFE haemochromatosis?" *Br J Hosp Med* (Lond) 77(6): C91–95.

Mathers, J. C. (2019). "Paving the way to better population health through personalised nutrition." *EFSA J* 17: e170713.

McAuley, E., H. McNulty, et al. (2016). "Riboflavin status, MTHFR genotype and blood pressure: current evidence and implications for personalised nutrition." *Proc Nutr Soc* 75(3): 405–414.

Mendes-Soares, H., T. Raveh-Sadka, et al. (2019). "Assessment of a personalized approach to predicting postprandial glycemic responses to food among individuals without diabetes." *JAMA Netw Open* 2(2): e188102.

Merritt, D. C., J. Jamnik, et al. (2018). "FTO genotype, dietary protein intake, and body weight in a multiethnic population of young adults: a cross-sectional study." *Genes Nutr* 13: 4.

Montonen, J., R. Jarvinen, et al. (2007). "Consumption of sweetened beverages and intakes of fructose and glucose predict type 2 diabetes occurrence." *J Nutr* 137(6): 1447–1454.

Nascimento, M., D. Silva, et al. (2016). "Effect of a nutritional intervention in athlete's body composition, eating behaviour and nutritional knowledge: A comparison between adults and adolescents." *Nutrients* 8(9).

Nicholson, J. K., E. Holmes, et al. (2012). "Host-gut microbiota metabolic interactions." *Science* 336(6086): 1262–1267.

Norris, J. M., X. Yin, et al. (2007). "Omega-3 polyunsaturated fatty acid intake and islet autoimmunity in children at increased risk for type 1 diabetes." *JAMA* 298(12): 1420–1428.

Olsen, S. F., T. I. Halldorsson, et al. (2007). "Milk consumption during pregnancy is associated with increased infant size at birth: prospective cohort study." *Am J Clin Nutr* 86(4): 1104–1110.

Ordovas, J. M., L. R. Ferguson, et al. (2018). "Personalised nutrition and health." *BMJ* 361: bmj.k2173.

Palatini, P., G. Ceolotto, et al. (2009). "CYP1A2 genotype modifies the association between coffee intake and the risk of hypertension." *J Hypertens* 27(8): 1594–1601.

Palidis, D. J., P. A. Wyder-Hodge, et al. (2017). "Distinct eye movement patterns enhance dynamic visual acuity." *PLoS One* 12(2): e0172061.

Pantopoulos, K., S. K. Porwal, et al. (2012). "Mechanisms of mammalian iron homeostasis." *Biochemistry* 51(29): 5705–5724.

Price, N. D., A. T. Magis, et al. (2017). "A wellness study of 108 individuals using personal, dense, dynamic data clouds." *Nat Biotechnol* 35(8): 747–756.

Reardon, T. F. and D. G. Allen (2009). "Iron injections in mice increase skeletal muscle iron content, induce oxidative stress and reduce exercise performance." *Exp Physiol* 94(6): 720–730.

Recalcati, S., G. Minotti, et al. (2010). "Iron regulatory proteins: from molecular mechanisms to drug development." *Antioxid Redox Signal* 13(10): 1593–1616.

Rucker, M. (2019). "In pursuit of personalized nutrition for disease prevention: using nutrigenetics and nutrigenomics in your practice." from https://www.verywellhealth.com/nutrigenetics-and-nutrigenomics-in-healthcare-4153579.

Sabate, J., H. Harwatt, et al. (2016). "Environmental nutrition: a new frontier for public health." *Am J Public Health* 106(5): 815–821.

Sekirov, I., S. L. Russell, et al. (2010). "Gut microbiota in health and disease." *Physiol Rev* 90(3): 859–904.

Shaw, G., A. Lee-Barthel, et al. (2017). "Vitamin C-enriched gelatin supplementation before intermittent activity augments collagen synthesis." *Am J Clin Nutr* 105(1): 136–143.

Siroux, V., L. Agier, et al. (2016). "The exposome concept: a challenge and a potential driver for environmental health research." *Eur Respir Rev* 25(140): 124–129.

Soares, R. N., A. Schneider, et al. (2018). "The influence of CYP1A2 genotype in the blood pressure response to caffeine ingestion is affected by physical activity status and caffeine consumption level." *Vascul Pharmacol* 106: 67–73.

Soranzo, N., T. D. Spector, et al. (2009). "A genome-wide meta-analysis identifies 22 loci associated with eight hematological parameters in the HaemGen consortium." *Nat Genet* 41(11): 1182–1190.

Thomas, D. T., K. A. Erdman, et al. (2016). "American College of Sports Medicine joint position statement: Nutrition and athletic performance." *Med Sci Sports Exerc* 48(3): 543–568.

Vandeloo, M. J., L. M. Bruckers, et al. (2007). "Effects of lifestyle on the onset of puberty as determinant for breast cancer." *Eur J Cancer Prev* 16(1): 17–25.

Vega-Lopez, S., L. M. Ausman, et al. (2007). "Interindividual variability and intra-individual reproducibility of glycemic index values for commercial white bread." *Diabetes Care* 30(6): 1412–1417.

Vidoni, M. L., K. Pettee Gabriel, et al. (2018). "Relationship between homocysteine and muscle strength decline: The Baltimore longitudinal study of aging." *J Gerontol A Biol Sci Med Sci* 73(4): 546–551.

Wang, T. J., F. Zhang, et al. (2010). "Common genetic determinants of vitamin D insufficiency: A genome-wide association study." *Lancet* 376(9736): 180–188.

Williams, R. J. (1950). "Concept of genetotrophic disease." *Nutr Rev* 8(9): 257–260.

Yang, A., A. A. Palmer, et al. (2010). "Genetics of caffeine consumption and responses to caffeine." *Psychopharmacology* 211(3): 245–257.

Yucha, C. and D. Guthrie (2003). "Renal homeostasis of calcium." *Nephrol Nurs J* 30(6): 621–626; quiz 627–628.

Zeevi, D., T. Korem, et al. (2015). "Personalized nutrition by prediction of glycemic responses." *Cell* 163(5): 1079–1094.

Zeisel, S. H. (2006). "Choline: critical role during fetal development and dietary requirements in adults." *Annu Rev Nutr* 26: 229–250.

Zeisel, S. H., K. A. Da Costa, et al. (1991). "Choline, an essential nutrient for humans." *FASEB J* 5(7): 2093–2098.

3 Personalized Nutrition in Cardiovascular Disease
From Concept to Realization

Marcella O'Reilly, Sarina Kajani,
Sean Curley, Sarah Mahayni, Helen M.
Roche and Fiona C. McGillicuddy

CONTENTS

3.1 EPIDEMIOLOGY

Cardiovascular disease (CVD) is the leading cause of death globally (2017) and is associated with numerous lifestylemodifiable risk factors including obesity/diabetes, metabolic dyslipidemia, hypertension, insulin resistance, lack of physical activity, smoking and alcohol consumption (Turner, Millns et al. 1998; Mackay and Mensah 2004). It is estimated that 17.7 million people died from CVD in 2015, equating to 31% of all global deaths. CVDs are a group of disorders of the heart and blood vessels and include: coronary heart disease (CHD), cerebrovascular disease, peripheral arterial disease, rheumatic heart disease, congenital heart disease, deep vein thrombosis and pulmonary embolisms (2017). In developed countries including North America, Europe, Australia, at least one-third of all CVD is attributable to five lifestyle-modifiable risk factors: tobacco use, alcohol use, high blood pressure (BP), high cholesterol and obesity (Mackay and Mensah 2004).

3.2 PERSONALIZED NUTRITION (PN) IN CVD

PN has been defined by Gibney and Goosens as an approach that assists individuals in achieving a lasting dietary behavior change that is beneficial for health (Gibney and Goosens 2016). There are multiple levels to PN (PN1–3) increasing in complexity and individualization and in turn, feasibility to implement (Gibney and Walsh 2013). The first level (PN1) that has seen widespread success is the use of technology to record and analyze imputed food intake and physical activity levels and tailor advice to the end-user to improve health (Gibney and Walsh 2013). There is a multitude of software apps currently available that have transformed PN1 from an ideological concept that could only be accessed by those with the means, to widespread dissemination to anyone with access to mobile phone technology and an interest in PN/health. In terms of PN in the management of CVD, PN1 certainly provides a means to educate patients on implementation and adherence to a healthy diet which is a critical lifestyle-modifiable factor that can profoundly reduce CVD risk.

PN2 involves tailoring nutritional advice based on a person's phenotype/metabotype. In the management of CVD, nutrition can and should be tailored to manage nutrient-modifiable risk factors such as weight, blood pressure, sodium intake and lipoprotein profiles as appropriate to the individual. There is growing interest in the use of dried blood spots (DBS) in PN2 as a means to derive biochemical/metabolic profiles of individuals and adapt a diet according to these parameters (Celis-Morales, Livingstone et al. 2015). The feasibility of using DBS in a large population, using blood samples taken remotely, to direct PN has recently been established in the European Union Seventh Framework Programme integrated project, Food4Me (Celis-Morales, Livingstone et al. 2015). This study analyzed a wide variety of biochemical measurements from DBS including glucose, total cholesterol, carotenoids, omega-3 (n-3) fatty acid index, fatty acids and vitamin D (Celis-Morales, Livingstone et al. 2015). DBS can also be used for lipidomic (Gao, McDaniel et al. 2017), metabolomic (Wang, Sun et al. 2016) and proteomic (Martin, Bunch et al. 2013) analyses and thus a vast amount of data can be derived from a small sample size. Indeed, the expansive information that is acquired from DBS, using more sensitive analytical platforms, provides the power to group individuals into metabotypes wherein individuals are grouped into clusters with similar metabolic profiles (Gibney and Goosens 2016). Metabotype subgroups may respond differently to nutrients contributing to interindividual variability that exists during nutrition interventions. A recent study, for example, hypothesized that the interindividual variability to the lipid lowering properties of pomegranate may be due to variability in polyphenol metabolism (Gonzalez-Sarrias, Garcia-Villalba et al. 2017). Participants were grouped according to the microbially derived ellagitannin-metabolizing phenotypes including urolithin metabotypes A, B and 0 (UM-A, UM-B and UM-0). The UM-B cluster exhibited the greatest CVD risk at baseline but also exhibited the greatest lipid-lowering response to phenolics (Gonzalez-Sarrias, Garcia-Villalba et al. 2017). This is a prime example of how PN may be implemented to prescribe the right diet to the right responder and move away from a one-size-fits-all approach to nutrition.

PN3 involves tailoring nutritional advice to a person's genotype. This is the premium level for PN but also the most challenging to implement. In the era of genome-wide association studies (GWAS) and post-the Human Genome Project, it has been widely envisioned that both personalized medicine and PN would be largely based on the ultimate blueprint of a person, their genetic code (Gibney and Walsh 2013). One of the first single nucleotide polymorphisms (SNPs) to be identified that contributes to dysbetalipoproteinemia was in the Apolipoprotein (Apo) E gene which resulted in accumulation of triglyceride (TG)-rich lipoproteins in circulation (Utermann, Hees et al. 1977). There are three common isoforms of ApoE (apoE2, apoE3 and apoE4) that are encoded for by a gene on chromosome 19—approximately 55–60% of individuals are homozygous for the E3 allele (Davignon, Gregg et al. 1988; Minihane, Khan et al. 2000). Individuals carrying the ApoE4 allele have the highest levels of low-density lipoprotein (LDL)-cholesterol and ApoB, intermediate levels are observed in individuals homozygous for ApoE3 and the lowest levels are evident in those with the ApoE2 allele (Davignon, Gregg et al. 1988). Individuals with the ApoE2 allele also display reduced postprandial TG response to fish-oil intervention relative to non-E2 carriers, indicative of better response to fish-oils in individuals with this genotype (Minihane, Khan et al. 2000). This is one example of how dietary advice could be tailored to genotype in that individuals with the ApoE2 allele should be encouraged to consume fish-oil enriched diets due to their propensity to yield beneficial effects from these nutrients.

Another example of a nutrient modifiable genetic variant (C677T) is within the gene encoding the enzyme methylenetetrahydrofolate reductase (MTHFR). Individuals carrying the TT allele of the MTHFR gene exhibit reduced activity of the enzyme, reduced flavin adenine dinucleotide (FAD) co-factor and are at increased risk of hypertension (Heux, Morin et al. 2004). In turn, individuals with the TT variant responded to riboflavin intervention with lowered homocysteine levels (McNulty, Dowey et al. 2006) and reduced BP (Wilson, Ward et al. 2012) while those carrying the C allele exhibited no such response. These gene-nutrient intervention studies were carried out pre-GWAS and indeed SNPs in MTHFR have not reached the significance criteria or been reproduced across GWAS for selection as one of the primary SNPs associated with CVD (Clarke, Bennett et al. 2012). However, they do provide the fundamental basis of PN3 in that understanding one's genetic code can translate to tailored nutritional advice for optimal health. There are a multitude of other gene-nutrient intervention studies, many in the pre-GWAS era that have been excellently reviewed (Ordovas and Corella 2004; Corella and Ordovas 2009; Ordovas, Ferguson et al. 2018).

3.3   PN1: A HEALTHY REFERENCE DIET FOR CVD

Implementation of PN1 for the management of CVD is contingent upon our under-standing of the optimal diet for the management of CVD. The current recommended dietary guidelines for CVD management include consuming a diet rich in fruits and vegetables, grains, low-fat or non-fat dairy products, fish, legumes, poultry and lean meats (Krauss, Eckel et al. 2000; Eckel, Jakicic et al. 2014). In turn, recommenda-tions are sub-divided into guidelines for body weight (match energy intake to energy needs), cholesterol profiles (limit foods high in saturated fat and cholesterol; and substitute unsaturated fats from vegetables, fish, legumes and nuts) and blood-pres-sure management (limit salt and alcohol, maintain a healthy body weight, increase intake of vegetables, fruits and low-fat/non-fat dairy products). In this section of the chapter we will review the literature about the importance of a healthy diet in CVD management, the evidence for the current dietary recommendations and the current controversies surrounding macronutrient quality and CVD (Figure 3.1).

FIGURE 3.1   Macronutrient quality is an important consideration to incorporate into dietary guidelines for CVD patients.

3.4   DIET AND CVD: LESSONS FROM AND LIMITATIONS OF OBSERVATIONAL STUDIES

Epidemiological studies provide the fundamental basis of our current understanding of the relationship between dietary macronutrient composition and CVD risk. In turn, the outcomes of epidemiological studies have been the major cornerstone for dietary advice currently prescribed. That notwithstanding, such observational studies do not prove causality and there are some major discrepancies in findings between studies, most notably about the relationship between saturated fat (SFA) and carbohydrate with CVD (Hu and Willett 2002; Siri-Tarino, Sun et al. 2010; Grasgruber, Sebera et al. 2016; Dehghan, Mente et al. 2017). There are of course fundamental limitations to epidemiological studies which likely contribute to conflicting data entering the public domain. Firstly, methods used for nutritional data collection vary widely from study to study and are often reliant on self-reported food frequency questionnaires (FFQ)/24h food recalls which are inherently flawed (Kristal, Peters et al. 2005). Biomarkers that accurately reflect nutritional status and provide an unbiased and quantifiable surrogate marker of dietary intake such as vitamin C/carotenoids for fruit and vegetable intake (Baldrick, Woodside et al. 2011) should be included in such studies to support/validate future epidemiology study outcomes. Finally, many studies are based solely on the breakdown of macronutrients into three main subgroups – protein, carbohydrate and fat. This is problematic, as macronutrient quality is most likely the critical determinant of health and is not captured in studies that subgroup macronutrients into three simplistic categories. For instance, carbohydrates are comprised of fiber/whole-grains which are considered protective against CVD (Hu and Willett 2002), as well as refined carbohydrates which are considered detrimental for CVD (Grasgruber, Sebera et al. 2016). Hence grouping of carbohydrates into one diet-group to infer a relationship to CVD is inherently flawed. The relationship between dietary macronutrient composition/quality and relationship to CVD will be discussed below.

While epidemiology studies are important to explore potential associations between dietary composition and disease, intervention studies are required to justify the benefit of one diet over another and prove causation. A major challenge for nutritional interventions to manage CVD is the time taken for the precipitation of cardiovascular events to interpret success of the intervention in question. Indeed, in the 1960s it was widely acclaimed that individuals should consume a low-SFA diet to reduce CVD risk. However, the large-scale replacement of SFA with refined carbohydrate has coincided with greater energy intake to precipitate a global obesity and type 2 diabetes epidemic. Many nutritional intervention studies determine the success of an intervention by measuring changes in traditional risk factors for CVD including LDL-C, triglycerides (TG) and HDL-C (Wilson, D'Agostino et al. 1998; Miller, Stone et al. 2011). While such an approach is logical, it is evident from pharmaceutical trials aimed at lowering TGs (Group 1975; Canner, Berge et al. 1986) and raising HDL-C (Group, Landray et al. 2014) that these risk factors are more biomarkers of disease, as opposed to causal, and therefore over-interpretation of the clinical relevance of changes in these biomarkers in response to nutritional interventions should be cautioned.

3.5   SATURATED FAT (SFA) INTAKE AND CVD

The association between serum cholesterol and CVD was demonstrated by Keys et al. in the 1960s (Keys, Taylor et al. 1963). The relationship between dietary fat consumption and serum cholesterol levels was discovered as far back as the 1950s (Keys, Mickelsen et al. 1950; Bronte-Stewart, Antonis et al. 1956). The Seven Countries Study demonstrated a positive correlation between SFA intake and CVD in 12,000 men with no association to polyunsaturated fatty acids (PUFA), proteins or carbohydrates (Keys, Menotti et al. 1986). The Nurse's Health Study demonstrated that diets high in SFA and refined carbohydrate were positively associated with CVD, while diets rich in fiber, whole-grains and unsaturated fats were associated with reduced CVD (Hu, Stampfer et al. 1997; Liu, Stampfer et al. 1999; Hu and Willett 2002; Yu, Rimm et al. 2016). Similarly, a large meta-analysis by Hu et al., which incorporated 147 original investigations and reviews highlighted three dietary strategies that are effective in preventing CVD including the replacement of SFA and trans-fat by unsaturated fat, increasing consumption of n-3 fatty acids and consuming a diet high in fruits, vegetables, nuts and whole-grains and low in refined grain products (Hu and Willett 2002). Greater understanding about the contribution of lifestyle factors to CVD and better medical management of associated risk factors ultimately culminated in a dramatic fall in deaths from CVD (Sytkowski, Kannel et al. 1990; Jones and Greene 2013). The Framingham study demonstrated an astonishing 60% reduction in mortality in the 1970 cohort compared to their 1950 cohort which was coincident with improvements in serum cholesterol levels and better medical management of blood pressure (Sytkowski, Kannel et al. 1990).

However, in more recent years the association of SFA with CVD has been called into question. A meta-analysis of prospective cohort studies (21 studies with 347,747 subjects of which 11,006 developed CVD or stroke) by SiriTarino et al. evaluated the association of SFA with CVD (Siri-Tarino, Sun et al. 2010) and demonstrated that dietary SFA was not associated with increased risk of CVD. More recently Dehghan et al. (Dehghan, Mente et al. 2017) found no association between dietary fat consumption and CVD in the Prospective Urban and Rural Epidemiological study (PURE). The PURE study investigated the association between fat and carbohydrate intake with CVD and mortality in 18 countries over five continents (135,000 subjects). Total, saturated, and unsaturated fat intake was not significantly associated with myocardial infarction risk or CVD mortality while high carbohydrate was associated with increased total mortality, but not CVD-related mortality (Dehghan, Mente et al. 2017). This study was specifically designed to include low-, middle- and high-income countries to ascertain whether findings from North America/Europe translate across the different demographics (Dehghan, Mente et al. 2017). However, more than half of participants (52%) within the PURE study consumed a high-carbohydrate diet (at least 60% of total kcal), a trend that was particularly prevalent in low-income countries (Dehghan, Mente et al. 2017). An astounding 80% of CVD deaths occur in low- and middle-income countries despite cardiovascular risk factors being significantly higher in highincome countries (Yusuf, Rangarajan et al. 2014). Lack of access to and affordability of appropriate medical management for both hypercholesterolemia and BP was identified as a significant factor causing increased

CVD deaths in lower-income countries, which coincided with a high-carbohydrate diet. On the contrary, in Western society the aggressive medical management of patients at high risk of CVD, particularly with statins, may have reduced the association between SFA-intake and CVD, particularly given the purported mechanism of enhanced risk is via increased LDL-C, the primary target of statins.

A recent prospective study in the US, where overnutrition predominates, assessed individual SFA and risk of CHD (73,147 women and 42,635 men), and found that higher intake of longer chain SFA was associated with increased CHD risk (Zong, Li et al. 2016). In 2015, the Cochrane Collaboration published a systematic review and meta-analysis of over 15 randomized controlled trials and concluded that those at risk of CVD should include a permanent reduction, but not elimination, of dietary SFA by replacing them with unsaturated fats (Hooper, Martin et al. 2015). Remarkably, only a 1% daily energy intake replacement of long chain fatty acids (LCFA) with PUFA, whole-grain carbohydrates or plant protein resulted in a 6–8% reduced risk of CHD (Zong, Li et al. 2016). The Scientific Advisory Committee on Nutrition (SACN) recently reviewed the role of SFA in health, given the extensive new literature in this area since their last review in 1994. A draft review of their findings was published in May 2018 and the recommendations remained the same as their previous review; saturated fat consumption should be no more than 10% of total calories (<30g for men, <20g for women) and dietary saturated fats should be substituted with unsaturated fats (Nutrition 2018). Perhaps most importantly this report concluded that there is no evidence that reducing intake of SFA increases risk of any of the health-related parameters assessed including CVD, blood lipids, BP, diabetes, dementia and some cancers.

3.6   FAT QUALITY AND CVD

It is important to note that recent findings from the PURE study and Siri-Tarino et al. are observational and ultimately interventional studies are required to prove causality. While reductions in SFA are recommended it is critical to highlight that dietary replacement of SFA for refined carbohydrates does not lower cardiovascular risk [31, 47–49]. In this section we will discuss the evidence supporting the substitution of saturated fats for unsaturated fats.

SFA are made up of carbons joined by single bonds, and can be short (1–6 carbons), medium (7–12 carbons, MCFA), or long-chained (13 or more carbons) (Briggs, Petersen et al. 2017). The predominant SFA in Western diets include palmitic acid (16 carbons) and stearic acid (18 carbons). Sources of SFA include red meat, whole milk, cheese, butter, processed food and confectionary (Figure 3.2). Monounsaturated fatty acids (MUFA) have a single double bond in the fatty acid chain while the remainder are single-bonded. Sources of MUFAs include red meat, olive oil, sunflower oil and avocados (Figure 3.2). In structure, polyunsaturated fatty acids (PUFA) contain more than one double bond in the fatty acid chain (Russo 2009). There are two major classes of PUFA that lead nutritional investigation; n-3 and n-6 PUFA, socalled for the position of their double bond (Poudyal and Brown 2015). Linoleic acid (C18:2n-6) is an n-6 PUFA and is the precursor to arachidonic acid (AA), a 20 carbon fatty acid while α-linoleic acid (ALA), an n-3 PUFA (C18:3n-3) is

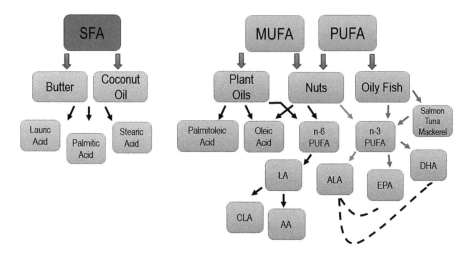

FIGURE 3.2 Saturated, mono- and polyunsaturated fat sources.

the precursor to both eicosapentenoic acid (EPA) (20:5n-3) and docosahexenoic acid (DHA) (C22:6n-3) (also n-3 PUFAs) (Simopoulos 2008). Both linoleic acid and ALA are considered essential fatty acids, as they cannot be synthesized in the body (Das 2006). Sources of PUFA are primarily from fish and vegetable oils (Figure 3.3).

Kafatos et al. noted in a paper that although Crete was an area of low CHD prevalence and incidence, consumption of fat was high, particularly MUFA (Kafatos, Diacatou et al. 1997). The Mediterranean diet is rich in the MUFA oleic acid (olive oil) and is the most widely accredited diet that is associated with reduced CVD risk (de Lorgeril, Salen et al. 1999; Sofi, Cesari et al. 2008; Fung, Rexrode et al. 2009; Estruch, Ros et al. 2013). However, it is difficult to extrapolate whether these improvements are just attributable to dietary fat quality, as a range of other factors within this diet may be equally or more important (fruit/vegetables/nuts/fish-rich also). In a meta-analysis of over 12 studies, those consuming higher amounts of MUFA (>12% total energy) exhibited lower fat mass and reduced hypertension

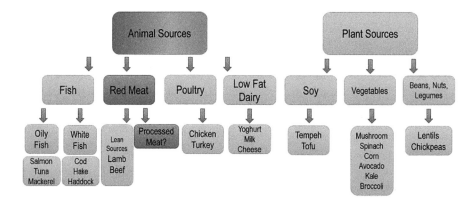

FIGURE 3.3 Animal and plant sources of protein.

(Schwingshackl, Strasser et al. 2011). Agreeing with these findings, Yu et al. analyzed over 18 studies and found that MUFA elicited a hypocholesterolemic effect on total cholesterol and LDL-C while increasing HDL-C (Yu, Derr et al. 1995). A previous meta-analysis examining isocaloric replacement of carbohydrate with MUFA similarly found augmented HDL-C in response to MUFA (Mensink, Zock et al. 2003). That notwithstanding, a pooled analysis of 11 cohort studies which included over 340,000 patients, by Jakobsen et al., found that there was no benefit to replacing SFA with MUFA by computer modeling (Jakobsen, O'Reilly et al. 2009). This may be in part due to the fact that MUFA-sources in the Western diet come primarily from animal sources (as opposed to plant) and this may confound results. The PREDIMED trial, for instance, exhibited a 30% reduction of CVD events in people who ate a diet higher in MUFA derived from olive oil/mixed nuts compared to a low-fat diet (Estruch, Ros et al. 2013).

Replacing SFA in the diet with PUFA has yielded the most promising results in terms of CVD prevention and management as it has been proven to lower circulating LDL-C (Hu and Willett 2002; Mozaffarian, Micha et al. 2010). A meta-analysis of the literature by Hu et al. has previously demonstrated that dietary replacement of 5% of carbohydrate with PUFA has the greatest impact on serum lipoprotein profiles compared to both SFA and MUFA alternatives with elevation in HDL-C and reductions in LDL-C predicted (Hu and Willett 2002). A combined analysis of the Nurses' Health Study and the Health Professionals Follow-up Study found a 20% CVD risk reduction with those in the highest quintile of PUFA intake compared with those in the lowest (Li, Hruby et al. 2015).

3.7 PROTEIN INTAKE AND CVD RISK

High-protein diets are particularly renowned for their satiating properties and in turn are effective in aiding weight-loss. The Atkin's diet, where protein and fat intake are increased while carbohydrate intake is reduced, is the classic example of a high-protein diet (Atallah, Filion et al. 2014). While reducing weight no doubt reduces CVD risk (Brown, Buscemi et al. 2016), high-protein diets are currently not recommended for CVD due to the potential that higher fat intake might have adverse effects (Hu, Stampfer et al. 1997; Hooper, Summerbell et al. 2001). In addition, there is a lack of understanding of long-term health consequences of high protein consumption. However, recent evidence from the PURE study, along with other recent epidemiological studies, has demonstrated no association or inverse association between dietary protein intake and CVD (Haring, Gronroos et al. 2014; Grasgruber, Sebera et al. 2016). The Nurse's Health Study (80,082 women), for instance, found that total protein (both animal and plant) inversely correlated with ischemic heart disease (Hu, Stampfer et al. 1999). Furthermore, randomized control trials have consistently shown that low-carbohydrate/high-protein diets reduce body weight and CVD risk (Dansinger, Gleason et al. 2005; Atallah, Filion et al. 2014; Hu and Bazzano 2014). Increasing dietary protein has also been shown in many clinical trials to increase HDL-C, and to a lesser extent, reduce TG levels (Foster, Wyatt et al. 2003; Dansinger, Gleason et al. 2005; Atallah, Filion et al. 2014; Hu and Bazzano 2014). While weight-loss likely contributes to these improved lipid profiles there is evidence of beneficial effects of high-protein diets beyond weight-loss effects. A study recruited 100 women who were assigned an isocaloric high-protein or high-carbohydrate diet. While weight loss was comparable between groups, those on the high-protein diet were found to have greater reduction in plasma TG than high-carbohydrate diet. Reductions observed in fasting LDL-C, HDL-C and CRP in both groups correlated to weightloss (Noakes, Keogh et al. 2005).

With respect to BP a number of randomized control trials found that high-protein diets decrease systolic and diastolic BP, in a weight-loss dependent manner, to a similar extent as isocaloric low-fat diets (Yancy, Olsen et al. 2004; Sacks, Bray et al. 2009; Hu and Bazzano 2014). However, high-protein diets have also been associated with exacerbated hypertension and renal damage but the data is conflicting (Juraschek, Appel et al. 2013). High protein intake was found to correlate with presence of microalbuminuria, a marker of renal dysfunction, in those with either type 1 and type 2 diabetes and hypertension compared to those with isolated diabetes or hypertension (Wrone, Carnethon et al. 2003). Overfeeding amino acids increases cellular exposure to advanced glycation end products, which is a potential mechanism linking protein intake to worsening diabetic kidney disease (Meek, LeBoeuf et al. 2013). Coincident with this finding, high-protein diets can increase glomerular cell death and drive inflammation (Meek, LeBoeuf et al. 2013). More recent findings indicate that patients with type 2 diabetes on a high-protein diet had improved metabolic outcome without negatively affecting renal function after 12 months (Pedersen, Jesudason et al. 2014). These findings indicate that high-protein diets can exert beneficial effects in certain populations (particularly obese/type 2 diabetes) who are exhibiting normal renal function, but care should be taken in patients with progressive renal damage and hypertension. Interestingly an eight-week study by Lopez-Legarrea et al. found in 96 participants that a high animal protein diet was positively associated with inflammatory markers CRP, IL-6, and TNF-α, but neither vegetable- nor fish-derived protein was found to influence inflammatory status (Lopez-Legarrea, de la Iglesia et al. 2014). These findings indicate the importance of dietary protein quality in CVD risk. Indeed in 2010, the Dietary Guidelines Advisory Committee's primary recommendation was to shift food intake to a more plant-based diet (Richter, Skulas-Ray et al. 2015).

3.8 CARBOHYDRATE INTAKE AND CVD RISK

Carbohydrates are found in a variety of foods and can be divided into four chemical subgroups: monosaccharides, disaccharides, oligosaccharides and polysaccharides. While mono- and disaccharides are commonly referred to as sugars, polysaccharides can be anything from an energy source like starch but also structural components like cellulose found in plants. As with protein, the quality of carbohydrate must be investigated to understand its influence on CVD risk factors; measures include glycemic index, dietary fiber, and glucose/fructose content. Wholegrain carbohydrate is associated with reduced CHD risk (Anderson, Hanna et al. 2000; Jacobs and Gallaher 2004); however, it is difficult to decipher what portion of the wholegrain is eliciting these beneficial responses as whole-grains are high in fiber, are a significant source of antioxidants (Chandrasekara and Shahidi 2011) and also often have a low glycemic load (GL).

3.8.1 GLYCEMIC INDEX/LOAD (GI/GL)

The glycemic index (GI) refers to the change in blood glucose levels in response to carbohydrate-containing food and is highly correlated with insulin response. Low-value GI foods are more slowly digested, absorbed and metabolized and as such cause a lower increase in blood glucose and a lower insulin response. Glucose Load (GL) is a ranking system for the amount of carbohydrate in food. While GI indicates how rapidly the carbohydrate is digested and released as glucose, it does not take into account the amount of carbohydrate in food and thus how a carbohydrate-based food will affect blood sugar (Siri-Tarino, Chiu et al. 2015). In the Nurse's Health Study, it was found that high GL was associated with an increased risk of CHD (Liu, Willett et al. 2000) with other meta-analyses confirming this finding, specifically in women (Dong, Zhang et al. 2012; Mirrahimi, de Souza et al. 2012). Concurrent with these findings, the OmniCarb study (finished in 2010, n=163) assessed the effect of low vs. high GI diets in the context of low and moderate carbohydrate diets. The increase in TGs and decrease in HDL-C seen in these patients was attributable to higher carbohydrate rather than higher GI (Sacks, Carey et al. 2014). While higher GI/GL foods have a negative impact on CVD and its associated risk factors, it appears low GI foods may correlate with lower disease risk (Jenkins, Wolever et al. 1985). Dietary fiber has a marked influence on the glycemic response to a carbohydrate meal. Low GI foods tend to have a higher proportion of fiber and it may be the fiber content influencing CVD outcomes.

3.8.2 FIBER

Dietary fiber is regarded as the non-digestible part of a plant and can be further classified as soluble and non-soluble. Soluble sources are legumes, oats and root vegetables while insoluble sources are wholegrain foods such as wheat and corn (Slyper 2013). Fiber can contribute to satiety and regularity and can also reduce overall fat and protein absorption and as a result, total caloric intake (Howarth, Saltzman et al. 2001). A study looking at 89,432 European adults found that there was an inverse association of total and cereal fiber intake and weight and waist circumference (Du, van der A et al. 2010). An analysis of ten cohort studies in 2004 found that for every 10g/day increment of dietary fiber, there was a 27% reduction in CVD mortality but also a 14% reduction in risk of CVD events (Pereira, O'Reilly et al. 2004). Corroborating this, a meta-analysis covering 14 studies found dietary fiber intake was moderately protective against CVD (Mente, de Koning et al. 2009). Specifically, water-soluble fibers (beta-glucan, psyllium) were the most effective at reducing LDL-C levels, without affecting HDL-C, with a greater effect seen in diabetic patients (Whitehead, Beck et al. 2014). Furthering this, the positive effects of fiber on CVD extends to glycemic control and management in both men and women (Hopping, Erber et al. 2010).

3.8.3 GLUCOSE/FRUCTOSE

The implications of sugar consumption on health remain a controversial subject. The low-fat era of the 1960s gave way to the rise of sugar-sweetened drinks and added sugars and coincided with the rise in obesity. It is important to note that sucrose and high fructose corn syrup (HFCS) (both frequently used sweeteners) are composed of glucose and fructose; sucrose is equal in its ratio of glucose and fructose while HFCS is composed of 55% fructose and 45% glucose (Kahn and Sievenpiper 2014). Increased sugar intake promotes a positive energy balance which leads to both weight gain and fat gain, leading to dysregulation in metabolism and with that the associated comorbidities, i.e. CVD and diabetes. In the absence of positive energy balance, Te Morenga et al. found no evidence of weight change in a meta-analysis of 12 studies that isoenergetic exchanges of dietary sugars with other carbohydrate sources or other macronutrients (Te Morenga, Mallard et al. 2012). However, despite no changes in weight, in the absence of positive energy balance, sugar consumption directly increases risk factors for metabolic disease. Using a cohort of over 10,000 adults from the National Health and Nutrition Examination Survey (NHANES), Yang et al. found those that consumed 10–25% calories from added sugar had a 30% higher risk of CVD mortality compared to those who consumed <10% calories from added sugar and those that regularly consumed sugar sweetened beverages ($\geq$7 servings/week) were at a higher risk of CVD, independent of other risk factors (de Koning, Malik et al. 2012; Yang, Zhang et al. 2014). Adding to this, another analysis of the same cohort noted an association of added sugars with lipid profiles, with incremental increases in added sugar correlating to decreases in HDL-C and increases in TGs (Welsh, Sharma et al. 2010). An intervention study in 48 adults (BMI 18–35kg/m^2) who received 25% energy requirements from glucose, fructose or HFCS in a beverage over two weeks demonstrated increased fasting non-HDL-C and ApoB concentrations after consumption of fructose and HFCS but not glucose (Stanhope, Bremer et al. 2011). Furthering this, a study using 47 overweight and obese patients who consumed a sucrose-sweetened soft drink over six months were found to have increased liver lipid, visceral adipose tissue (VAT) depot size and circulating TGs (Maersk, Belza et al. 2012; Bruun, Maersk et al. 2015). It is important to note that glucose and fructose are metabolized differently and it is these differences that may be associated with poor metabolic outcome (Softic, Gupta et al. 2017). Glucose metabolism is slower than fructose metabolism but is utilized by every cell in the body while fructose is mainly metabolized by the liver and strongly increases fatty acid synthesis enzymes (Softic, Gupta et al. 2017). Fructose has been shown to promote dyslipidemia, decrease insulin sensitivity and increase VAT, all risk factors for CVD (Stanhope, Schwarz et al. 2009; Stanhope, Bremer et al. 2011). Thus, sugar quality is also a likely contributor to modulation of traditional risk factors for CVD.

High-carbohydrate diets, which are associated with lower LDL-C, exhibit a positive correlation with proatherogenic small dense (sd)LDL particles in randomized controlled clinical trials (Hoogeveen, Gaubatz et al. 2014; Tsai, Steffen et al. 2014; Mora, Caulfield et al. 2015; Siri-Tarino, Chiu et al. 2015). One such study by Krauss et al. looked at a group of 178 mildly obese and overweight men who consumed a 26, 39 or 54% kcal from carbohydrate diet with 7–9% energy intake as SFA (low-SFA diet) and one group that consumed a 26% carbohydrate diet and 15% energy intake as SFA (high-SFA diet) for three weeks. They found a linear relationship of increased carbohydrate intake with increased prevalence of sdLDL particles. Those on the 26% carbohydrate, low-SFA diet had reduced TGs, ApoB, sdLDL particles and total HDL cholesterol compared to those on a 54% carbohydrate, low-SFA diet. Those on the 26% carbohydrate diet with high-SFA exhibited increased concentrations of medium- and large-sized LDL particles (Krauss, Blanche et al. 2006).

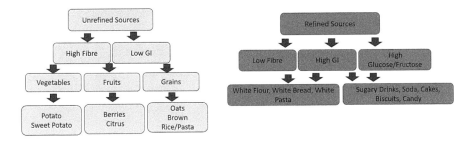

FIGURE 3.4 Carbohydrate sources, refined and unrefined.

As discussed above there are multiple different types of carbohydrates—some considered healthy (wholegrains, fiber) and others unhealthy (refined, fructose) and this makes it difficult to interpret epidemiology studies that pool carbohydrates into one category (Figure 3.4). A previous meta-analysis study by Hu et al. across 147 original research articles, however, summarized the association between major dietary factors, including fiber/wholegrains and CHD risk (Hu and Willett 2002). There has been a consistent inverse correlation observed between nut intake, fruit and vegetable intake and whole-grain intake and CHD across a multitude of epidemiological studies (Hu and Willett 2002). The largest study (84,251 women and 42,148 men) demonstrated that those in the highest quintile of fruit and vegetable intake had a relative risk of CHD of 0.8 compared to those in the lowest quintile (Joshipura, Hu et al. 2001). The Nurse's Health Study (75,521) similarly reported a significant reduction in CHD risk in the highest quintile of whole-grain intake (relative risk 0.67) compared to lowest quintile (Liu, Stampfer et al. 1999). These findings demonstrate the importance of accounting for carbohydrate quality within epidemiological studies and not just focusing on carbohydrate quantity.

3.9 DIETARY CHOLESTEROL AND PLANT STEROLS AND STANOLS AND CVD RISK

3.9.1 Dietary Cholesterol

Systemic cholesterol levels unarguably correlate to CVD; however, whether dietary cholesterol impacts on lipoprotein parameters has been rather elusive. It has been found that relative to fatty acid composition, dietary cholesterol has minimal influence on systemic levels of cholesterol (Oliver 1976; Pyorala 1987; Adams 2011). Cholesterol *de novo* lipogenesis which occurs in the liver (under the control of enzyme HMG-CoA reductase) is primarily responsible for systemic cholesterol levels (Brown and Goldstein 1984; Brown and Goldstein 1986).

3.9.2 Plant Sterols and Stanols

While dietary cholesterol contributes minimally to systemic cholesterol concentrations, cholesterol re-absorption in the gut is extremely efficient with up to 98% reabsorption of bile-salts (derived from hepatic cholesterol) and 25–80% reabsorption

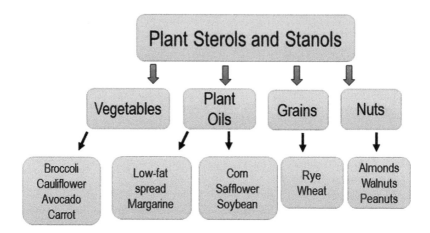

FIGURE 3.5   Plant sterol and stanol sources.

of free cholesterol within the lumen (Cohen 2008). Sterols and stanols are bioactive components found in plants that function similarly to cholesterol in mammals (Figure 3.5). Plant sterols are steroid alkaloids and are structurally different from cholesterol at their side chain, whereas plant stanols are 5α-saturated derivatives of plant sterols (Gylling, Plat et al. 2014). Taking 2g/day of plant sterols/stanols inhibits cholesterol absorption and lowers LDLC levels up to 10%, additive to statin treatment (Blair, Capuzzi et al. 2000; Musa-Veloso, Poon et al. 2011). In the intestine sterols are transported into the enterocyte (intestinal wall) via Niemann–Pick C1-Like1 transporter (NPC1L1). Plant sterols and stanols are not taken up by NPC1L1 and are pumped back to the intestine by ATP-binding cassette (ABC) G5/8. Plant sterols and stanols are believed to inhibit cholesterol absorption in the intestine but the exact mechanisms are not fully understood (Gylling, Plat et al. 2014). The mechanism of plasma LDL-C lowering is by blocking absorption of cholesterol and allows more to be excreted into the feces and also increases the hepatic expression of the LDL-receptor (LDLR) (Gylling, Plat et al. 2014). There is some concern over the accumulation of plant sterols and stanols in tissues, particularly arterial tissue and large, long-term clinical studies are needed to assess if they can modulate plaque regression or progression. Plant sterols and stanols should not be added to the diet of those with phytosterolemia who have a loss of function mutation in ABCG5/ABCG8 due to accumulation of plant sterols in circulation (Kidambi and Patel 2008).

3.9.3 Salt Intake and CVD Risk

Limiting salt intake, particularly in patients with hypertension, remains an important target for optimal nutrition to prevent CVD. In 1988, the international study INTERSALT (n=10,079) analyzed salt intake and BP from 52 centers around the world and showed that across centers, high salt intake was associated with increased BP with age (p<0.001) while within-center analysis revealed a significant association between salt intake and median BP (p<0.001) (1988). Re-analysis of the data in 1996 confirmed significant association between salt intake and BP within populations and revealed that increased sodium intake by over 100 mmol/24 h is associated with a mean increase of between 3/0 and 6/3 mmHg (Elliott, Stamler et al. 1996). While these epidemiologic studies reinforce the association of salt intake with hypertension, multiple interventions have shown a causal effect. The Dietary Approaches to Stop Hypertension (DASH)-Sodium trial (n=412) showed reduction in systolic BP (-2.1 mmHg, p<0.001) when patients were shifted from a high sodium diet (150 mmol/day) to a normal sodium diet (100 mmol/day) and a further reduction (−4.6 mmHg, p<0.001) when shifted from the normal sodium diet to a low sodium diet (50 mmol/day) (Sacks, Svetkey et al. 2001). The Trials of Hypertension Prevention (TOHP) phase I (n=2,182) aimed to determine the effects of lifestyle changes on BP with an 18 month follow up. Participants in the sodium reduction group had decreased urinary sodium excretion (−44 mmol/24h, p<0.01) after 18 months, which resulted in significantly lower diastolic (−0.9 mmHg, p<0.05) and systolic (1.7 mmHg, p<0.01) BP (1992). A TOHP phase II trial (n=2,382) was also conducted with the sodium reduction group displaying significantly reduced systolic BP (−1.2 mmHg, p<0.05) after 36 months (1997). Not only does lowering salt intake have beneficial effects on BP, it also decreases the risk of cardiovascular events, as shown by a 10–15 year follow-up study of TOHP I and II where the intervention group had a 25% risk reduction for cardiovascular events (p<0.05) (Cook, Cutler et al. 2007). The above studies are consistent with the current WHO guidelines that suggest reducing salt intake from 9–12 g/day to less than 5 g/day.

3.10 PN1: HEALTHY DIET RECOMMENDATIONS FOR PATIENTS WITH, OR AT RISK OF, CVD

The findings from both epidemiological and interventional studies indicate that macronutrient quality over macronutrient quantity is a critical factor to consider when prescribing a diet to manage CVD. Based on the findings from the epidemiological/interventional studies discussed, several conclusive points emerge.

- Reduce (not eliminate) saturated fat in the diet and replace with unsaturated, non-hydrogenated forms. Keep saturated fat consumption below 10% of total kcal.
- Increase consumption of n-3 PUFA from plant and fish sources.
- Increase consumption of fruits and vegetables.
- Carbohydrate consumption should be of high quality: wholegrain, high in fiber and low in refined products.
- Include functional foods with meals such as plant sterol/stanol-containing spreads and drinks (unless genetically deficient for ABCG5/8).
- Consume foods low in salt content.
- Avoid a positive energy balance by adhering to recommended caloric intake.
- In overweight/obese patients encourage weight-loss promoting diets with beneficial effects evident even at 5–10% weight-loss.

- Encourage routine physical activity to prevent a positive energy balance and promote a healthy lifestyle.
- Quit tobacco smoking and reduce/avoid alcohol consumption.

3.11   PN2: TAILORING DIETARY ADVICE TO MANAGE CVD BASED ON A PERSON'S PHENOTYPE

Our ability to implement PN2 to prevent CVD based on traditional risk factors and a person's phenotype is quite advanced at this point. Appropriate advice is surmised in Table 3.1 and should be prescribed in addition to appropriate pharmacological therapy. It is anticipated that other phenotypic traits including metabotype and proteomic profile may also be incorporated into personalization of nutritional advice but more research into this area is necessary to guide appropriate nutritional advice.

3.12   PN3: TAILORING NUTRITIONAL ADVICE TO GENETIC COMPOSITION: FROM CONCEPT TO REALIZATION

PN3 is the ultimate level of personalization in that a person's dietary advice is tailored to their individual genotypic data (Gibney and Walsh 2013). It is hoped through the increased understanding of disease-related SNPs that we can identify at risk

TABLE 3.1
Action Table of Proposed Dietary Advice

Traditional Risk Factor	Primary Target	Secondary Target	Nutritional Intervention
High LDL-C	<3 mmol/L	<2.5 mmol/L	↓ Total fat <30% (Kcal); SFA ≤10%, <1% Trans-fat. Replace SFA for MUFA or PUFA (Omega-3 >1g/d.). Consume 2–3 g/d of phytosterols with meals (Except in those with defective ABCG8). ↓ Sugar intake particularly from fructose and HFCS.
High Triglycerides	<2.5 mmol/L	<1.7 mmol/L	↓ Carbohydrate <40–50% (Kcal), particularly from simple/ refined sources. Replace simple carbohydrates (sugar, fructose and HFCS) for whole-grain sources.
Low HDL-C		>1 mmol/L (M) >1.3 mmol/L (F)	↓Simple carbohydrate intake and replace SFA for MUFA and PUFA in diet (<30% kcal from fat).
BP	<140/90 mm Hg		↓ Salt intake from sodium to <5 g/day, use potassium instead and ↑ fiber intake to 30–45 g; ≥ 200 g fruit (2–3) + vegetables.
Overweight/ obese	BMI <25 and/or 10% weight reduction		Avoid a positive energy balance, increase consumption of high-protein foods to promote satiety. Reduce saturated fat and refined carbohydrate consumption. Increase fruit and vegetable intake.

individuals and intervene accordingly. Furthermore, certain genotypes may exhibit greater predisposition to respond to one diet over another. A classic example of personalization based on genotype is in patients who have familial hypercholesterolemia (FH), caused by carrying one (heterozygous FH) or two (homozygous FH) mutant alleles of the LDLR gene (MIM 606945) (Cuchel, Bruckert et al. 2014). FH is a common hereditary disorder, thought to occur between 1/500 and 1/200 in most countries (Rader, Cohen et al. 2003; Brice, Burton et al. 2013). It is caused by a defect in the hepatic LDL receptor gene resulting in severely elevated LDL-C at a young age and enhanced atherosclerosis (Austin, Hutter et al. 2004). Early diagnosis and therapy with lipid-lowering medications significantly reduces risk of CVD events (van der Graaf, Avis et al. 2011). Adherence to a low-fat diet is also critical for patients with FH, who are recommended to consume <3% saturated fat in their diet (Connor and Connor 1989; Cuchel, Bruckert et al. 2014). In this section of the book chapter we will review our current understanding of genetics and CVD and some examples of gene-nutrient interactions that are relevant for CVD.

3.13   GENETICS AND CVD: TESTING THE NATURE VERSUS NURTURE HYPOTHESIS

The intricate link between multiple lifestyle-associated risk factors and CVD indicate the importance of maintaining a healthy diet for CVD management but there is also an irrefutable genetic component to CVD (Khera and Kathiresan 2017). Clinical observation from 1950s have supported the opinion that CVD is heritable (Gertler, Garn et al. 1951). A study of ~20,000 twins in Sweden confirmed this and showed increased CVD risk for those with close relatives affected by the disease and estimated that heritable factors account for 30–60% of the interindividual variation in CVD risk (Marenberg, Risch et al. 1994; Zdravkovic, Wienke et al. 2002). The Framingham Heart Study similarly showed that having a parent or sibling with CVD was a strong predictor of the incidence of disease (Lloyd-Jones, Nam et al. 2004; Murabito, Nam et al. 2004). However, to date it has been poorly understood whether poor lifestyle further exacerbates CVD in high-risk individuals. In updated GWAS, it is estimated that heritability of CVD is 40–50% (Won, Natarajan et al. 2015). Many researchers once assumed that genetic predisposition is deterministic (White 1957) but in fact an individual's risk of developing CVD is a dynamic interplay between genetic and lifestyle factors (Khera, Emdin et al. 2016). Thus, lifestyle may accentuate genetic susceptibility, or not, as the case may be. A cornerstone study by Khera et al. investigated to what extent increased genetic risk can be offset by healthy lifestyle (Khera, Emdin et al. 2016). The study involved 55,685 participants across three prospective cohorts where adherence to healthy lifestyle was determined by using a scoring system consisting of four factors: no current smoking, no obesity (BMI <30), a physical activity at least once weekly and consumption of a healthy diet. The polygenic risk score was derived from analysis of 50 SNPs that achieved genome-wide significance for association with CVD (Consortium, Deloukas et al. 2013). A healthy diet pattern was determined on the basis of adherence to at least half of the following (Mozaffarian 2016); high consumption of fruits, nuts, vegetables, whole grains, fish and dairy products and low intake of refined carbohydrates, processed meats, unprocessed red meats, sugar-sweetened beverages, trans-fats and sodium. The relative risk of CVD events was ~91% higher among participants at high genetic risk than among those at low genetic risk. Each of the four healthy lifestyle factors was associated with decreased risk of CVD events. An unfavorable lifestyle was associated with a higher risk of CVD than a favorable lifestyle (adjusted hazard ratio 1.71). Adherence to a favorable lifestyle, as compared to an unfavorable lifestyle, was associated with an ~45% reduction in CVD events among participants at low, intermediate and high genetic risk. This study showed that within any genetic risk category, adherence to a healthy lifestyle significantly decreased risk of both CVD events and subclinical burden of CVD. This is evidence that lifestyle factors may powerfully modify CVD risk regardless of a patient's genetic predisposition (Khera, Emdin et al. 2016).

3.14   GWAS AND CVD

The scientific community is only at the beginning of understanding the complexity of gene regulation. In turn, the nutrition field is in its infancy in terms of understanding how environmental factors, including diet, are overlaid on genotypic data to affect disease outcomes. GWAS is a method that searches for small variations in the genome called SNPs that occur at a higher frequency in people with a particular disease than those without the disease (Bush and Moore 2012). SNPs are single base-pair changes in the DNA sequence (Genomes Project, Abecasis et al. 2010) and occur at high frequency in the human genome, although not all have a biological impact (Genomes Project, Abecasis et al. 2010). In general, in rare disorders that are inherited, the affected gene tends to have a large effect size or biological effect; however, common disorders are likely influenced by genetic variation that is common in the population. Thus, the effect size of common genetic variants is small, relative to those found in rare disorders (Bush and Moore 2012). This common disease/common variant hypothesis led to the use of polygenetic risk scores (GRS) whereby a considerable proportion of phenotypic variation can be explained by grouping together SNPs (Dudbridge 2013). Modest effect sizes of SNPs have the potential to hinder prediction of risk using a single SNP, therefore grouping SNPs together give greater predictability of an individual's risk of a disease (Lewis and Vassos 2017). Boyle et al., have further hypothesized that for complex traits, association signals are spread out across most of the genome including near many genes that have no known connection to the disease. They have established an 'omnigenic' model whereby it is hypothesized that gene regulatory networks are sufficiently interconnected that all genes expressed in disease-relevant cells are liable to affect the functions of core disease-related genes (Boyle, Li et al. 2017). Hence, the multitude of SNPs that have been identified as associating with CVD, but not with any known risk factor, may be affecting the function of disease-related genes.

Before GWAS studies, Mendelian diseases provided insights into the causal genes/markers involved in CVD and helped with understanding the traditional risk factors for CVD. Mendelian diseases are caused by a single variant in a single gene and include FH, cardiomyopathies, congenital heart disease and primary arrhythmias (Kalayinia, Goodarzynejad et al. 2017). The findings from GWAS have both identified/confirmed some known risk SNPs but also identified many new SNPs that are not associated with traditional risk factors (Consortium, Deloukas et al.

2013). Meta-analysis was carried out on GWAS from multiple consortia including the Myocardial Infarction Genetics Consortium (MIGen) (Myocardial Infarction Genetics, Kathiresan et al. 2009), the Coronary Artery Disease Genetics Consortium (C4D) (Coronary Artery Disease Genetics 2011) and the Coronary Artery Disease Genome-Wide Replication and Meta-analysis (Cardiogram) (Schunkert, Konig et al. 2011) and multiple other meta-analyses with the UK Biobank (Klarin, Zhu et al. 2017; Nelson, Goel et al. 2017; Verweij, Eppinga et al. 2017) to identify SNPs that significantly associated with CVD and replicated across consortia. Ninety-seven common and reproducible variants were identified that were associated with CVD. One of the first identified was on chromosome region 9p21.3 which contained multiple SNPs with a strong association with CVD (Helgadottir, Thorleifsson et al. 2007; McPherson, Pertsemlidis et al. 2007; Samani, Erdmann et al. 2007). Surprisingly these SNPs were not associated with traditional risk factors such as plasma lipids, BP, diabetes, older age or obesity and were located in a non-coding region so the mechanism underlying enhanced risk remains unclear. The minor risk allele of 9p21.3 occurs at high frequency, with up to 50% frequency in European populations, and confers an ~1.3-fold increase in risk (Assimes, Knowles et al. 2008; Schunkert, Gotz et al. 2008). Of the 97 common variants, many are linked to lipoprotein metabolism and hypertension; however, such as in the case of 9p21.3, a large number are within gene regions that have not been previously linked to CVD risk or pathogenesis. A further 64 novel loci were identified in 2018, bringing the number to 161, so this number is constantly evolving (van der Harst and Verweij 2018). Table 3.2 encompasses known SNPs associated with CVD to date.

3.15   PN3: PROOF OF CONCEPT FROM LIPGENE

There is major potential for nutrition research and primary care to use GWAS data for more in-depth PN approaches to treat CVD; however, there are multiple hurdles that must be overcome to translate this into a clinical reality. A greater understanding of potential gene-nutrient interactions on newly identified SNPs is essential to implement PN3 and therefore much research is required in this field. Gene-nutrient interactions and implications for CVD have been comprehensively reviewed by Corella et al. (Corella, Coltell et al. 2017). A proof of concept study that has explored diet-nutrient interactions in patients with the metabolic syndrome was the pan-European LIPGENE study. LIPGENE was a multicenter, single-blind, randomized controlled trial that aimed to: 1) determine the effect of reducing SFA consumption by altering quality of dietary fat and reducing quantity of dietary fat on the metabolic syndrome and its associated risk factors; and 2) determine if common genetic polymorphisms affect an individual's responsiveness to dietary therapy. The study included eight European centers and 486 volunteers with the metabolic syndrome who were assigned to one of four dietary interventions: 1) high-fat (38% energy) SFA-rich diet; 2) high-fat (38% energy), MUFA-rich diet; 3) isocaloric low-fat (28% energy), high complex carbohydrate diet and 4) isocaloric low-fat (28% energy), high complex carbohydrate diet with 1g/d LC n-3 PUFA for 12 weeks (Shaw, Tierney et al. 2009; Tierney, McMonagle et al. 2011). LIPGENE attempted to use genotype information of individuals to investigate responsiveness of singular SNPs to dietary

TABLE 3.2
Known SNPs Associated with CVD

Gene	Lead SNP	Gene	Lead SNP	Gene	Lead SNP	Gene	Lead SNP	Gene	Lead SNP	Gene	Lead SNP	Gene	Lead SNP
SORT1	rs599839	ZNF827	rs35879803	GOSR2	rs17608766	CYP17A1	rs12413409	MAP3K7CL, BACH1	rs2832227	UMPS–ITGB5	rs142695226	CDKN1A, PI16	rs1321309
TDRKH	rs11810571	GUCY1A3	rs7692387	PCNX3	rs12801636	FURIN/FES	rs17514846	CALCRL, TFPI	rs840616	ARHGEF26	rs12493885	PLEKHG1, IYD	rs17080091
ATP1B1	rs1892094	CFDP1	rs3851738	HOXC4	rs11170820	PECAM1	rs1867624	CDC123, NUDT5, OPTN	rs61848342	CDKN2BAS	rs3217992	CTTNBP2, CFTR-ASZ1	rs975722
KCNE2	rs9982601	CXCL12	rs2047009	C1S	rs11838267	FNDC3B	rs12897	MAP3K1, MIER3	rs3936511	DAB2IP	rs885150	TMEM106B, THSD7A	rs11509880
LMOD1	rs2820315	CXCL12	rs501120	SWP70	rs10840293	SMG6	rs216172	ALS2CL, RTP3	rs7633770	HSD17B12	rs7116641	PRDM8-FGF5	rs10857147
PPAP2B	rs17114036	ANKS1A	rs17609940	PDGFD	rs974819	UBE2Z	rs46522	HNRNPD, RASGEF1B	rs11099493	N4BP2L2, PDS5B	rs9591012	N4BP2L2, PDS5B	rs9591012
ZFPM2	rs10093110	ApoE-ApoC1	rs2075650, rs445925	ZNF831	rs260020	ZNF507-LOC400684	rs12976411	KLF4	rs944172	STAG1, MSL2, NCK1, PPP2R3A	rs667920	DNAJC13, NPHP3, ACAD11, UBA5	rs10512861
IL6R	rs4845625	PHACTR1	rs12526453	APOA5	rs964184	BCAS3	rs7212798	CCNL1, TIPARP	rs4266144	HGFAC, RGS12, MSANTD1	rs16844401	SEMA5A, TAS2R1	rs1508798
MIA3	rs17465637	TCF21	rs12190287	RASD1	rs12936587	ADAMTS7	rs3825807	COPRS, RAB11FIP4	rs769954792	RP11-664H17.1	rs10841443	AGT, CAPN9, GNPAT	rs699
FIGN	rs12999907	ARNTL	rs3993105	DHX38	rs1050362	ADAMTS7	rs7173743	C12orf43/HNF1A	rs2258287	MAD1L1	rs10267593	MAD2L1, PDE5A	rs11723436
ABCG8	rs6544713	LPA	rs379822	SCARB1	rs11057830	SNRPD2	rs1964272	PLCG2, CENPN	rs7199941	FAM46A	rs4613862	MFGE8-ABHD2	rs8042271
FN1	rs1250229	LPA	rs10455872	HNF1A	rs2244608	POM121L9P-ADORA2A	rs180803	SLC22A3/LPAL2/LPA	rs2048327	ZNF259-APOA5-APOA1	rs7173743	CORO6,§ ANKRD13B, GIT1, SSH2, EFCAB5	rs13723
TNS1	rs2571445G	PLG	rs4252120	PMAIP1-MC4R	rs663129	HDGFL1	rs6909752	TRIM5, TRIM22, TRIM6, OR52N1OR52B6	rs11601507	STN1, SH3PXD2A	rs4918072	BMP1, SFTPC, DMTN, PHYHIP§DOK2, XPO7	rs6984210

(Continued)

TABLE 3.2 (CONTINUED)
Known SNPs Associated with CVD

Gene	Lead SNP	Gene	Lead SNP	Gene	Lead SNP	Gene	Lead SNP	Gene	Lead SNP	Gene	Lead SNP	Gene	Lead SNP
APOB	rs515-35	PARP12	rs10237377	HNF1A	rs2259816	PROCR	rs867186	PALLD, DDX60L	rs7696431	HTRA1, PLEKHA1	rs4752700	SERPINA2, SERPINA1	rs112635299
COL6A3	rs11677932	HDAC9	rs2023938	SH2B3	rs3184504	SERPINH1	rs590121	PRDM16, PEX10, PLCH2	rs2493298	ARHGAP26	rs246600	CDC25A, SPINK8, MAP4, ZNF589	rs7617773
WDR12	rs6725887	BCAP29	rs10953541	FLT1	rs9319428	EDNRA	rs1878406	PDE5A–MAD2L1	rs7678555	ARHGAP42	rs7947761	ALS2CL, RTP3	rs7633770
FN1	rs175-7928	ZC3HC1	rs11556924	COL4A1	rs4773144	CDH13	rs7500448	DX59/CAMSAP2	rs6700559	CCM2, MYO1G	rs2107732	KIAA1462	rs2505083
RHOA	rs7623687	NOS3	rs3918226	PCSK9	rs11206510	MORN1,SKI	rs36096196	GGCX/VAMP8	rs1561198	ARID4A, PSMA3	rs2145598	LIPA	rs2246833
LDLR	rs1122608	LPL	rs264	HHIPL1	rs2895811	SGEF, DHX36	rs433903	ZEB2-AC074093.1	rs2252641	NDUFA12, FGD6	rs7306455		rs17581137
UNC5C	rs3775058	TRIB1	rs2954029	OAZ2, RBPMS2	rs6494488	HHATSERTAD4, DIEXF	rs60154123	MCF2L,*§ PCID2,‡ CUL4A‡	rs1317507	VEGFA, MRPL14, TMEM63B	rs6905288	TSPAN14, MAT1A, FAM213A	rs17680741
MRAS	rs9818870	TGFB1	rs8108632	SMAD3	rs56062135	NAT2	rs6997340	PRKCE, TMEM247	rs582384	ACAA2, RPL17	rs9964304	PCIF1, ZNF335, NEURL2, PLTP	rs3827066
FGD5	rs748-31	CDKN2A	rs4977574	CENPW	rs1591805	COL4A1/COL4A2	rs9515203	FHL3, TP11, SF3A3, MANEAL, INPP5B	rs61776719	MAP1S, FCHO1, COLGALT1	rs73015714	PRIM2, RAB23, DST, BEND6	rs9367716
KCNK5	rs109-7789	ABO	rs579459	FOXC1	rs9501744	NGF, CASQ2	rs11806316	SHROOM3, SEPT11, FAM47E, STBD1	rs12500824	ZHX3, PLCG1, TOP1	rs6102343	DHX58, AT2A, RAB5C, NKIRAS2, DNAJC7, KCNH4, HCRT, GHDC	rs2074158

Adapted from Khera et al. (2016) *N Eng J Med* 375: 24, Argam and Kathiresan *Genetics of Coronary Atherosclerosis* 2017 and van der Harst, P. and N. Verweij (2018) "Identification of 64 Novel Genetic Loci Provides an Expanded View on the Genetic Architecture of Coronary Artery Disease." *Circ Res* 122(3): 433–434.

fat on markers of the metabolic syndrome. For example, individuals with the minor allele at rs1799983 in the nitric oxide synthase 3 (NOS3) gene were shown to have a greater TG-lowering response to n-3 PUFA than major allele carriers (Ferguson, Phillips et al. 2010). Individuals with this genotype in turn were projected to yield the greatest benefit from increasing PUFA intake. By contrast, individuals with the minor alleles of rs266729 and rs10920533 SNPs in ADIPOQ and in ADIPOR1 were particularly sensitive to insulin resistance in response to SFA consumption (Ferguson, Phillips et al. 2010). It was therefore speculated that individuals carrying these minor alleles would be particularly responsive to SFA-lowering to improve insulin sensitivity. These findings are excellent proof of concept of how PN3 might be realized. However, a lot of SNPs studied in previous nutrigenomic studies have not been validated in recent GWAS for CVD and go beyond the scope of this book chapter. Table 3.3 is a review of some nutrigenomics studies carried out on validated SNPs related to traditional risk factors for CVD.

3.16   PN3: THE CHALLENGE OF KEEPING UP WITH GWAS

A major challenge and indeed opportunity for the nutrition field is the catch-up phase after the identification and validation of CVD-related SNPs from GWAS. What research design will be implemented for such studies? Do we go for a simple approach assessing one SNP at a time with a few different diets? Do we standardize such interventions globally so cross-comparisons can be made between studies? What will our primary outcomes be? Do we continue to use nutrient-modifiable risk factors such as TG and HDL-C as read-outs for CVD risk despite lack of evidence proving CVD causality? Do we conduct much longer interventions and actually measure CVD events? If so, how will we ensure compliance and what biomarkers will we use? If we were to implement multiple dietary intervention studies with each SNP currently validated, there would be too many permutations for quite unpredictable gain. It is also noteworthy that many of these SNPs occur simultaneously in the same individual which further complicates this reductionist SNP by SNP approach. Indeed, the presentation of multiple SNPs within an individual form the basis of calculating a GRS which in turn could be exploited for PN3. Precedence for this approach come from pharmacogenomics studies—Natarajan et al., for example, demonstrated that individuals at highest genetic risk (top quintile for GRS) yielded the greatest CVD risk reduction in response to statin therapy than lower risk groups (Natarajan, Young et al. 2017). Using GRS as the selection criteria for nutrition intervention studies is certainly attractive and offers the advantage of analyzing the contribution of individual SNPs to phenotypes observed by *post hoc* analysis. On the other hand, there is a risk that data relating to specific gene-nutrient interactions will be diluted/underpowered by such an approach. It is critical that nutritional intervention studies are conducted in this postGWAS era, as despite the large number of SNPs/GRS that have been identified, they only explain small percentage of heritability. It is suggested that gene-environment interactions could explain the 'missing heritability' and this is a critical question to explore in order to understand disease pathogenesis. In addition, greater understanding of gene-nutrient interactions is an important precedent for the implementation of PN3 (Corella, Coltell et al. 2017).

TABLE 3.3
Example of Nutrigenomic Studies Carried Out on GWAS-Validated SNPs That Are Associated with CVD

Gene	Identified Study	Number of Subjects	Dietary Intervention and Results
Apolipoprotein A1 (APOA1)	Polyunsaturated fatty acids modulate the effects of the APOA1 G-A polymorphism on HDL-cholesterol concentrations in a sex-specific manner: the Framingham Study (Ordovas, Corella et al. 2002).	1577	The Framingham Study provided data on lipid phenotypes, dietary intake, DNA genotypes and CVD risk factors of the participants. Significant gene-nutrient interactions were only observed in women GG carriers with a PUFA intake of <4% of energy had a 14% greater HDL-C than A allele carriers. A allele carriers with a PUFA intake of >8% had 13% higher HDL-C compared to GG carriers.
Apolipoprotein E (APOE)	Effect of apolipoprotein E polymorphism on serum lipid response to the separate modification of dietary fat and dietary cholesterol (Sarkkinen, Korhonen et al. 1998).	75	3 APOE genotype groups (3/3, 3/4, 4/4) were established (n=15 mild hypercholesterolemic subjects in each group). Three separate diets were administered to all participants in identical order; 1) baseline diet A: 38% fat, 300 mg cholesterol), 2) diet B: 34% fat, 264 mg cholesterol and 3) diet C: 34% fat, 566 mg cholesterol. The greatest reduction in total cholesterol was in Apo E 4/4 group that adhered to diet B (−14.1%). The greatest increase in total cholesterol was observed Apo E 4/4 group that adhered to diet C (10.4%).
Lipoprotein lipase (LPL)	Genetic variations at the lipoprotein lipase gene influence plasma lipid concentrations and interact with plasma n-6 polyunsaturated fatty acids to modulate lipid metabolism (Garcia-Rios, Delgado-Lista et al. 2011).	1754	The plasma fatty acid composition and plasma lipid concentrations of metabolic syndrome patients from the European LIPGENE were examined. Carriers of the rs328variant of the LPL gene with low plasma n-6 PUFA had higher non-esterfied fatty acid concentrations than major allele homozygotes.

(Continued)

TABLE 3.3 (CONTINUED)
Example of Nutrigenomic Studies Carried Out on GWAS-Validated SNPs
That Are Associated with CVD

Gene	Identified Study	Number of Subjects	Dietary Intervention and Results
LDLR	Reducing saturated fat intake is associated with increased levels of LDL receptors on mononuclear cells in healthy men and women (Mustad, Etherton et al. 1997).	25	Participants were randomly assigned to three diets: 34% fat (15% saturated fat), 29% fat (9% saturated fat) and 25% fat (6% saturated fat). LDLR were measured in mononuclear cells after six to eight weeks. LDLRs increased by 10.5% after the diet lowest in saturated fat.
PCSK9	Marine n-3 polyunsaturated fatty acids lower plasma proprotein convertase subtilisin kexin type 9 levels in pre- and postmenopausal women: A randomised study (Graversen, Lundbye-Christensen et al. 2016).	92	Women were randomly assigned to a diet containing either: 2.2 g marine n-3 PUFA or a control oil (thistle oil) for three months. Adipose tissue and blood samples were measured pre- and post-intervention. The premenopausal and postmenopausal women supplemented with 2.2 g marine oil showed reduced PCSK9 plasma levels by 11.4% and 9.8% respectively compared with the women supplemented with the control oil.
PCSK9	Docosahexaenoic Acid Attenuates Cardiovascular Risk Factors via a Decline in Proprotein Convertase Subtilisin/Kexin Type 9 (PCSK9) Plasma Levels (Rodriguez-Perez, Ramprasath et al. 2016).	54	Metabolic syndrome patients were randomly assigned diets containing beverages supplemented with three different oils: canola oil, high DHA canola oil and high oleic acid canola oil. Participants who consumed high oleic acid oil showed a reduction in plasma concentrations while those who consumed high DHA oil showed a reduction in plasma PCSK9 and TG levels.

A futuristic model of PN is depicted in Figure 3.6 which currently has a gene-centric approach; however, as our understanding of genetics and disease evolves, so too will this model of PN. Indeed, it is quite plausible that implementation of PN2 based on metabotype may prove more feasible than PN3 based on genotype. Ideally, a patient's genetic risk for CVD should be calculated early in life to allow

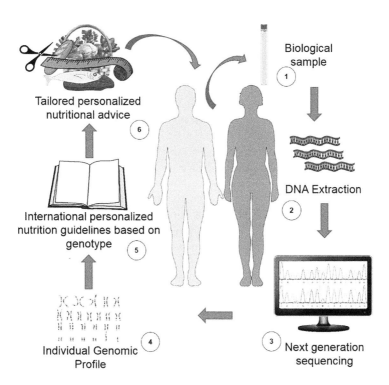

FIGURE 3.6 The future potential for the role of personalized nutrition advice for management/prevention of CVD. 1) collect a biological sample from the population early in life; 2) Extract DNA; 3) Perform next generation sequencing to identify disease-related SNPs; 4) Map out an individual's SNP/genome profile; 5) Refer to a set of international guidelines for nutritional advise on SNPs; 6) Prescribe a personalized diet plan based on individual SNP profile. Disseminate this information from a technological platform to the end-user.

appropriate lifestyle intervention therapies to be implemented to reduce disease burden in later life. Such a population-based approach is currently limited by financial resources (Phillips, Pletcher et al. 2015), infrastructure limitations and lack of staff with appropriate training to interpret the resulting data. If such testing is made available to everyone, will the benefits outweigh the cost, by promoting healthier eating, improving patient outcomes and reducing national expenditures on healthcare? There is huge potential and scope for the use of PN3 for the management/prevention of CVD; however, there is a critical need for well-designed clinical trials, to test the interactions of individual nutrients on the SNPs associated with CVD, on hard clinical endpoints and associated risk factors.

3.17   PERSPECTIVE ON THE FUTURE OF PN FOR CVD: IS SOCIETY READY TO EMBRACE CHANGE?

Prior to delving into expensive large-scale nutritional intervention studies to better tailor PN3 advice, it is critical to understand if the information acquired by these

trials will be harnessed by the general population and motivate them to change/
adapt their eating patterns to improve health. In Food4Me, individuals (n=1269)
were randomized to follow conventional dietary advice or PN based on individual
baseline diet (PN1), individual baseline diet plus phenotype (anthropometry and
blood biomarkers) (PN2) or individual baseline diet plus phenotype plus genotype
(five dietresponsive genetic variants) (PN3). Following six months' intervention, participants randomized to PN consumed less red meat, salt and saturated fat, increased
folate intake and had a higher Healthy Eating Index score compared to the control
arm (Celis-Morales, Livingstone et al. 2015). However, the addition of phenotypic
and genotypic information did not enhance the effectiveness of PN. That notwithstanding, the population in question were not selected due to predisposition to any
given disease and were selected from the general population who volunteered and
therefore may have been biased towards motivated people who did not gain any
additional motivation by PN2/3. Similarly, a personalized intervention trial (PRO-
FIT) in individuals with FH (n=349) demonstrated no additional benefit of personalized intervention (web-based tailored lifestyle advice and personal counseling)
on plasma lipids, blood pressure, glucose, body mass index or waist circumference
after 12 months compared to standard care group (Broekhuizen, van Poppel et al.
2010). Contrary to a healthy population, individuals with FH already have height-
ened awareness of their disease which in turn needs intensive medical management.
Hence, lack of efficacy, particularly on plasma lipids, may have been due to lack of
additional benefit of lifestyle intervention beyond medical management. In addition,
the control arm of this study would have been acutely aware of their extremely high-
risk status and in turn likely consumed a low-saturated fat diet.

The greatest benefit for PN is likely to be harnessed by individuals who are currently unaware of their genetic risk status but who with such awareness, could adapt
appropriate lifestyle changes to yield positive health benefits. There is much research
to be conducted in the nutrition field to fully understand specific gene-nutrient interactions that will both enhance our understanding of disease pathogenesis as well as
better inform dietary advice to high-risk individuals. In the interim, knowledge of
risk status in early life coupled with PN1 platforms has great potential to revolutionize CVD management.

ABBREVIATIONS

Apo:	Apolipoprotein
ABC G5/G8 (ABCG5/ABCG8):	ATP-binding cassette
BP:	Blood pressure
CVD:	Cardiovascular disease
CHD:	Coronary heart disease
CRP:	C-Reactive Protein
DHA:	Docosahexenoic acid
EPA:	Eicosapentenoic acid
FH:	Familial hypercholesterolemia
GRS:	Genetic risk score
GWAS:	Genome wide association studies

GI:	Glycemic index
GL:	Glycemic load
HDL-C:	High density lipoprotein cholesterol
HFCS:	High fructose corn syrup
IL:	Interleukin
LDL-R:	LDL-Receptor
LDL-C:	Low density lipoprotein cholesterol
MTHFR:	Methylenetetrahydrofolate reductase
MUFA:	Monounsaturated fatty acid
NHANES:	National Health and Nutrition Examination Survey
NPC1L1:	Niemann–Pick C1-Like 1 transporter
PN:	Personalized nutrition
PUFA:	Polyunsaturated fatty acid
PURE:	Prospective Urban and Rural Epidemiological study
SFA:	Saturated fatty acid
SNPs:	Single nucleotide polymorphisms
TG:	Triglycerides
UM:	Urolithin metabotypes
VAT:	Visceral adipose tissue
ALA:	α-linoleic acid

REFERENCES

(1988). "Intersalt: an international study of electrolyte excretion and blood pressure. Results for 24 hour urinary sodium and potassium excretion. Intersalt Cooperative Research Group." *BMJ* 297(6644): 319–328.

(1992). "The effects of nonpharmacologic interventions on blood pressure of persons with high normal levels: results of the Trials of Hypertension Prevention, phase I." *JAMA* 267(9): 1213–1220.

(1997). "Effects of weight loss and sodium reduction intervention on blood pressure and hypertension incidence in overweight people with high-normal blood pressure: the Trials of Hypertension Prevention, phase II. The Trials of Hypertension Prevention Collaborative Research Group." *Arch Intern Med* 157(6): 657–667.

(2017). *Cardiovascular Diseases (CVD) Factsheet*, WHO.

Adams, D. D. (2011). "The great cholesterol myth; unfortunate consequences of Brown and Goldstein's mistake." *QJM* 104(10): 867–870.

Anderson, J. W., T. J. Hanna, et al. (2000). "Whole grain foods and heart disease risk." *J Am Coll Nutr* 19(3 Suppl): 291S–299S.

Assimes, T. L., J. W. Knowles, et al. (2008). "Susceptibility locus for clinical and subclinical coronary artery disease at chromosome 9p21 in the multi-ethnic ADVANCE study." *Hum Mol Genet* 17(15): 2320–2328.

Atallah, R., K. B. Filion, et al. (2014). "Long-term effects of 4 popular diets on weight loss and cardiovascular risk factors: a systematic review of randomized controlled trials." *Circ Cardiovasc Qual Outcomes* 7(6): 815–827.

Austin, M. A., C. M. Hutter, et al. (2004). "Genetic causes of monogenic heterozygous familial hypercholesterolemia: a HuGE prevalence review." *Am J Epidemiol* 160(5): 407–420.

Baldrick, F. R., J. V. Woodside, et al. (2011). "Biomarkers of fruit and vegetable intake in human intervention studies: a systematic review." *Crit Rev Food Sci Nutr* 51(9): 795–815.

Blair, S. N., D. M. Capuzzi, et al. (2000). "Incremental reduction of serum total cholesterol and low-density lipoprotein cholesterol with the addition of plant stanol ester-containing spread to statin therapy." *Am J Cardiol* 86(1): 46–52.

Boyle, E. A., Y. I. Li, et al. (2017). "An expanded view of complex traits: from polygenic to omnigenic." *Cell* 169(7): 1177–1186.

Brice, P., H. Burton, et al. (2013). "Familial hypercholesterolaemia: a pressing issue for European health care." *Atherosclerosis* 231(2): 223–226.

Briggs, M. A., K. S. Petersen, et al. (2017). "Saturated fatty acids and cardiovascular disease: replacements for saturated fat to reduce cardiovascular risk." *Healthcare (Basel)* 5(2).

Broekhuizen, K., M. N. M. van Poppel, et al. (2010). "A tailored lifestyle intervention to reduce the cardiovascular disease risk of individuals with familial hypercholesterolemia (FH)." *Psychol Health* 25: 168–168.

Bronte-Stewart, B., A. Antonis, et al. (1956). "Effects of feeding different fats on serum-cholesterol level." *Lancet* 270(6922): 521–526.

Brown, J. D., J. Buscemi, et al. (2016). "Effects on cardiovascular risk factors of weight losses limited to 5–10." *Transl Behav Med* 6(3): 339–346.

Brown, M. S. and J. L. Goldstein (1984). "How LDL receptors influence cholesterol and atherosclerosis." *Sci Am* 251(5): 58–66.

Brown, M. S. and J. L. Goldstein (1986). "A receptor-mediated pathway for cholesterol homeostasis." *Science* 232(4746): 34–47.

Bruun, J. M., M. Maersk, et al. (2015). "Consumption of sucrose-sweetened soft drinks increases plasma levels of uric acid in overweight and obese subjects: a 6-month randomised controlled trial." *Eur J Clin Nutr* 69(8): 949–953.

Bush, W. S. and J. H. Moore (2012). "Chapter 11: genome-wide association studies." *PLoS Comput Biol* 8(12): e1002822.

Canner, P. L., K. G. Berge, et al. (1986). "Fifteen year mortality in Coronary Drug Project patients: long-term benefit with niacin." *J Am Coll Cardiol* 8(6): 1245–1255.

Celis-Morales, C., K. M. Livingstone, et al. (2015). "Design and baseline characteristics of the Food4Me study: a web-based randomised controlled trial of personalised nutrition in seven European countries." *Genes and Nutrition* 10(1).

Chandrasekara, A. and F. Shahidi (2011). "Bioactivities and antiradical properties of millet grains and hulls." *J Agric Food Chem* 59(17): 9563–9571.

Clarke, R., D. A. Bennett, et al. (2012). "Homocysteine and coronary heart disease: meta-analysis of MTHFR case-control studies, avoiding publication bias." *PLoS Med* 9(2): e1001177.

Cohen, D. E. (2008). "Balancing cholesterol synthesis and absorption in the gastrointestinal tract." *J Clin Lipidol* 2(2): S1–S3.

Connor, W. E. and S. L. Connor (1989). "Dietary treatment of familial hypercholesterolemia." *Arteriosclerosis* 9(1 Suppl): I91–I105.

Consortium, C. A. D., P. Deloukas, et al. (2013). "Large-scale association analysis identifies new risk loci for coronary artery disease." *Nat Genet* 45(1): 25–33.

Cook, N. R., J. A. Cutler, et al. (2007). "Long term effects of dietary sodium reduction on cardiovascular disease outcomes: observational follow-up of the trials of hypertension prevention (TOHP)." *BMJ* 334(7599): 885–888.

Corella, D., O. Coltell, et al. (2017). "Utilizing nutritional genomics to tailor diets for the prevention of cardiovascular disease: a guide for upcoming studies and implementations." *Expert Rev Mol Diagn* 17(5): 495–513.

Corella, D. and J. M. Ordovas (2009). "Nutrigenomics in cardiovascular medicine." *Circ Cardiovasc Genet* 2(6): 637–651.

Coronary Artery Disease Genetics, C. (2011). "A genome-wide association study in Europeans and South Asians identifies five new loci for coronary artery disease." *Nat Genet* 43(4): 339–344.

Cuchel, M., E. Bruckert, et al. (2014). "Homozygous familial hypercholesterolaemia: new insights and guidance for clinicians to improve detection and clinical management. A position paper from the Consensus Panel on Familial Hypercholesterolaemia of the European Atherosclerosis Society." *Eur Heart J* 35(32): 2146–2157.

Dansinger, M. L., J. A. Gleason, et al. (2005). "Comparison of the Atkins, Ornish, Weight Watchers, and Zone diets for weight loss and heart disease risk reduction: a randomized trial." *JAMA* 293(1): 43–53.

Das, U. N. (2006). "Essential fatty acids: biochemistry, physiology and pathology." *Biotechnol J* 1(4): 420–439.

Davignon, J., R. E. Gregg, et al. (1988). "Apolipoprotein-E polymorphism and atherosclerosis." *Arteriosclerosis* 8(1): 1–21.

de Koning, L., V. S. Malik, et al. (2012). "Sweetened beverage consumption, incident coronary heart disease, and biomarkers of risk in men." *Circulation* 125(14): 1735–1741, S1731.

de Lorgeril, M., P. Salen, et al. (1999). "Mediterranean diet, traditional risk factors, and the rate of cardiovascular complications after myocardial infarction: final report of the Lyon Diet Heart Study." *Circulation* 99(6): 779–785.

Dehghan, M., A. Mente, et al. (2017). "Associations of fats and carbohydrate intake with cardiovascular disease and mortality in 18 countries from five continents (PURE): a prospective cohort study." *Lancet* 390(10107): 2050–2062.

Dong, J. Y., Y. H. Zhang, et al. (2012). "Meta-analysis of dietary glycemic load and glycemic index in relation to risk of coronary heart disease." *Am J Cardiol* 109(11): 1608–1613.

Du, H., D. van der A, et al. (2010). "Dietary fiber and subsequent changes in body weight and waist circumference in European men and women." *Am J Clin Nutr* 91(2): 329–336.

Dudbridge, F. (2013). "Power and predictive accuracy of polygenic risk scores." *PLoS Genet* 9(3): e1003348.

Eckel, R. H., J. M. Jakicic, et al. (2014). "2013 AHA/ACC guideline on lifestyle management to reduce cardiovascular risk: a report of the American College of Cardiology/American Heart Association Task Force on Practice Guidelines." *J Am Coll Cardiol* 63(25 Pt B): 2960–2984.

Elliott, P., J. Stamler, et al. (1996). "Intersalt revisited: further analyses of 24 hour sodium excretion and blood pressure within and across populations. Intersalt Cooperative Research Group." *BMJ* 312(7041): 1249–1253.

Estruch, R., E. Ros, et al. (2013). "Primary prevention of cardiovascular disease with a Mediterranean diet." *N Engl J Med* 368(14): 1279–1290.

Ferguson, J. F., C. M. Phillips, et al. (2010a). "NOS3 gene polymorphisms are associated with risk markers of cardiovascular disease, and interact with omega-3 polyunsaturated fatty acids." *Atherosclerosis* 211(2): 539–544.

Ferguson, J. F., C. M. Phillips, et al. (2010b). "Gene-nutrient interactions in the metabolic syndrome: single nucleotide polymorphisms in ADIPOQ and ADIPOR1 interact with plasma saturated fatty acids to modulate insulin resistance." *Am J Clin Nutr* 91(3): 794–801.

Foster, G. D., H. R. Wyatt, et al. (2003). "A randomized trial of a low-carbohydrate diet for obesity." *N Engl J Med* 348(21): 2082–2090.

Fung, T. T., K. M. Rexrode, et al. (2009). "Mediterranean diet and incidence of and mortality from coronary heart disease and stroke in women." *Circulation* 119(8): 1093–1100.

Gao, F., J. McDaniel, et al. (2017). "Dynamic and temporal assessment of human dried blood spot MS/MSALL shotgun lipidomics analysis." *Nutr Metab* 14(1): 28.

Garcia-Rios, A., J. Delgado-Lista, et al. (2011). "Genetic variations at the lipoprotein lipase gene influence plasma lipid concentrations and interact with plasma n-6 polyunsaturated fatty acids to modulate lipid metabolism." *Atherosclerosis* 218(2): 416–422.

Genomes Project, C., G. R. Abecasis, et al. (2010). "A map of human genome variation from population-scale sequencing." *Nature* 467(7319): 1061–1073.

Gertler, M. M., S. M. Garn, et al. (1951). "Young candidates for coronary heart disease." *J Am Med Assoc* 147(7): 621–625.

Gibney, M. J. and W. M. Goosens (2016). "Chapter 5.1: personalized nutrition: paving the way to better population health." *Good Nutrition: Perspectives for the 21st Century*: 234–248.

Gibney, M. J. and M. C. Walsh (2013). "The future direction of personalised nutrition: my diet, my phenotype, my genes." *Proc Nutr Soc* 72(2): 219–225.

Gonzalez-Sarrias, A., R. Garcia-Villalba, et al. (2017). "Clustering according to urolithin metabotype explains the interindividual variability in the improvement of cardiovascular risk biomarkers in overweight-obese individuals consuming pomegranate: a randomized clinical trial." *Mol Nutr Food Res* 61(5).

Grasgruber, P., M. Sebera, et al. (2016). "Food consumption and the actual statistics of cardiovascular diseases: an epidemiological comparison of 42 European countries." *Food Nutr Res* 60: 31694.

Graversen, C. B., S. Lundbye-Christensen, et al. (2016). "Marine n-3 polyunsaturated fatty acids lower plasma proprotein convertase subtilisin kexin type 9 levels in pre- and post-menopausal women: a randomised study." *Vasc Pharmacol* 76: 37–41.

Group, C. D. P. R. (1975). "Clofibrate and niacin in coronary heart disease." *JAMA* 231(4): 360–381.

Group, H. T. C., M. J. Landray, et al. (2014). "Effects of extended-release niacin with laropiprant in high-risk patients." *N Engl J Med* 371(3): 203–212.

Gylling, H., J. Plat, et al. (2014). "Plant sterols and plant stanols in the management of dyslipidaemia and prevention of cardiovascular disease." *Atherosclerosis* 232(2): 346–360.

Haring, B., N. Gronroos, et al. (2014). "Dietary protein intake and coronary heart disease in a large community based cohort: results from the Atherosclerosis Risk in Communities (ARIC) study [corrected]." *PLoS One* 9(10): e109552.

Helgadottir, A., G. Thorleifsson, et al. (2007). "A common variant on chromosome 9p21 affects the risk of myocardial infarction." *Science* 316(5830): 1491–1493.

Heux, S., F. Morin, et al. (2004). "The methylentetrahydrofolate reductase gene variant (C677T) as a risk factor for essential hypertension in Caucasians." *Hypertens Res* 27(9): 663–667.

Hoogeveen, R. C., J. W. Gaubatz, et al. (2014). "Small dense low-density lipoprotein-cholesterol concentrations predict risk for coronary heart disease: the Atherosclerosis Risk In Communities (ARIC) study." *Arterioscler Thromb Vasc Biol* 34(5): 1069–1077.

Hooper, L., N. Martin, et al. (2015). "Reduction in saturated fat intake for cardiovascular disease." *Cochrane Database Syst Rev* (6): CD011737.

Hooper, L., C. D. Summerbell, et al. (2001). "Dietary fat intake and prevention of cardiovascular disease: systematic review." *BMJ* 322(7289): 757–763.

Hopping, B. N., E. Erber, et al. (2010). "Dietary fiber, magnesium, and glycemic load alter risk of type 2 diabetes in a multiethnic cohort in Hawaii." *J Nutr* 140(1): 68–74.

Howarth, N. C., E. Saltzman, et al. (2001). "Dietary fiber and weight regulation." *Nutr Rev* 59(5): 129–139.

Hu, F. B., M. J. Stampfer, et al. (1997). "Dietary fat intake and the risk of coronary heart disease in women." *N Engl J Med* 337(21): 1491–1499.

Hu, F. B., M. J. Stampfer, et al. (1999). "Dietary protein and risk of ischemic heart disease in women." *Am J Clin Nutr* 70(2): 221–227.

Hu, F. B. and W. C. Willett (2002). "Optimal diets for prevention of coronary heart disease." *JAMA* 288(20): 2569–2578.

Hu, T. and L. A. Bazzano (2014). "The low-carbohydrate diet and cardiovascular risk factors: evidence from epidemiologic studies." *Nutr Metab Cardiovasc Dis* 24(4): 337–343.

Jacobs, D. R., Jr. and D. D. Gallaher (2004). "Whole grain intake and cardiovascular disease: a review." *Curr Atheroscler Rep* 6(6): 415–423.

Jakobsen, M. U., E. J. O'Reilly, et al. (2009). "Major types of dietary fat and risk of coronary heart disease: a pooled analysis of 11 cohort studies." *Am J Clin Nutr* 89(5): 1425–1432.

Jenkins, D. J., T. M. Wolever, et al. (1985). "Low glycemic index carbohydrate foods in the management of hyperlipidemia." *Am J Clin Nutr* 42(4): 604–617.

Jones, D. S. and J. A. Greene (2013). "The decline and rise of coronary heart disease: understanding public health catastrophism." *Am J Public Health* 103(7): 1207–1218.

Joshipura, K. J., F. B. Hu, et al. (2001). "The effect of fruit and vegetable intake on risk for coronary heart disease." *Ann Intern Med* 134(12): 1106–1114.

Juraschek, S. P., L. J. Appel, et al. (2013). "Effect of a high-protein diet on kidney function in healthy adults: results from the OmniHeart trial." *Am J Kidney Dis* 61(4): 547–554.

Kafatos, A., A. Diacatou, et al. (1997). "Heart disease risk-factor status and dietary changes in the Cretan population over the past 30 y: the Seven Countries Study." *Am J Clin Nutr* 65(6): 1882–1886.

Kahn, R. and J. L. Sievenpiper (2014). "Dietary sugar and body weight: have we reached a crisis in the epidemic of obesity and diabetes?: we have, but the pox on sugar is overwrought and overworked." *Diabetes Care* 37(4): 957–962.

Kalayinia, S., H. Goodarzynejad, et al. (2017). "Next generation sequencing applications for cardiovascular disease." *Ann Med*: 1–19.

Keys, A., A. Menotti, et al. (1986). "The diet and 15-year death rate in the seven countries study." *Am J Epidemiol* 124(6): 903–915.

Keys, A., O. Mickelsen, et al. (1950). "The relation in man between cholesterol levels in the diet and in the blood." *Science* 112(2899): 79–81.

Keys, A., H. L. Taylor, et al. (1963). "Coronary heart disease among Minnesota business and professional men followed fifteen years." *Circulation* 28: 381–395.

Khera, A. V., C. A. Emdin, et al. (2016). "Genetic risk, adherence to a healthy lifestyle, and coronary disease." *N Engl J Med* 375(24): 2349–2358.

Khera, A. V. and S. Kathiresan (2017). "Genetics of coronary artery disease: discovery, biology and clinical translation." *Nat Rev Genet* 18(6): 331–344.

Kidambi, S. and S. B. Patel (2008). "Sitosterolaemia: pathophysiology, clinical presentation and laboratory diagnosis." *J Clin Pathol* 61(5): 588–594.

Klarin, D., Q. M. Zhu, et al. (2017). "Genetic analysis in UK Biobank links insulin resistance and transendothelial migration pathways to coronary artery disease." *Nat Genet* 49(9): 1392–1397.

Krauss, R. M., P. J. Blanche, et al. (2006). "Separate effects of reduced carbohydrate intake and weight loss on atherogenic dyslipidemia." *Am J Clin Nutr* 83(5): 1025–1031; quiz 1205.

Krauss, R. M., R. H. Eckel, et al. (2000). "AHA Dietary Guidelines: revision 2000: a statement for healthcare professionals from the Nutrition Committee of the American Heart Association." *Circulation* 102(18): 2284–2299.

Kristal, A. R., U. Peters, et al. (2005). "Is it time to abandon the food frequency questionnaire?" *Cancer Epidemiol Biomarkers Prev* 14(12): 2826–2828.

Lewis, C. M. and E. Vassos (2017). "Prospects for using risk scores in polygenic medicine." *Genome Med* 9(1): 96.

Li, Y., A. Hruby, et al. (2015). "Saturated fats compared with unsaturated fats and sources of carbohydrates in relation to risk of coronary heart disease: a prospective cohort study." *J Am Coll Cardiol* 66(14): 1538–1548.

Liu, S., M. J. Stampfer, et al. (1999). "Whole-grain consumption and risk of coronary heart disease: results from the Nurses' Health Study." *Am J Clin Nutr* 70(3): 412–419.

Liu, S., W. C. Willett, et al. (2000). "A prospective study of dietary glycemic load, carbohydrate intake, and risk of coronary heart disease in US women." *Am J Clin Nutr* 71(6): 1455–1461.

Lloyd-Jones, D. M., B. H. Nam, et al. (2004). "Parental cardiovascular disease as a risk factor for cardiovascular disease in middle-aged adults: a prospective study of parents and offspring." *JAMA* 291(18): 2204–2211.

Lopez-Legarrea, P., R. de la Iglesia, et al. (2014). "The protein type within a hypocaloric diet affects obesity-related inflammation: the RESMENA project." *Nutrition* 30(4): 424–429.

Mackay, J. and G. A. Mensah (2004). *The Atlas of Heart Disease and Stroke*, World Health Organization.

Maersk, M., A. Belza, et al. (2012). "Sucrose-sweetened beverages increase fat storage in the liver, muscle, and visceral fat depot: a 6-mo randomized intervention study." *Am J Clin Nutr* 95(2): 283–289.

Marenberg, M. E., N. Risch, et al. (1994). "Genetic susceptibility to death from coronary heart disease in a study of twins." *N Engl J Med* 330(15): 1041–1046.

Martin, N. J., J. Bunch, et al. (2013). "Dried blood spot proteomics: surface extraction of endogenous proteins coupled with automated sample preparation and mass spectrometry analysis." *J Am Soc Mass Spectrom* 24(8): 1242–1249.

McNulty, H., L. R. C. Dowey, et al. (2006). "Riboflavin lowers homocysteine in individuals homozygous for the MTHFR 677C->T polymorphism." *Circulation* 113(1): 74–80.

McPherson, R., A. Pertsemlidis, et al. (2007). "A common allele on chromosome 9 associated with coronary heart disease." *Science* 316(5830): 1488–1491.

Meek, R. L., R. C. LeBoeuf, et al. (2013). "Glomerular cell death and inflammation with high-protein diet and diabetes." *Nephrol Dial Transplant* 28(7): 1711–1720.

Mensink, R. P., P. L. Zock, et al. (2003). "Effects of dietary fatty acids and carbohydrates on the ratio of serum total to HDL cholesterol and on serum lipids and apolipoproteins: a meta-analysis of 60 controlled trials." *Am J Clin Nutr* 77(5): 1146–1155.

Mente, A., L. de Koning, et al. (2009). "A systematic review of the evidence supporting a causal link between dietary factors and coronary heart disease." *Arch Intern Med* 169(7): 659–669.

Miller, M., N. J. Stone, et al. (2011). "Triglycerides and cardiovascular disease: a scientific statement from the American Heart Association." *Circulation* 123(20): 2292–2333.

Minihane, A. M., S. Khan, et al. (2000). "ApoE polymorphism and fish oil supplementation in subjects with an antherogenic lipoprotein phenotype." *Arterioscler Thromb Vasc Biol* 20(8): 1990–1997.

Mirrahimi, A., R. J. de Souza, et al. (2012). "Associations of glycemic index and load with coronary heart disease events: a systematic review and meta-analysis of prospective cohorts." *J Am Heart Assoc* 1(5): e000752.

Mora, S., M. P. Caulfield, et al. (2015). "Atherogenic lipoprotein subfractions determined by ion mobility and first cardiovascular events after random allocation to high-intensity statin or placebo: the justification for the use of statins in prevention: an intervention trial evaluating rosuvastatin (JUPITER) trial." *Circulation* 132(23): 2220–2229.

Mozaffarian, D. (2016). "Dietary and policy priorities for cardiovascular disease, diabetes, and obesity: a comprehensive review." *Circulation* 133(2): 187–225.

Mozaffarian, D., R. Micha, et al. (2010). "Effects on coronary heart disease of increasing polyunsaturated fat in place of saturated fat: a systematic review and meta-analysis of randomized controlled trials." *PLoS Med* 7(3): e1000252.

Murabito, J. M., B. H. Nam, et al. (2004). "Sibling cardiovascular disease as a risk factor for cardiovascular disease in middle-aged adults: the Framingham Offspring Study." *Circulation* 110(17): 764–764.

Musa-Veloso, K., T. H. Poon, et al. (2011). "A comparison of the LDL-cholesterol lowering efficacy of plant stanols and plant sterols over a continuous dose range: results of a meta-analysis of randomized, placebo-controlled trials." *Prostaglandins Leukot Essent Fatty Acids* 85(1): 9–28.

Mustad, V. A., T. D. Etherton, et al. (1997). "Reducing saturated fat intake is associated with increased levels of LDL receptors on mononuclear cells in healthy men and women." *J Lipid Res* 38(3): 459–468.

Myocardial Infarction Genetics, C., S. Kathiresan, et al. (2009). "Genome-wide association of early-onset myocardial infarction with single nucleotide polymorphisms and copy number variants." *Nat Genet* 41(3): 334–341.

Natarajan, P., R. Young, et al. (2017). "Polygenic risk score identifies subgroup with higher burden of atherosclerosis and greater relative benefit from statin therapy in the primary prevention setting." *Circulation* 135(22): 2091–2101.

Nelson, C. P., A. Goel, et al. (2017). "Association analyses based on false discovery rate implicate new loci for coronary artery disease." *Nat Genet* 49(9): 1385–1391.

Noakes, M., J. B. Keogh, et al. (2005). "Effect of an energy-restricted, high-protein, low-fat diet relative to a conventional high-carbohydrate, low-fat diet on weight loss, body composition, nutritional status, and markers of cardiovascular health in obese women." *Am J Clin Nutr* 81(6): 1298–1306.

Nutrition, S. A. C. o. (2018). "Saturated fats and health draft report."

Oliver, M. (1976). "Dietary cholesterol, plasma cholesterol and coronary heart disease." *Br Heart J* 38(3): 214–218.

Ordovas, J. M. and D. Corella (2004). "Genes, diet and plasma lipids: the evidence from observational studies." *World Rev Nutr Diet* 93: 41–76.

Ordovas, J. M., D. Corella, et al. (2002). "Dietary fat intake determines the effect of a common polymorphism in the hepatic lipase gene promoter on high-density lipoprotein metabolism: evidence of a strong dose effect in this gene-nutrient interaction in the Framingham Study." *Circulation* 106(18): 2315–2321.

Ordovas, J. M., L. R. Ferguson, et al. (2018). "Personalised nutrition and health." *BMJ* 361: bmj k2173.

Pedersen, E., D. R. Jesudason, et al. (2014). "High protein weight loss diets in obese subjects with type 2 diabetes mellitus." *Nutr Metab Cardiovasc Dis* 24(5): 554–562.

Pereira, M. A., E. O'Reilly, et al. (2004). "Dietary fiber and risk of coronary heart disease: a pooled analysis of cohort studies." *Arch Intern Med* 164(4): 370–376.

Phillips, K. A., M. J. Pletcher, et al. (2015). "Is the '$1000 Genome' really $1000? Understanding the full benefits and costs of genomic sequencing." *Technol Health Care* 23(3): 373–379.

Poudyal, H. and L. Brown (2015). "Should the pharmacological actions of dietary fatty acids in cardiometabolic disorders be classified based on biological or chemical function?" *Prog Lipid Res* 59: 172–200.

Pyorala, K. (1987). "Dietary cholesterol in relation to plasma cholesterol and coronary heart disease." *Am J Clin Nutr* 45(5 Suppl): 1176–1184.

Rader, D. J., J. Cohen, et al. (2003). "Monogenic hypercholesterolemia: new insights in pathogenesis and treatment." *J Clin Invest* 111(12): 1795–1803.

Richter, C. K., A. C. Skulas-Ray, et al. (2015). "Plant protein and animal proteins: do they differentially affect cardiovascular disease risk?" *Adv Nutr* 6(6): 712–728.

Rodriguez-Perez, C., V. R. Ramprasath, et al. (2016). "Docosahexaenoic acid attenuates cardiovascular risk factors via a decline in proprotein convertase subtilisin/kexin type 9 (PCSK9) plasma levels." *Lipids* 51(1): 75–83.

Russo, G. L. (2009). "Dietary n-6 and n-3 polyunsaturated fatty acids: from biochemistry to clinical implications in cardiovascular prevention." *Biochem Pharmacol* 77(6): 937–946.

Sacks, F. M., G. A. Bray, et al. (2009). "Comparison of weight-loss diets with different compositions of fat, protein, and carbohydrates." *N Engl J Med* 360(9): 859–873.

Sacks, F. M., V. J. Carey, et al. (2014). "Effects of high vs low glycemic index of dietary carbohydrate on cardiovascular disease risk factors and insulin sensitivity: the OmniCarb randomized clinical trial." *JAMA* 312(23): 2531–2541.

Sacks, F. M., L. P. Svetkey, et al. (2001). "Effects on blood pressure of reduced dietary sodium and the Dietary Approaches to Stop Hypertension (DASH) diet. DASH-Sodium Collaborative Research Group." *N Engl J Med* 344(1): 3–10.

Samani, N. J., J. Erdmann, et al. (2007). "Genomewide association analysis of coronary artery disease." *N Engl J Med* 357(5): 443–453.

Sarkkinen, E., M. Korhonen, et al. (1998). "Effect of apolipoprotein E polymorphism on serum lipid response to the separate modification of dietary fat and dietary cholesterol." *Am J Clin Nutr* 68(6): 1215–1222.

Schunkert, H., A. Gotz, et al. (2008). "Repeated replication and a prospective meta-analysis of the association between chromosome 9p21.3 and coronary artery disease." *Circulation* 117(13): 1675–1684.

Schunkert, H., I. R. Konig, et al. (2011). "Large-scale association analysis identifies 13 new susceptibility loci for coronary artery disease." *Nat Genet* 43(4): 333–338.

Schwingshackl, L., B. Strasser, et al. (2011). "Effects of monounsaturated fatty acids on cardiovascular risk factors: a systematic review and meta-analysis." *Ann Nutr Metab* 59(2–4): 176–186.

Shaw, D. I., A. C. Tierney, et al. (2009). "LIPGENE food-exchange model for alteration of dietary fat quantity and quality in free-living participants from eight European countries." *Br J Nutr* 101(5): 750–759.

Simopoulos, A. P. (2008). "The importance of the omega-6/omega-3 fatty acid ratio in cardiovascular disease and other chronic diseases." *Exp Biol Med (Maywood)* 233(6): 674–688.

Siri-Tarino, P. W., S. Chiu, et al. (2015). "Saturated fats versus polyunsaturated fats versus carbohydrates for cardiovascular disease prevention and treatment." *Annu Rev Nutr* 35: 517–543.

Siri-Tarino, P. W., Q. Sun, et al. (2010). "Meta-analysis of prospective cohort studies evaluating the association of saturated fat with cardiovascular disease." *Am J Clin Nutr* 91(3): 535–546.

Slyper, A. H. (2013). "The influence of carbohydrate quality on cardiovascular disease, the metabolic syndrome, type 2 diabetes, and obesity - an overview." *J Pediatr Endocrinol Metab* 26(7–8): 617–629.

Sofi, F., F. Cesari, et al. (2008). "Adherence to Mediterranean diet and health status: meta-analysis." *BMJ* 337: a1344.

Softic, S., M. K. Gupta, et al. (2017). "Divergent effects of glucose and fructose on hepatic lipogenesis and insulin signaling." *J Clin Invest* 127(11): 4059–4074.

Stanhope, K. L., A. A. Bremer, et al. (2011). "Consumption of fructose and high fructose corn syrup increase postprandial triglycerides, LDL-cholesterol, and apolipoprotein-B in young men and women." *J Clin Endocrinol Metab* 96(10): E1596–E1605.

Stanhope, K. L., J. M. Schwarz, et al. (2009). "Consuming fructose-sweetened, not glucose-sweetened, beverages increases visceral adiposity and lipids and decreases insulin sensitivity in overweight/obese humans." *J Clin Invest* 119(5): 1322–1334.

Sytkowski, P. A., W. B. Kannel, et al. (1990). "Changes in risk factors and the decline in mortality from cardiovascular disease. The Framingham Heart Study." *N Engl J Med* 322(23): 1635–1641.

Te Morenga, L., S. Mallard, et al. (2012). "Dietary sugars and body weight: systematic review and meta-analyses of randomised controlled trials and cohort studies." *BMJ* 346: e7492.

Tierney, A. C., J. McMonagle, et al. (2011). "Effects of dietary fat modification on insulin sensitivity and on other risk factors of the metabolic syndrome—LIPGENE: a European randomized dietary intervention study." *Int J Obes (Lond)* 35(6): 800–809.

Tsai, M. Y., B. T. Steffen, et al. (2014). "New automated assay of small dense low-density lipoprotein cholesterol identifies risk of coronary heart disease: the Multi-ethnic Study of Atherosclerosis." *Arterioscler Thromb Vasc Biol* 34(1): 196–201.

Turner, R. C., H. Millns, et al. (1998). "Risk factors for coronary artery disease in non-insulin dependent diabetes mellitus: United Kingdom Prospective Diabetes Study (UKPDS: 23)." *BMJ* 316(7134): 823–828.

Utermann, G., M. Hees, et al. (1977). "Polymorphism of apolipoprotein-E and occurrence of dys-beta-lipoproteinemia in man." *Nature* 269(5629): 604–607.

van der Graaf, A., H. J. Avis, et al. (2011). "Molecular basis of autosomal dominant hypercholesterolemia: assessment in a large cohort of hypercholesterolemic children." *Circulation* 123(11): 1167–1173.

van der Harst, P. and N. Verweij (2018). "Identification of 64 novel genetic loci provides an expanded view on the genetic architecture of coronary artery disease." *Circ Res* 122(3): 433–443.

Verweij, N., R. N. Eppinga, et al. (2017). "Identification of 15 novel risk loci for coronary artery disease and genetic risk of recurrent events, atrial fibrillation and heart failure." *Sci Rep* 7(1): 2761.

Wang, Q. J., T. Sun, et al. (2016). "A dried blood spot mass spectrometry metabolomic approach for rapid breast cancer detection." *Oncotargets Ther* 9: 1389–1398.

Welsh, J. A., A. Sharma, et al. (2010). "Caloric sweetener consumption and dyslipidemia among US adults." *JAMA* 303(15): 1490–1497.

White, P. D. (1957). "Genes, the heart and destiny." *N Engl J Med* 256(21): 965–969.

Whitehead, A., E. J. Beck, et al. (2014). "Cholesterol-lowering effects of oat beta-glucan: a meta-analysis of randomized controlled trials." *Am J Clin Nutr* 100(6): 1413–1421.

Wilson, C. P., M. Ward, et al. (2012). "Riboflavin offers a targeted strategy for managing hypertension in patients with the MTHFR 677TT genotype: a 4-y follow-up." *Am J Clin Nutr* 95(3): 766–772.

Wilson, P. W., R. B. D'Agostino, et al. (1998). "Prediction of coronary heart disease using risk factor categories." *Circulation* 97(18): 1837–1847.

Won, H. H., P. Natarajan, et al. (2015). "Disproportionate contributions of select genomic compartments and cell types to genetic risk for coronary artery disease." *PLoS Genet* 11(10): e1005622.

Wrone, E. M., M. R. Carnethon, et al. (2003). "Association of dietary protein intake and microalbuminuria in healthy adults: third National Health and Nutrition Examination Survey." *Am J Kidney Dis* 41(3): 580–587.

Yancy, W. S., Jr., M. K. Olsen, et al. (2004). "A low-carbohydrate, ketogenic diet versus a low-fat diet to treat obesity and hyperlipidemia: a randomized, controlled trial." *Ann Intern Med* 140(10): 769–777.

Yang, Q., Z. Zhang, et al. (2014). "Added sugar intake and cardiovascular diseases mortality among US adults." *JAMA Intern Med* 174(4): 516–524.

Yu, E., E. Rimm, et al. (2016). "Diet, lifestyle, biomarkers, genetic factors, and risk of cardiovascular disease in the nurses' health studies." *Am J Public Health* 106(9): 1616–1623.

Yu, S., J. Derr, et al. (1995). "Plasma cholesterol-predictive equations demonstrate that stearic acid is neutral and monounsaturated fatty acids are hypocholesterolemic." *Am J Clin Nutr* 61(5): 1129–1139.

Yusuf, S., S. Rangarajan, et al. (2014). "Cardiovascular risk and events in 17 low-, middle-, and high-income countries." *N Engl J Med* 371(9): 818–827.

Zdravkovic, S., A. Wienke, et al. (2002). "Heritability of death from coronary heart disease: a 36-year follow-up of 20 966 Swedish twins." *J Intern Med* 252(3): 247–254.

Zong, G., Y. Li, et al. (2016). "Intake of individual saturated fatty acids and risk of coronary heart disease in US men and women: two prospective longitudinal cohort studies." *BMJ* 355: i5796.

4 Nutraceuticals for Hypertension Control

Balázs Varga, Mariann Bombicz,
Andrea Kurucz and Béla Juhász

CONTENTS

4.1 HYPERTENSION

Blood pressure (BP) is necessary to maintain organ perfusion, oxygen and nutrient supply, and is created by the force of blood pushing against the walls of blood vessels (arteries). It can be determined by the following equation:

$$\text{Blood Pressure } (BP) = \text{Cardiac Output } (CO)$$

$$\times \text{Systemic Vascular Resistance } (SVR)$$

Hypertension, also known as high or raised blood pressure, is a condition where there is persistently raised pressure in the blood vessels. The higher the pressure, the harder the heart must pump. Since 1980, the American College of Cardiology (ACC) and the American Heart Association (AHA) have published numerous clinical practice guidelines with recommendations to improve cardiovascular health (Reboussin, Allen et al. 2018). Between 1934 and 1954, a strong direct relationship was noted between the level of BP and the risk of clinical complications and death. In the 1960s, a series of reports from the Framingham Heart Study proved these findings (Mahmood, Levy et al. 2014). The first profound guideline for detection, evaluation and management of high BP was published in 1977, sponsored by the NHLBI (1977). In the following years, a series of Joint National Committee (JNC) BP guidelines were published to assist the practice community and improve prevention, awareness, treatment and control of high BP. The American College of Cardiology/American Heart Association (ACC/AHA) Guideline for the Prevention, Detection, Evaluation and Management of High Blood Pressure in Adults updated the Seventh Report of the Joint National Committee, on Prevention, Detection, Evaluation and Treatment of High Blood Pressure (JNC7), which was published in 2003 (Chobanian, Bakris et al. 2003). Compared with the JNC7 guideline, the 2017 ACC/AHA guideline recommends using lower systolic blood pressure (SBP) and diastolic blood pressure (DBP) levels to define hypertension Table 4.1.

TABLE 4.1
Blood Pressure Levels to Define Hypertension and Recommended Antihypertensive Medication According to the 2017 ACC/AHA Guideline and JNC7 Guideline

Guideline Definition of Hypertension	2017 ACC/AHA	JNC7
Systolic BP, mm HG		
General population	≥130	≥140
Diastolic BP, mm HG		
General population	≥80	≥90
Guideline-Recommended Antihypertensive Medication		
Systolic BP, mm HG		
General population	≥140	≥140
Diabetes or CKD	≥130	≥130
High cardiovascular disease risk	≥130	–
Age ≥65 yrs	≥130	–
Diastolic BP, mm HG		
General population	≥90	≥90
Diabetes or CKD	≥80	≥80
High cardiovascular disease risk	≥80	–

ACC/AHA = American College of Cardiology/American Heart Association; BP = blood pressure; CKD = chronic kidney disease; DBP = diastolic blood pressure; JNC7 = Seventh Report of the Joint National Committee on Prevention, Detection, Evaluation and Treatment of High Blood Pressure; SBP = systolic blood pressure. No specific blood pressure threshold is given by the JNC7 guideline, the other thresholds, provided in the group should be applied.

TABLE 4.2
Blood Pressure Categories in Adults

Blood Pressure Category	Systolic Blood Pressure	Diastolic Blood Pressure
Normal	<120 mm Hg and	<80 mm Hg
Elevated	120–129 mm Hg and	<80 mm Hg
Hypertension		
Stage 1	130–139 mm Hg and	80–89 mm Hg
Stage 2	≥140 mm Hg and	≥90 mm Hg

According to the 2017 ACC/AHA guideline, BP is categorized into four levels on the basis of average BP measured in a healthcare setting (office pressures): normal, elevated and stage 1 or 2 hypertension (Reboussin, Allen et al. 2018 Table 4.2).

The high risk of cardiovascular diseases (CVDs) with stage 2 hypertension is wellknown. However, there is growing evidence (individual studies and meta-analyses of observational data) that the risk of CVDs progressively increases from normal to elevated BP and stage 1 hypertension (Guo, Zhang et al. 2013; Huang, Cai et al. 2014; Huang, Cai et al. 2014).

The diagnosis of hypertension requires integration of home or ambulatory blood pressure monitoring (ABPM), with great emphasis on multiple measurements.

The following diagnostic criteria were suggested by the 2017 ACC/AHA guidelines; meeting one or more of these criteria using ABPM can be of diagnostic value for hypertension.

- A 24-hour mean of 125/75 mm Hg or above.
- Daytime (awake) mean of 130/80 mm Hg or above.
- Night-time (asleep) mean of 110/65 mm Hg or above.

4.2 EPIDEMIOLOGY

The number of individuals with systolic blood pressure (SBP) of at least 110 to 115 mm Hg increased from 1.87 billion in 1990 to 3.47 billion in 2015, and the associated annual number of projected deaths increased from 7.2 million in 1990 to 10.7 million in 2015. The projected number of individuals with SBP of 140 mm Hg or higher increased from 442 million in 1990 to 874 million in 2015, and the associated annual number of projected deaths in 2015 (7.8 million) or 14.0% of total deaths were related to SBP of 140 mm Hg or higher. SBP of at least 110 to 115 mm Hg was associated with all hypertensive heart disease deaths, 68.7% of chronic kidney disease deaths, 54.4% of cerebrovascular disease deaths (50.0% of ischemic stroke and 58.3% of hemorrhagic stroke deaths) and 54.5% of ischemic heart disease deaths (Forouzanfar, Liu et al. 2017).

4.3 HYPERTENSION CAN BE CLASSIFIED INTO TWO FORMS: PRIMARY AND SECONDARY

4.3.1 PRIMARY HYPERTENSION

The pathogenesis of primary hypertension (formerly called 'essential' hypertension) is poorly understood, but is most likely the result of a large number of genetic and environmental factors.

4.3.1.1 Genetic Factors

Hypertension is a complex polygenic disorder in which many genes or gene combinations influence BP (Lifton, Gharavi et al. 2001). Several monogenic forms of hypertension and multiple single nucleotide polymorphisms that can contribute to hypertension have been identified (e.g. Liddle's syndrome, Gordon's syndrome, glucocorticoid-remediable aldosteronism). However, these diseases are rare. The collective effect of all BP loci identified through genome-wide association studies accounts for only about 3.5% of BP variability (Whelton, Carey et al. 2018). Future studies will need to better elucidate genetic expression, epigenetic effects, transcriptomics and proteomics that link genotypes with underlying pathophysiological mechanisms.

4.3.1.2 Environmental Factors

Poor diet (i.e. excess intake of sodium and insufficient intake of potassium, magnesium, calcium, proteins from vegetables, fibers and fish fats), aging, persistent stress, physical inactivity and excess intake of alcohol, alone or in combination, are responsible for a large proportion of hypertension (Savica, Bellinghieri et al. 2010). Gut microbiota has also been linked to hypertension, especially in experimental animals (Tang, Kitai et al. 2017).

4.3.1.2.1 *Overweight and Obesity*

Several studies, including the Framingham Heart Study and the Nurses' Health Study identified a direct, strong and almost linear correlation between obesity and hypertension. The most frequently examined parameter is the body mass index (BMI), but waist-to-hip ratio and central fat distribution measured with computed tomography (CT) are better prognostic markers (Garrison, Kannel et al. 1987; Forman, Stampfer et al. 2009; Juonala, Magnussen et al. 2011).

4.3.1.2.2 *Sodium Intake*

Several cohort and cross-sectional studies have shed light on the positive association between excessive sodium intake and increased blood pressure (1988; Takase, Sugiura et al. 2015). Salt sensitivity differs among different individuals, being very prominent in elder people, black people, those who have high BP or those with comorbidities like chronic kidney disease (CKD), diabetes or metabolic syndrome, and can be associated with elevated risk of severe complications like CVD or stroke (Weinberger 1996; Morimoto, Uzu et al. 1997).

4.3.1.2.3 *Potassium*

Different epidemiological studies have demonstrated that potassium intake is inversely related to high blood pressure, and a lower sodium–potassium ratio decreases the risk of CVD (Cook, Obarzanek et al. 2009; Zhang, Cogswell et al. 2013).

4.3.1.2.4 *Physical Activity*

It has long been known that regular physical exercise plays a protective role in human health, and that is the case regarding blood pressure as well. Hundreds of studies have shown that even moderate physical activity is inversely associated with high blood pressure and cardiovascular risk (Lesniak and Dubbert 2001; Carnethon, Gidding et al. 2003).

4.3.1.2.5 *Alcohol Consumption*

There is a direct relationship between alcohol consumption and increased blood pressure, and it was first demonstrated in 1915. In the US alcohol is responsible for 10% of high blood pressure cases. However, it also has to be mentioned that moderate alcohol (especially red wine) intake increases the level of high-density lipoprotein (Klatsky 2010).

4.3.1.3 Childhood Risk for Hypertension

A meta-analysis of 50 studies dealing with the relationship of childhood BP and adult BP showed that there was a correlation coefficient of about 0.38 for SBP and 0.28 for DBP. Childhood obesity, genetic factors, premature birth and low birth weight can increase the possibility of adult high blood pressure (Chen and Wang 2008; de Jong, Monuteaux et al. 2012; Juhola, Oikonen et al. 2012).

4.3.2 SECONDARY HYPERTENSION

A huge number of medical conditions (i.e. renal parenchymal or renovascular diseases, coarctation of aorta, obstructive sleep apnea, drug and alcohol abuse, adrenal hyperplasia, primary hyperaldosteronism, phaeochromocytoma, hyperparathyroidism, hyperthyroidism, acromegaly and Cushing's disease), sometimes coexisting with risk factors of primary hypertension can lead to elevated blood pressure. If the underlying cause can be identified and cured, the patient can experience a significant reduction in blood pressure. Therefore, in a patient newly presenting with hypertension, it is essential to perform a precise examination to discover the cause of it (Hirsch, Haskal et al. 2006; Funder, Carey et al. 2008; Nieman, Biller et al. 2008; Pedrosa, Drager et al. 2011; Katznelson, Laws et al. 2014; Lenders, Duh et al. 2014).

Prescriptions and over-the-counter drugs can also lead to secondary hypertension. These drugs include (Grossman and Messerli 2012):

- Oral contraceptives.
- Non-steroidal anti-inflammatory drugs (NSAIDs).
- Antidepressants (tricyclic antidepressants, selective serotonin reuptake inhibitors and monoamine oxidase inhibitors).
- Corticosteroids.
- Decongestants (phenylephrine, pseudoephedrine).
- Sodium-containing antacids.
- Weight-loss medications.
- Erythropoietin.
- Cyclosporine or tacrolimus.
- Stimulants (methylphenidate and amphetamine).
- Atypical antipsychotics (olanzapine and clozapine).
- Bevacizumab.
- Tyrosine kinase inhibitors (sorafenib, sunitinib).

4.3.3 Complications of Hypertension

Hypertension can be associated with significantly increased risk of adverse cardio-vascular and renal complications, like left ventricular hypertrophy, heart failure, acute myocardial infarction, ischemic stroke, intracerebral hemorrhage and chronic kidney disease (Wilson 1994; Levy, Larson et al. 1996; Thrift, McNeil et al. 1996; Lorell and Carabello 2000; Coresh, Wei et al. 2001).

Hypertension is the most important modifiable risk factor for premature cardio-vascular disease, and it often coexists with other factors like smoking, dyslipidemia, obesity, diabetes and physical inactivity, that can potentiate each other's effect. The likelihood of having any cardiovascular problems increases in parallel with blood pressure above 115/75 mm Hg (Lewington, Clarke et al. 2002). The systolic pressure and the pulse pressure are greater predictors of risk in patients over the age of 50 to 60 years (Franklin, Larson et al. 2001). Under the age of 50 years, diastolic blood pressure is a better predictor of mortality than systolic readings (Taylor, Wilt et al. 2011).

TABLE 4.3
Nonpharmacological Interventions in the Treatment of Hypertension

Nonpharmacological Intervention	Dose	Impact on Hypertension
Weight loss	Best goal is ideal body weight, but aim for at least a 1 kg reduction in body weight for most adults who are overweight.	About 1 mm Hg for every 1 kg reduction in body weight.
Healthy diet, e.g. DASH diet	Consume a diet rich in fruits, vegetables, whole-grains, and low-fat dairy products, with reduced content of saturated and total fat.	−11 mm Hg
Reduced intake of dietary sodium	Optimal goal is <1500 mg/d, but aim for at least a 1,000 mg/d reduction in most adults.	−5/6 mm Hg
Enhanced intake of dietary potassium	Aim for 3,500–5,000 mg/d, preferably by consumption of a diet rich in potassium.	−4/5 mm Hg
Aerobicactivity	90–150 min/wk 65–75% heart rate reserve	−5/8 mm Hg
Dynamic resistance activity	90–150 min/wk 50–80% 1 rep maximum 6 exercises, 3 sets/exercise, 10 repetitions/set	−4 mm Hg
Isometric resistance activity	4×2 min (hand grip), 1 min rest between exercises, 30%–40% maximum voluntary contraction, 3 sessions/wk, 8–10 wk	−5 mm Hg
Moderation in alcohol intake	Reduce alcohol to: Men: ≤2 drinks daily Women: ≤1 drink daily	−4 mm Hg

DASH=Dietary Approaches to Stop Hypertension is a diet rich in fruits, vegetables, whole grains and low–fat dairy with reduced intake of sodium, saturated and total fat (Saneei, Salehi-Abargouei et al. 2014) (http.//dashdiet.org/dash_diet_tips.asp.). In the United States, one 'standard' drink contains roughly 14 g of pure alcohol, which is typically found in 12 oz (1 oz=30 ml) of regular beer (usually about 5% alcohol), 5 oz of wine (usually about 12% alcohol), and 1.5 oz of distilled spirits (usually about 40% alcohol).

4.3.4 TREATMENT OF HYPERTENSION

In the treatment of high blood pressure, we can apply nonpharmacological and pharmacological tools (Reboussin, Allen et al. 2018).

According to the 2017 ACC/AHA Guideline, the most effective nonpharmacological interventions are based on behavioral strategies aimed at lifestyle change. These factors are summarized in Table 4.3.

With these nonpharmacological interventions, we may be able to prevent the evaluation of hypertension and could normalize blood pressure in patients with stage 1 hypertension, and they should be an integral part of the management of patients with stage 2 hypertension.

The decision to initiate drug therapy should be individualized and involve shared decision-making between patient and provider (Figure 4.1). According to the 2017 ACC/AHA Guideline, the following patients should receive pharmacological treatment as complementary to the nonpharmacological one (Reboussin, Allen et al. 2018):

- Patients with out-of-office daytime blood pressure ≥135 mmHg systolic or ≥85 mmHg diastolic.
- Patients with an out-of-office blood pressure (mean home or daytime ambulatory) ≥130 mmHg systolic or ≥80 mmHg diastolic together with one of the following conditions:
  - Established clinical cardiovascular disease (e.g. stable ischemic heart disease, heart failure, carotid disease, previous stroke or peripheral arterial disease).
  - Type 2 diabetes mellitus.
  - Chronic kidney disease.
  - Aged 65 years or older.
  - An estimated 10-year risk of atherosclerotic cardiovascular disease of at least 10%.

The appropriate choice of pharmacological agent must depend on the individual characteristics of the patient, i.e. age, comorbidities, concurrent medication, drug adherence, drug interactions, the overall treatment regimen and out-of-pocket costs should be considered. Many patients can be started on a single agent; however, in the case of stage 2 hypertensive patients, two or more drugs with different mechanisms of action may be necessary and more useful (Gradman, Basile et al. 2011). Those antihypertensive drugs that are scientifically proven to decrease clinical events (i.e. CVD, cerebrovascular events or death) should be used preferentially (1967). The primary agent used in the treatment of high blood pressure must be one of the following: thiazide diuretics, angiotensin-converting enzyme (ACE) inhibitors, angiotensin receptor blockers (ARBs) or calcium-channel blockers (CCBs) (Law, Morris et al. 2009). Drug regimens with complementary activity, where a second antihypertensive agent is used to block compensatory responses to the initial agent or affect a different pressor mechanism, can result in additive lowering of BP (Reboussin, Allen et al. 2018). For example, thiazide diuretics may stimulate the renin-angiotensin-aldosterone

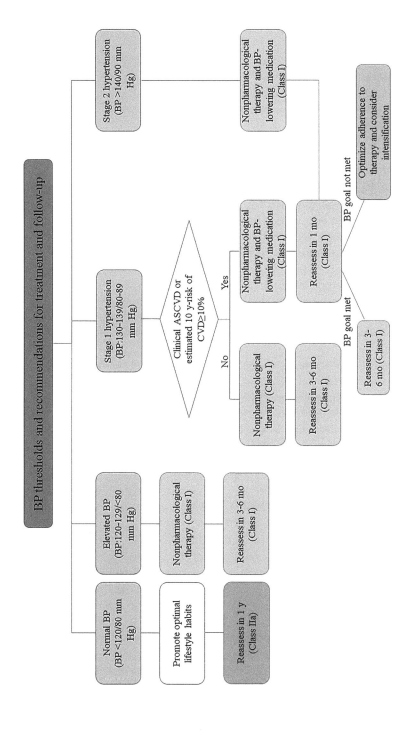

FIGURE 4.1 High BP treatment recommendations. ASCVD=atherosclerotic cardiovascular disease.

system. By adding an ACE inhibitor or ARB to the thiazide, an additive BP-lowering effect may be obtained (Doumas, Tsioufis et al. 2014).

The different antihypertensive agents are listed below:

- **Thiazide or thiazide-type diuretics**: Chlorthalidone, Hydrochlorothiazide, Indapamide, Metolazone.
- **ACE-inhibitors:** Benazepril, Captopril, Enalapril, Fosinopril, Lisinopril, Moexipril, Perindopril, Perindopril, Quinapril, Ramipril, Trandolapril.
- **ARBs:** Azilsartan, Candesartan, Eprosartan, Irbesartan, Losartan, Olmesartan, Telmisartan, Valsartan.
- **CCB-dihydropyridines**: Amlodipine, Felodipine, Isradipine, Nicardipine SR, Nifedipine LA, Nisoldipine.
- **CCB-nondihydropyridines**:Diltiazem ER, Verapamil IR, Verapamil SR, Verapamil-delayed onset ER.
- **Loop diuretics**: Bumetamide, Furosemide, Torsemide.
- **Potassium sparing diuretics**: Amiloride, Triamteren.
- **Aldosterone antagonist diuretics**: Eplerenone, Spironolactone.
- **Cardioselective β-blockers**: Atenolol, Betaxolol, Bisoprolol, Metoprolol tartrate, Metoprolol succinate.
- **Cardioselective and vasodilatory β-blockers**: Nebivolol.
- **Noncardioselective β-blockers**: Nadolol, Propranolol LA, Propranolol IR.
- **β-blockers with intrinsic sympatomimetic activity**: Acebutol, Pindolol, Penbutol.
- **β-blockers—combined alpha- and beta-receptor**: Carvedilol, Carvedilol phosphate, Labetalol.
- **Direct renin inhibitor**: Aliskiren.
- **α-1 blockers**: Doxazosin, Prazosin, Terazosin.
- **Central α-2-agonist and other centrally acting drugs**: Clonidine oral, Clonidine patch, Methyldopa, Guanfacine.
- **Direct vasodilators**: Hydralazine, Minoxidil.

4.4   MINERALS AND HYPERTENSION

4.4.1   SODIUM

Sodium chloride is an essential component of our diet; however, the amount of intake should be kept under strict control. An ever-growing body of evidence informs us about the many different risks related to high sodium intake, including higher blood pressure, as well as an increased risk for cardiovascular diseases such as coronary heart disease, myocardial infarction, left ventricular hypertrophy or cerebrovascular accidents (stroke), renal insufficiency and others (Kotchen and McCarron 1998; Yang, Liu et al. 2011).

Many human experiments, studies and surveys have been carried out to assess the effects of dietary sodium intake, including the INTERSALT study (1988), the National Health and Nutrition Examination Surveys (NHANES) (Alderman, Cohen et al. 1998), the Trials of Hypertension Prevention (TOHP) (Cook, Cutler et al. 2007) or the Dietary Approaches to Stop Hypertension (DASH) experiment (Sacks, Appel et al. 1999), each tending to confirm the importance of salt reduction.

In many countries, including the low- or middle-income countries, (e.g. Hungary) or even the high-income US, the average salt intake is very high. According to the most recent (i.e. 2014) data, sodium intake of men in Hungary is an average of 6.2 g (which equals 15.9 g sodium chloride), while these numbers in women are 4.4 g (=11.2 g NaCl) (Nagy, NagyLorincz et al. 2017). Unfortunately, in the case of men, the highest sodium intake was characteristic of the youngest age class, while the lowest value was detectable in the oldest. In the case of sodium intake of women, age group differences were not observable. A little more favorable is the situation in the US, where there is a tendency of a continuously decreasing sodium intake based on the NHANES carried out on a regular two-year basis. However, according to the most recent data (2014), for Americans two years old or older, the average daily sodium intake is higher than 3,400 mg (=8,640 mg of NaCl) (Jackson, Cogswell et al. 2018). This, indeed, is still an improvement given the fact that formerly this number was an average 5,000 mg/day, while in some areas of the country the average sodium intake even reached the 15,000–20,000 mg/day levels (Kotchen and McCarron 1998).

It is not only kitchen salt used at the table or during cooking that contributes to the high sodium intake. Such 'discretionary' salt is a considerable 36% of our salt intake. However, the other part, the non-discretionary intake, consists of the salt present in the food itself (circa 10%) and the sodium used for preservation purposes and to enhance the flavor of food. The latter constitutes the remaining 54%, which is a very large amount of our intake as well (Spagnolo, Giussani et al. 2013). Within the latter group of processed foods, much of the sodium comes from bakery products and meat products (e.g. in Hungary 40% and 30%, respectively) (Nagy, Nagy-Lorincz et al. 2017). Food preservatives and flavor enhancers also often contain sodium in the form of acid-salts such as sodium benzoate and sodium sorbate, or sodium-glutamate. Thus, as our nutrition shifted over the ages to a diet including more processed and/or precooked foods with additives (such as sugars, fats and sodium), our sodium intake increased as well (Subasinghe, Arabshahi et al. 2016).

The daily recommendation of sodium intake in the United States for the population over 19 years of age is 2,300 mg/day. While this reduction may reduce cases of high blood pressure by 11 million annually (Palar and Sturm 2009), in some countries in Europe the recommendation may get even stricter—as in Germany, where the German Nutrition Society set a goal of 1,500 mg/day (Nagy, Nagy-Lorincz et al. 2017). In Hungary, the recommendation is set to the WHO value of 2,000 mg/day sodium intake, which equals practically 5.0 g (just under a teaspoon) of NaCl per day (Nagy, Nagy-Lorincz et al. 2017). Beside the apparent health gains, reducing the amount of salt intake also has economic benefits; according to a published calculation, reduction of average sodium intake of the US population to 2,300 mg/day may save as much as $18 billion dollars in health care (Palar and Sturm 2009).

According to 2014 NHANES measurements, there is a clear correlation between sodium intake and blood pressure: the higher the urinary sodium excretion, the higher the systolic and diastolic blood pressure is. In numbers: each 1,000 mg difference means 4.58 mmHg and 2.25 mmHg of systolic and diastolic blood pressure, respectively (Jackson, Cogswell et al. 2018). Some data analysing review studies demonstrate a lesser degree of decrease (−2.08 mmHg/g systolic blood pressure) in the case of salt reduction (Mente, O'Donnell et al. 2016); others have shown even more than a 5.5 mmHg-decrease (−5.51/−2.9 mmHg/g systolic/diastolic blood pressure) in case of white people with hypertension (Graudal, Hubeck-Graudal et al. 2017). Nevertheless, a linear association between hypertension and progressively higher sodium intake is established (Subasinghe, Arabshahi et al. 2016).

The mechanism of action of the above-mentioned correlation is well-studied, but nevertheless, it is still not fully understood. Important factors include endothelial cells, endothelial nitric oxide synthase (eNOS) and nitric oxide (NO), as well as kidney and aldosterone.

Sodium homeostasis is regulated by the kidneys. Here, in the case of excess sodium intake, its reabsorption results in a loss of potassium. The reabsorbed sodium prevents water molecules being excreted from the systemic circulation, resulting in a direct blood pressure increasing effect. Nevertheless, in the case of an even higher sodium intake, reabsorption cannot keep pace with the increment and eventually, beside the pre-existing increased blood pressure, the sodium concentration in the urine will increase as well. This is the basis of measuring dietary sodium from urinary sodium excretion, which can account for 95–98% of its intake (Bentley 2006).

Vascular endothelial cells are directly affected by increased sodium levels; their size and volume decreases, their deformability reduces, which results in stiffness, higher sensitivity to blood flow turbulences and shear-stress, and their balance between oxidative and antioxidant mediators is lost, resulting in increased oxidative stress, probably due to lower activity of eNOS for example (Oberleithner, Riethmuller et al. 2007; Kanbay, Chen et al. 2011). Vascular smooth muscle cells are also affected by the high sodium and/or low potassium levels; by inhibiting eNOS, and thus the production of NO, vasoconstriction and reduced vasodilation will develop (Kanbay, Chen et al. 2011).

All the aforementioned vascular events are further deteriorated if aldosterone is present (Fels, Oberleithner et al. 2010). The mechanism behind the aldosterone-sensitivity of endothelial cells may be related to epithelial sodium channels located on the surface of endothelial cells activated by aldosterone (Kusche-Vihrog, Callies et al. 2010). Through these channels, endothelial cells react to increased serum sodium by altering the viscosity of their plasma membrane and the submembranous actin network (Oberleithner, Kusche-Vihrog et al. 2010), which eventually results in reduction of physiological vasodilation and in increased stiffness (Kusche-Vihrog and Oberleithner 2012). Difference in salt-sensitivity or insensitivity, i.e. individual responses to sodium overload, may be attributable to expression or activity of such protein products of genes as epithelial sodium channels, among probable multiple others.

Although the exact mechanisms underlying the blood pressure elevating effects of sodium are not totally clear, and even though the relationship between these two is questioned by some researchers, from the high number of published comments of different cardiological-related authors on the *Lancet* article (Mente, O'Donnell et al. 2016), we can conclude lowering sodium intake is recommended—not mindlessly, as there are also articles about the elevated risks of cardiovascular diseases in case of too low salt intake (O'Donnell, Mente et al. 2014; Pfister, Michels et al. 2014; Saulnier, Gand et al. 2014)—but to a circa 2 g sodium per day limit.

4.4.2 POTASSIUM

The role of potassium is fundamentally intertwined with sodium: the balance between these two ions is not only important in blood pressure-control, but also in prevention of cardiovascular and cerebrovascular events (O'Donnell, Mente et al. 2014). By increasing the sodium/potassium ratio it is shown that the risk of cardiovascular disease and mortality also increases (Messerli, Schmieder et al. 1997).

According to recent data (NHANES 2014), dietary intake of potassium in the US is an average 2.0 g/day (Jackson, Cogswell et al. 2018), and some sources report even lower values (1.75 g/day (Houston 2011)), while in lower-income countries such as Hungary, it is an average 3.0 g/day (Nagy, Nagy-Lorincz et al. 2017). Such differences may be attributed to income-related dietary habits such as possible differences in meat and vegetable intake, based on the lower cost and easier access to the latter in lower-income countries (Bailey, Akabas et al. 2017), and it is well known that vegetables and fruits are the main sources of potassium, while meat contains lower values (Weaver 2013). For example, in Hungary about half of the potassium-intake comes from plant-derived foods—vegetables, fruits and cereals—while meat is responsible for circa 20% and dairy products account for 15% (Nagy, Nagy-Lorincz et al. 2017).

According to NHANES 2014 survey, a 1,000 mg-increase in potassium intake will result in a 3.72 mm Hg decrease (Jackson, Cogswell et al. 2018) and thus many articles emphasize the importance of increased potassium intake (Aburto, Hanson et al. 2013; McDonough, Veiras et al. 2017). What seems to be even more important and clinically more relevant is the relationship between sodium to potassium (Na/K) ratio and cardiovascular risk: findings of many articles including analyses of randomized controlled trials, showed that Na/K ratio appears to be more strongly related to blood pressure in hypertensive adult populations than either sodium or potassium alone (Perez and Chang 2014; Mirmiran, Bahadoran et al. 2018). Accordingly, sodium-to-potassium ratio is linearly associated with developed hypertension and a 0.5-unit difference is directly proportional to systolic blood pressure (1.72 mm Hg, (Jackson, Cogswell et al. 2018)). Some reviews showed even greater differences per one-unit increase of Na/K ratio (4.33–4.39 mm Hg (Yin, Deng et al. 2018)).

The exact mechanism of potassium to interfere with blood pressure is well-studied but not fully known. Sodium and potassium together play an important role in endotheliumdependent vasodilation; while sodium promotes a reduction in the synthesis of the vasodilator nitric oxide (Fujiwara, Osanai et al. 2000), potassium works against it by decreasing cytosolic calcium in vascular smooth-muscle cells, hindering contraction (Haddy, Vanhoutte et al. 2006). Dietary potassium may also inhibit sodium-sensitivity in a dosedependent manner (Getz and Reardon 2007). Other articles report other proposed mechanisms as well by which potassium may influence blood pressure including natriuresis, modulation of baroreceptor sensitivity or even altering DNA synthesis and thus decreasing enzyme levels such as NADPH oxidase, causing oxidative stress and/or inflammation (Ando, Matsui et al. 2010; Houston 2011).

The recommended potassium-intake based on all the above-mentioned information is clearly more than present values; the guidelines of the American Heart Association suggest an increase in potassium intake to 120 mmol/day (4.7 g/day), which was also incorporated into the Dietary Approach to Stop Hypertension

(DASH) diet (Appel, Brands et al. 2006). These recommendations are further supported by European Society of Hypertension and WHO guidelines as well (Whitworth and Chalmers 2004; Mancia, Fagard et al. 2013). However, increased potassium levels alone are not always statistically enough to be associated with blood pressure decrease, mainly due to within-individual variabilities in urinary potassium versus sodium excretion: only 77–90% of consumed potassium is excreted in urine while this number in the case of sodium is 90–95% (Jackson, Cogswell et al. 2018). Sodium-to-potassium ratio, on the other hand, correlates well with blood pressure (Jackson, Cogswell et al. 2018). Unfortunately, at present the Na/K ratio of many populations is about 3:1; in the US, 3.2–3.4 (Jackson, Cogswell et al. 2018), in Hungary (central Europe), it is 2.0 (Nagy, Nagy-Lorincz et al. 2017), in China, it is 2.8 ± 0.8 (Yin, Deng et al. 2018). The recommended Na/K ratio based on the recommended sodium and potassium amounts is from 1:2 to 1:3, and there are also articles that suggest an even higher (1:5) ratio (Houston 2011).

4.4.3 MAGNESIUM

As is the case with potassium, magnesium is also inversely related to hypertension; in most studies, higher dietary intake of magnesium is linked with lower blood pressure, although the correlation is not so consistent as with sodium and potassium (Houston and Harper 2008; Schutten, Joosten et al. 2018).

Foods rich in magnesium include cereals, nuts and vegetables (Song and Liu 2012). Dietary intake of magnesium in the US is an average 265 mg per day (Rosanoff, Weaver et al. 2012; Zhang and Qiu 2018).There is some deviation from the average in male and female values, with female amounts being significantly lower than male, but even the higher values are insufficient to meet the magnesium requirements. The recommended dietary allowance is 300/400 mg per day (male/female) (1997). In a lower-income country such as Hungary, magnesium intake is an average 420 mg per day (Nagy, Nagy-Lorincz et al. 2017), probably due to the relative lower costs of magnesium-rich foods compared to other, magnesium-poor foods such as meat or many fast foods commonly eaten in the US (Bailey, Akabas et al. 2017).

Magnesium insufficiency may lead to chronic diseases such as cardiovascular diseases, including hypertension (de Baaij, Hoenderop et al. 2015). The mechanisms of action of magnesium include: decreasing the inflammation-related serum C-reactive protein levels (Zhang and Qiu 2018), increasing the synthesis of the vasodilator prostaglandin E_1 (PGE1), prostacyclin (PGI2) and nitric oxide (NO) and improving endothelial function (Houston 2013; Schutten, Joosten et al. 2018; Tangvoraphonkchai and Davenport 2018). Magnesium also acts as a direct vasodilator due to competition with sodium for binding sites on the surface of vascular smooth muscle cells (Houston 2014). It is also needed for ATP-generation and all ATP-dependent reactions that require hydrolization and transfer of phosphate groups (Tangvoraphonkchai and Davenport 2018).

According to recent data, a 100 mg/day increase in magnesium intake lowers the risk of hypertension by a significant 5% (Han, Fang et al. 2017), and a circa 370 mg/day magnesium-dose significantly reduces both systolic and diastolic blood pressure (–2.0 and 1.78 mmHg, respectively (Zhang, Li et al. 2016)). Even as high as 500–1,000 mg/day doses have been tried out in controlled trials and were found to produce a significant blood pressure decrease (Widman, Wester et al. 1993), based on which we can conclude that magnesium supplementation has a good safety profile and may have potential for the control of blood pressure. Of course, we must also emphasize that excessive elevation of magnesium intake may prove harmful. Caution is advised especially in renal insufficiency, or in the case of patients taking medications that induce magnesium retention.

4.4.4   CALCIUM

The balance between other nutrients and ions, such as sodium, potassium, magnesium and calcium, is important in decreasing cardiovascular and cerebrovascular events as well (Houston and Harper 2008); high calcium intake as well as high sodium or low potassium and low magnesium intake, has been linked with hypertension (Schutten, Joosten et al. 2018). However, data on the relationship between calcium intake and blood pressure is controversial and inconsistent.

Usual average calcium intake in the US is about 1,000 mg/day (Bailey, Akabas et al. 2017), while in a lower-income country as Hungary, it is about 700 mg/day (Balk, Adam et al. 2017). Most important sources of calcium are dairy products, and to a lesser extent, vegetables (Nagy, Nagy-Lorincz et al. 2017).

Some studies found no associations at all between calcium intake and risk of hypertension (Liu, Fang et al. 2018), whilst some found little blood pressure lowering effect of lower dietary calcium intake (−0.11/−0.08 mmHg, systolic and diastolic blood pressure, respectively) (Billington, Bristow et al. 2017; Skowronska-Jozwiak, Jaworski et al. 2017). Others have found, in complete contrast to this, that lower calcium intake may be associated with elevated blood pressure (Magalhaes, Pessoa et al. 2017; Skowronska-Jozwiak, Jaworski et al. 2017). These statistically significant, but practically small, differences in blood pressure values, and the contradictory results make it impossible to clearly draw a conclusion and make a definitive statement regarding calcium intake to control blood pressure.

4.4.5   ZINC

In the pathogenesis of hypertension, zinc may participate in blood pressure regulation; an inverse association was found between dietary zinc intake and systolic blood pressure (Kim 2013). Zinc is also known to induce taste acuity for salt (McDaid, Stewart-Knox et al. 2007), which may further contribute to blood pressure decrease through decreased sodium intake: people with zinc deficiency may tend to use more salt in their foods, which results in higher blood pressure (Saltman 1983).

Zinc is an essential trace biometal (Carpenter, Lam et al. 2013). Dietary zinc intake varies in the global population at about 7–10 mg/day based on geographical, gender and dietary habit differences (Kim 2013; Vance and Chun 2015; Nagy, Nagy-Lorincz et al. 2017).

Unfortunately, the relationship between zinc and blood pressure is not an easy task to determine, because the amount of zinc varies in different fluid compartments in different ways; zinc supplementation may cause an increase in plasma zinc concentration, but a reduction in erythrocyte zinc (Paz Matias, Costa e Silva et al. 2014). Intracellular zinc may cause contraction of muscular cells of vessels due to inhibiting ATP-dependent calcium pump and thus Ca^{2+}-efflux from cells, while extracellular zinc blocks the calcium channel (Tubek 2007). Also, there are contradictory articles about indirect blood pressure-elevating actions of excess zinc intake through impairing renal function (Yanagisawa, Miyazaki et al. 2014), but most articles assess zinc intake to be beneficial in lowering blood pressure (Bergomi, Rovesti et al. 1997; Canatan, Bakan et al. 2004; Carpenter, Lam et al. 2013). A suggested daily intake may be as high as 50 mg/day (Houston 2014).

4.4.6 COPPER

Copper, like zinc, is also an essential trace metal, and may have an effect on blood pressure. However, scientific data on the copper–hypertension relationship is contradictory, as both increased and decreased levels of copper have been found in hypertensive patients as well. An indirect effect on blood pressure was established in a paper where copper deficiency was shown to lead to anemia due to reduced hemoglobin synthesis, which contributes to increased blood pressure (Saltman 1983). Others also demonstrated a decreased level of copper in hypertensive patients (Taneja and Mandal 2007). In other articles, however, copper level was found to be increased or zinc/copper ratio was decreased in hypertension (Bergomi, Rovesti et al. 1997; Canatan, Bakan et al. 2004; Ghayour-Mobarhan, Shapouri-Moghaddam et al. 2009; Carpenter, Lam et al. 2013). One such study measured the level of daily dietary copper intake to be 1.3 mg/day (Lee, Lyu et al. 2015). However, one cannot infer a clear conclusion whether to elevate or to lower this value to decrease blood pressure.

4.4.7 CHROMIUM

A few articles can be found (Preuss, Gondal et al. 1995; Preuss, Grojec et al. 1997; Preuss, Jarrell et al. 1998; Perricone, Bagchi et al. 2008) regarding a possible blood pressurelowering effect of the trivalent ion of chromium, a trace element found only in micrograms per food serving (Thor, Harnack et al. 2011). In a recent double-blind, placebocontrolled randomised trial, however, no association between blood pressure and chrome supplementation (600ug) has been found (Kim, Kim et al. 2018), although it should be noted that the administered form of chromium was CrCl3 instead of other, more effective formulations such as niacin-bound chromium or chromium-picolinate (Preuss, Echard et al. 2008), though it also raises questions about the effectiveness of the non-chromium parts of these formulations. Formerly, there was also a proposed link between blood pressure change and the activity of chromium to improve glycemic control; however, even the latter effect of chromium could not be clearly demonstrated (Kleefstra, Houweling et al. 2006; Kleefstra, Houweling et al. 2007; Ali, Ma et al. 2011). Not to mention the risk of chromium overexposure: chromium may prove nephrotoxic in higher doses, especially in case of coexposure with lead and cadmium (Tsai, Kuo et al. 2017). Thus, currently chromium supplementation is scientifically unsupported.

4.4.8 SELENIUM

Deficiency or excessive intake of selenium, an important antioxidant trace element, may both lead to adverse cardiovascular effects (Wells, Goldman et al. 2012). Although in a normal nutritional range it may lower blood pressure (Nawrot, Staessen et al. 2007; Wells, Goldman et al. 2012), in other publications dietary selenium had inconsistent effects on blood pressure (Taittonen, Nuutinen et al. 1997; Kuruppu, Hendrie et al. 2014), and higher selenium levels were even associated with hypertension (Laclaustra, Navas-Acien et al. 2009; Berthold, Michalke et al. 2012). Based on the available scientific data, there is no conclusive evidence to support an association between selenium and hypertension.

4.4.9 MANGANESE

Despite manganese being a toxic heavy metal, it may play a role in blood pressure management. A study-group has found a significant negative correlation between manganese intake and blood pressure (Lee, Lyu et al. 2015); others confirmed the same from toenail measurements (Mordukhovich, Wright et al. 2012). Most recently, the same negative association of urinary manganese and blood pressure was also established based on a comprehensive national study in the US (Wu, Woo et al. 2017). Manganese dietary intake is about 2–3 mg/day (Nkwenkeu, Kennedy et al. 2002); however, manganese supplementation is not advised, as excessive exposure may lead to many toxicity symptoms, including neurological ones, as well as hypertension (Jiang and Zheng 2005; Cowan, Zheng et al. 2009; Chrissobolis and Faraci 2010; Lee and Kim 2011).

4.5 MACRONUTRIENTS AFFECTING HYPERTENSION

4.5.1 PROTEIN

Observational studies show an association between blood pressure decrease and high protein intake; at least a modest inverse association exists according to a review of cross-sectional studies, meaning a −0.20 mm Hg systolic decrease per 25 g total protein intake (Tielemans, Altorf-van der Kuil et al. 2013), but other studies show an even higher reduction such as −1.76 and −2.9 mm Hg systolic and 1.15 and 1.7 mm Hg diastolic blood pressure decrease (Rebholz, Friedman et al. 2012; Tielemans, Kromhout et al. 2014). Some differences exist between the sources of protein intake as well, based on recent, comprehensive analyses; in most cases plant-derived proteins produce more beneficial results than animal proteins in lowering the risk of hypertension (Elliott, Stamler et al. 2006; Wang, Yancy et al. 2008; Rebholz, Friedman et al. 2012; Tielemans, Kromhout et al. 2014). There are also some cases when animal proteins proved better in reducing blood pressure (Umesawa, Sato et al.

2009; Liu, Dang et al. 2013), a possible reason for which might be a higher proportion of fish in protein intake of animal origin.

We should not forget, however, that protein intake is high in energy, which means that a higher protein intake for the sake of blood pressure reduction can be detrimental to body mass index (BMI), especially in early childhood (Weber, Grote et al. 2014). Furthermore, scientific literature provides insufficient evidence for a consistent association between blood pressure and protein intake in children (Voortman, Vitezova et al. 2015; Lind, Larnkjaer et al. 2017), thus, increased protein intake is not advised in case of children. This is not the case with adults, however, where high protein diets are usually recommended for weight loss purposes (Sargrad, Homko et al. 2005; Santesso, Akl et al. 2012), although, in these cases one cannot rule out the effects of weight loss itself,and lower carbohydrate intake on blood pressure beside the presumed effect of higher intake of proteins. In such dietapproaches, protein intake in energy percentages can reach as high as 30% (Sargrad, Homko et al. 2005), without the practical elevation of protein intake as opposed to common diet; here the ratio between protein and carbohydrates is suggested to be changed for weight loss purposes, while retaining a daily recommended protein intake of 1–1.5 g/kg body weight (Houston 2014). However, for hypertensive patients with type 2 diabetes, higher protein intake and meat consumption is not recommended, as it may be associated with higher blood pressure values (Mattos, Viana et al. 2015). Other disadvantages of higher protein intake may include adverse gastrointestinal events as well (Santesso, Akl et al. 2012). A reason for this might be the source of the plant-derived proteins, as many of these vegetables and seeds may produce abdominal bloating and distension; sources of plant proteins include beans, lentils, peanuts, almonds and many kinds of different seeds, potato, spinach, corn, avocado, broccoli, Brussels sprouts, etc. (Richter, Skulas-Ray et al. 2015).

A reason for effectiveness of higher protein intake—apart from the lower intake of other food components with high risk for cardiovascular diseases, such as fats and carbohydrates—may also be associated with active ingredients in the ingested protein mix; many types of different proteins and peptides are shown to have angiotensin-converting enzyme (ACE)-inhibitor effect (FitzGerald, Murray et al. 2004; Aihara, Kajimoto et al. 2005; Ricci, Artacho et al. 2010; Cicero, Aubin et al. 2013), or to induce natriuresis or alter catecholamine responses (Kawasaki, Seki et al. 2000). Amino acids in proteins may also provide beneficial effects such as L-arginine, a potent source for generation of nitric oxide (NO) by nitric oxide synthase (NOS) (McRae 2016), or L-carnitine, an antioxidant and antiinflammatory aminoacid with antihypertensive effects (Wang, Liu et al. 2018). Taurine, a sulphur-containing beta-amino acid, may also act on the nitric oxide synthesis to produce a blood pressure-lowering effect through dilation of vascular smooth muscles (Chen, Guo et al. 2016; Sun, Wang et al. 2016).

4.5.2 Fat and Fatty Acids

In a study comparing different diets rich in either proteins (animal and plant ratio 1:1), or carbohydrates or unsaturated fat (predominantly monounsaturated fat, MUFA), the blood pressure-lowering effect of a protein-rich and monounsaturated diet proved to be in the same order of magnitude (Appel, Sacks et al. 2005). It is also well-known that omega-3 polyunsaturated fatty acids (PUFA) have cardioprotective effects (Mori 2006; Noreen and Brandauer 2012); their anti-inflammatory actions are also famous (Calder 2015). The difference between the two groups is only the number of double-bonds in the hydrocarbon side chain of the fatty acids: in case of PUFAs, two or more exist. From the end of the side chain, the double-bond starting from the 3rd, 6th or 9th carbon determines their name as omega-3, −6 or −9 fatty acids, omega-9 being mostly monounsaturated, while −3 and −6 usually being polyunsaturated (Houston 2014). According to recent studies, PUFAs decrease blood pressure significantly compared to saturated fatty acids (Livingstone, Givens et al. 2013), especially omega-3 fatty acids which are associated inversely with blood pressure, most specifically docosahexaenoic acid (DHA) and eicosapentaenoic acid (EPA), the latter producing a −4.7 mm Hg decrease per 0.25 g increase in intake (Ueshima, Stamler et al. 2007; O'Sullivan, Bremner et al. 2012). Eating fish three times a week, as they are rich in MUFAs and PUFAs, can be an effective countermeasure against high blood pressure, and fish-proteins may also have synergistic antihypertensive properties (Houston 2007).

4.5.3 CARBOHYDRATES

Low glycemic index and low glycemic load diets are both associated with reduced risk for cardiovascular diseases (Barclay, Petocz et al. 2008). Lowering of carbohydrate intake is usually not an isolated phenomenon; most commonly it is done in favor of protein intake (mainly for weight loss purposes), a topic on which a few scientific articles can be found. According to one such study where dietary intake of carbohydrates was exchanged for protein, significant systolic blood pressure reduction (−5.2 mm Hg) could be seen beside a somewhat higher fasting plasma glucose level (Hodgson, Burke et al. 2006). (The latter effect could also be considered healthy as—according to common belief—maintaining a blood glucose level without fluctuations produces less hunger; however, a recent paper demonstrated there is no relationship between the two (Schultes, Panknin et al. 2016).) In another study, where saturated fat was replaced with either carbohydrate-, protein- or unsaturated fat-rich diet, even the carbohydrates proved to lower blood pressure (−8.2 mm Hg), although a further blood pressure decrease (1.4 mm Hg) could be seen if a protein- or unsaturated fat-diet was used (Appel, Sacks et al. 2005).

The mechanism behind such blood pressure-increasing effects of carbohydrates may be related to the vasodilatory effects of insulin, and the consequential, sympathetic counterregulation and over-activity (Kopp 2005).

The contribution of carbohydrates to hypertension is not an easy matter to decipher. On the one hand there is their evident effect leading to obesity and consequently, higher blood pressure and diabetes; on the other hand some articles talk about insulin-sensitizing and the blood pressure-lowering effect of high carbohydrate intake (Sargrad, Homko et al. 2005; Zhu, Lin et al. 2014). The link between the seemingly controversial results might be found in the quality of carbohydrates; while the dietary glycemic index is inversely associated, the solid/total carbohydrate ratio, crude fibre intake and whole grain/total grain ratio are positively correlated with cardiovascular diseases, which was calculated by a recent article into a so-called carbohydrate quality index (Kim, Kim et al. 2018). According to this study, it absolutely does matter what kind of carbohydrate is consumed, as their carbohydrate quality index was associated with the risk of obesity as well as hypertension.

4.5.4 FIBERS

Beside their evident beneficial effects on changing the consistency of feces, fibers are associated with reduced cardiometabolic risks, including hypertension; in a most recent metaanalysis a significant −1.59/-0.39 mm Hg systolic/diastolic blood pressure decrease was proven (Khan, Jovanovski et al. 2018). The same result was reinforced in the case of diabetic patients as well in a cross-sectional study, with a mean decrease in blood pressure of 9.2 and 5.6 mm Hg, systolic and diastolic respectively (Beretta, Bernaud et al. 2018). Crude fiber intake was included in a calculated carbohydrate quality index as well by another study, where this index, and so the fiber intake, was inversely related to the prevalence of hypertension (Kim, Kim et al. 2018). Other trials also confirmed the same beneficial effects of fibers via improvement in blood pressure (He, Streiffer et al. 2004; Streppel, Arends et al. 2005; Whelton, Hyre et al. 2005; Aleixandre and Miguel 2016). Daily recommended fiber intake is 14 g fiber/1,000 kcal based on American Diabetes Association guidelines (Evert, Boucher et al. 2013).

4.6   MICRONUTRIENTS AND HYPERTENSION

4.6.1 VITAMINS

Blood pressure-lowering effects of many vitamins and vitamin supplements have been recorded; amongst the most important vitamins, vitamin C and D were included in the most studies (Ceriello, Motz et al. 2000; Puglisi 2013). The many effects of vitamin C may contribute to its blood pressure-lowering effect, including, but not limited to: inducing sodium diuresis, improving endothelial function, increasing nitric oxide and PGI2 synthesis, decreasing binding affinity of angiotensin II-receptor, activating potassium channels, reducing cytosolic calcium, etc. (Houston 2014). Based on systematic reviews and metaanalyses, a 500 mg/day dose of vitamin C may lower systemic and diastolic blood pressures by −3.84 and −1.48 mm Hg, respectively (Juraschek, Guallar et al. 2012).

As for vitamin D, results are not so unequivocal. Vitamin D is a key nutrient in regulating calcium and bone homeostasis (DeLuca 2004); it also has actions on the reninangiotensin-aldosterone system (Lee, O'Keefe et al. 2008) and improves endothelial functions (Sugden, Davies et al. 2008), and as such it also regulates blood pressure; metaanalyses and prospective studies demonstrate an inverse relationship between vitamin D and blood pressure (Kunutsor, Burgess et al. 2014). According to some studies, for a 1 ng/ml increase in plasma vitamin D level, a −0.2 mm Hg blood pressure decrease can be realized (Forman, Scott et al. 2013). However, there are also contradictory results existing in the literature showing the ineffectiveness of vitamin D supplementation (Beveridge, Struthers et al. 2015; Pilz, Gaksch et al.

2015; Cremer, Tambosco et al. 2018; Shu and Huang 2018), only a non-significant reduction in blood pressure (Kunutsor, Burgess et al. 2014) or efficacy only in case of hypertensive individuals with vitamin D insufficiency (Mirhosseini, Vatanparast et al. 2017). But taking into account its many other beneficial effects, vitamin D supplementation can be advised: the recommended daily dose is 600 IU, that is, 15 ug per day (EFSA 2016).

Vitamin E supplementation showed controversial results: a study demonstrated beneficial effects decreasing both systolic and diastolic blood pressure in patients with type 2 diabetes mellitus (Rafraf, Bazyun et al. 2012), while in another study carried out also with diabetic patients, vitamin E supplementation was proven to increase blood pressure significantly (Ward, Wu et al. 2007). A research group drew the conclusion based on their study, that vitamin E supplementation has no clinical relevance (Palumbo, Avanzini et al. 2000), but, according to another trial, vitamin E supplementation in a 200 IU/day-dose might be recommended for blood pressure reduction in mild hypertensive patients (Boshtam, Rafiei et al. 2002).

In case of Vitamin B supplementation, dietary intake of B12 may be related to lower blood pressure (Tamai, Wada et al. 2011), while if intravenously administered, it increases mean blood pressure by an average of 12 mm Hg (Ried, Travica et al. 2016). In another study, no assocations were observed between vitamin B12 and blood pressure (Husemoen, Skaaby et al. 2016). In the Healthy Lifestyle in Europe by Nutrition in Adolescence (HELENA) study, holotranscolbalamin, a marker of vitamin B12 status, was inversely related to blood pressure (de Moraes, Gracia-Marco et al. 2014), while levels of red blood cell folate (vitamin B9) and vitamin B6 in blood was found to be positively associated with it in adolescents. In contrast to this, oral pyridoxine (vitamin B6) supplementation significantly reduced systolic and diastolic blood pressure in many other studies (Aybak, Sermet et al. 1995; van Dijk, Rauwerda et al. 2001; Noori, Tabibi et al. 2013; Dakshinamurti and Dakshinamurti 2015), a proposed mechanism of action of which might be the reduction of homocysteine levels (van Dijk, Rauwerda et al. 2001). A review found a link between vitamin B2 and hypertension as well, but blood pressure rises only in individuals with a special genetic predisposition based on gene-polymorphism (Wilson, McNulty et al. 2010).

Vitamin A levels were consistently shown to be positively associated with blood pressure; in a cross-sectional study this was only proven in male adolescents (de Moraes, Gracia-Marco et al. 2014), although intracranial pressure increase was confirmed in many other studies as well (Libien and Blaner 2007; Perera, Sandamal et al. 2014; Mohammad, Raslan et al. 2016). A probable mechanism may seem to unfold in recent articles showing a link between serum retinol binding protein and the regulation of blood pressure (Chiba, Saitoh et al. 2010; Kraus, Sartoretto et al. 2015; Zhang, Zhu et al. 2017).

In case of vitamins, increasing intake cannot be clearly suggested to control hypertension; however, their many diverse effects and the fact that these molecules are essential for any living human, make them important—and indispensable—to consume, regardless of their particular effects on blood pressure.

4.6.2   COENZYME Q10 (UBIQUINONE)

Coenzyme Q10 (CoQ10) is a naturally occuring fat-soluble benzoquinon found in high concentrations in the mitochondria of aerobic organisms and it plays a key role in the respiratory transport chain. It is an electron carrier for oxidative phosphorylation and for adenosine triphosphate production, and has an important function in cell membrane stabilization (DiNicolantonio, Bhutani et al. 2015). There is evidence that insufficient contractility of myocardial cells seen in patients who suffer from heart failure is exacerbated by CoQ10 deficiency. Nevertheless, the clinical severity of heart failure seems to correlate with CoQ10 deficiency (Rasmussen, Glisson et al. 2012). Baggio et al. (1994) in a multicenter, postmarketing drug surveillance study, examined the safety and clinical efficacy of CoQ10 as an adjunctive treatment in patients with congestive heart failure, who were treated with conventional therapy for at least six months. Patients with congestive heart failure of NYHA II and III were followed over a three-month period. Compared to the baseline measurements, significant reductions were observed in respiratory rate, systolic blood ressure, diastolic blood pressure and in heart rate for 50–150 mg daily CoQ10 dosage (Chagan, Ioselovich et al. 2002). Despite the free radical scavenging and membrane stabilizing theories, it is still unclear how CoQ10 exactly lowers BP. Nonetheless, it has been suggested to increase nitric oxide bioavailability and/or boost the production of prostacyclin (Lonnrot, Porsti et al. 1998).

4.7   ANTIHYPERTENSIVE HERBS AND THEIR MECHANISM OF ACTION

A diverse range of plant or herbal extract and its metabolites can alter the cascade implicated in the pathogenesis of hypertension. The most favorably modulated cellular pathways by herbs or their extracts in the pathogenesis of hypertension are as follows: reactive oxygen species (ROS), nitric oxide (NO), hydrogen sulfide (H_2S), Angiotensin II and the ReninAngiotensin-Aldosterone (RAA) system (Figure 4.2) (Al Disi, Anwar et al. 2015).

4.7.1   ANTIOXIDANT DEFENSE

4.7.1.1   Flavonoids

There are thousands of different flavonoids with potent free radical scavenger activity that have been proven to protect cells from oxidative stress. In the past decades it has become apparent that the oxidation of lipids, also known as lipid peroxidation, is a crucial step in the pathogenesis of several diseases (Gaschler and Stockwell 2017). Via inhibition of harmful lipid peroxidation, flavonoids have long been used for a variety of cardiovascular conditions, especially to prevent atherosclerosis and hypertension.

4.7.1.2   *Vitis vinifera* (Grapevine)-Resveratrol

Vitis vinifera is one of the most widely consumed fruit in the world. Due to the fact that 90–95% of grape polyphenols exist in the seeds and skin, grape is widely

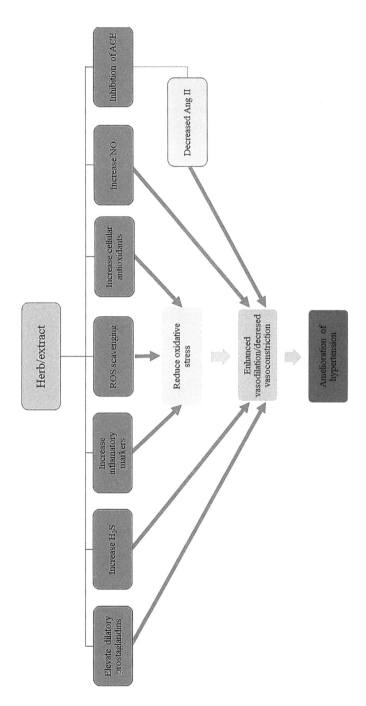

FIGURE 4.2 Most favorably modulated molecular or cellular pathways by herbs or their extracts in the pathogenesis of hypertension.

considered as one of the major sources of natural antioxidants, including flavanol, quercetin, catechins, procyanidins, anthocyanidins and last but not least, resveratrol (Nassiri-Asl and Hosseinzadeh 2009; Nassiri-Asl and Hosseinzadeh 2016). Polyphenols from the grape give health benefits, such as: improvement of cognition and neuronal function with aging and neurodegenerative disease, anti-inflamation, antimicrobe and antitumor properties and reducing the incidence of cardiovascular diseases. The resveratrol containing 250 mL of either regular or dealcoholized red wine, reduced augmentation index, impoved arterial compliance and lowered central arterial pressure in a human study (Hashimoto, Kim et al. 2001). There was a significant reduction in the aortic augmentation index with the dealcoholized (6.1%) and with the regular red wine (10.5%) as well. There was a significant decrease in central arterial pressure of 7.4 mmHg by dealcoholized red wine, and of 5.4 mmHg by regular red wine. Moreover, resveratrol is able to increase flow-mediated vasodilation in a dose-related manner. Uncoupling of eNOS and Angiotensin II blocking mechanism is supposed to be behind the vasodilatory and antihypertensive effects (Bhatt, Lokhandwala et al. 2011; Dolinsky, Chakrabarti et al. 2013). *Ex vivo* analyses demonstrated that resveratrol prevents AngIIinduced aortic contractions via AMPK activation (Cao, Luo et al. 2014). For adults, the recommended dose of trans-resveratrol is 1 g/day (Weiskirchen and Weiskirchen 2016).

4.7.1.3 *Lycopersicon esculentum* (Tomato)-Lycopene

Lycopene is a naturally occuring fat-soluble red carotenoid phytonutrient found in tomatoes, guava, pink grapefruit, apricots, papaya and watermelon. The large carotenoid family of plant pigments has two subtypes: hydrocarbon carotenoids and xanthophylls. Lycopene is a hydrocarbon type carotenoid, as it is composed entirely of hydrogen and carbon. Of the carotenoids, lycopene is the most effective singlet molecular oxygen scavenger *in vitro* (Sies and Stahl 1995). Some clinical trials have also shown a reduced effect in serum lipid peroxides and oxidized LDL of lycopene (Agarwal and Rao 1998; Hadley, Clinton et al. 2003). Moreover, a double-blind, placebo-controlled pilot study proved that a short-term treatment with tomato extract can reduce blood pressure in patients with grade-I hypertension, while no significant changes were found in lipid status (Engelhard, Gazer et al. 2006). These findings unanimously confirm the relevance of lycopene administration in the treatment of blood pressure.

4.7.1.4 *Camellia sinensis* (Green Tea)

There are many potential health benefits of habitual green tea consumption, namely antioxidative, anti-inflammatory, anti-proliferative, anti-thrombogenic, lipid lowering and last, but not least, antihypertensive. The most prominent effects of green tea on human health are mainly attributed to catechins (e.g. epigallocatechin-3-gallate), which belong to the flavonoids, like polyphenols or flavanols (Khan and Mukhtar 2007). A meta-analysis of 13 randomized control trials showed that green tea consumption significantly decreased BP in subjects with Stage 1 hypertension or prehypertension, and the beneficial effects of green tea on lowering BP were more prominent in subjects with grade-I hypertension (Peng, Zhou et al. 2014). Consistent with these findings, *in vivo* studies on hypertensive rats have revealed that green tea extract can significantly reduce their BP and enhance endothelial function (Negishi, Xu et al. 2004; Potenza, Marasciulo et al. 2007). The following mechanisms are predicted to underlie the beneficial effect of green tea to control blood pressure: 1) Green tea extract can balance vasoconstrictor molecules, including prostaglandins, endothelin-1, angiotensin II and vasodilating molecules, such as prostacyclin and endothelium-derived hyperpolarizing factors (Bhardwaj and Khanna 2013); 2) Green tea and green tea extract exert beneficial effects via increasing nitric oxide (NO) production from endothelium (Potenza, Marasciulo et al. 2007; Babu and Liu 2008); 3) As a catechin-rich plant, green tea reduces the oxidative stress by inhibiting the generation of ROS, the expression of nuclear factor-kappa B, redox-sensitive transcription factors, and activator protein-1 (Cabrera, Artacho et al. 2006; Agarwal, Prasad et al. 2010). In addition, green tea catechins can induce an anti-inflammatory effect by the suppression of several inflammatory factors, such as cytokines, and adhesion molecules (Lin and Lin 1997).

4.7.2 DIURETICS

4.7.2.1 *Taraxacum officinale* (Dandelion)

The first reference to its application is reflected in its Greek name 'taraxis,' meaning inflammation and 'akeomai' meaning curative. The commonly used French name, 'pissenlit' (bedwetter) refers to its diuretic action. In the following, scientific investigations supporting the diuretic properties ascribed to *Taraxacum* are reviewed. Firstly, a study performed on male rats proved that the aqueous extract of dandelion herb (*Taraxaci herba*) has consistently stronger diuretic effect than dandelion roots (*Taraxaci radix*) at a dose of 50 mL/kg body weight. Moreover, 8 g dried herb/kg body weight had comparable diuretic and saluretic indices to 80 mg/kg body weight of furosemide. In addition, the high potassium content of the leaf compensated for the potassium loss by urinary elimination. Thus, dandelion is devoid from the potassium loss-derived incidences such as hepatic coma and circulatory collapse, which are commonly occuring side-effects in furosemide therapy (Racz-Kotilla, Racz et al. 1974). Furtheremore, Grases et al. could not find significant differences in diuresis on the prevention of kidney stone formation in female Wistar rats receiving an aqueous dandelion root extract compared to the control group (Grases, Melero et al. 1994). The diuretic activity of dandelion may be due to its ascorbic acid, caffeic acid, chlorogenic acid, isoquercitrin, luteolin, magnesium, mannitol and inulin content via different diuretic and saluretic pathways (Duke 2005).

4.7.2.2 *Petroselinum crispus* (Parsley)

In folk medicine parsley has been used as a remedy for its powerful diuretic effect. Several studies ascribe additional hypotensive properties to the plant which could be considered a consequence of its diuretic effect (Farzaei, Abbasabadi et al. 2013). Kreydiyyeh and Usta (2002) investigated this evidence and aimed to uncover its underlying mechanism. Rats treated with aqueous parsley seed extract eliminated a significantly larger volume of urine as compared to the control group. These finding were supported by an *in situ* kidney perfusion technique with similar results as well.

According to this, they assume that the diuretic effect of extract is mediated through an increase in K^+ retention in the lumen. In conclusion, the mechanism of action of parsley seems to be mediated through an inhibition of the Na^+–K^+ pump that would lead to a decrease in Na^+ and K^+ reabsorption leading thus to an osmotic water flow into the lumen, and diuresis (Kreydiyyeh and Usta 2002).

4.7.2.3   *Urtica dioica* (Stinging Nettle)

The antihypertensive effect of the herb is a well-known fact from traditional medicine. Based on this, a number of studies examined, then published its acute diuretic, natriuretic, hypotensive, cardiovascular, immunostimulatory and anti-rheumatic effect (Wagner, Willer et al. 1989; Tahri, Yamani et al. 2000; Testai, Chericoni et al. 2002). Nettle is nutritionally high in vitamins A, C and D, and also in minerals iron, manganese, potassium and calcium, but not sodium (Bisht, Bhandari et al. 2012).

Tahri, Yamani et al. (2000) demonstrated a progressive reduction of the arterial blood pressure following a continuous intravenous perfusion of aqueous extract of *U. dioica* that depended on the dose perfused. Both diuresis and natriuresis increased in parallel to the dose perfused. Authors suggest that the diuretic and natriuretic effects is derived from a sodium surcharge independent pathway, due to the negligible electrolyte content of the plant (Tahri, Yamani et al. 2000). The observed acute hypotensive effect of the extract may assume a direct action on the cardiovascular system. Another study demonstrated antihypertensive effect by vasodilation under *in vitro* and *in vivo* circumstances as well. According to their results, they state NO release and Ca^{2+} inhibition as a dual underlying mechanism. Based on their earlier preliminary phytochemical analysis, the crude extract of *U. dioica* is rich in alkaloids that may have vasorelaxant properties (Qayyum, Qamar et al. 2016).

4.7.3   ACE-INHIBITORS

4.7.3.1   *Allium ursinum* (Ramsons)

Different extracts obtained from the fresh leaves of *Allium ursinum* were tested *in vitro* on human platelet aggregation. The results showed a significant inhibitory activity of the ethanol extract on ADP-induced aggregation. The mechanism of action was similar to that of a reference drug Clopidogrel (Hiyasat, Sabha et al. 2009). It was suggested that the active compounds exerting antiaggregatory effect are 1,2-d i-O-α-linolenoyl-3-O-β-D-galactopyranosyl-sn-glycerol (DLGG) and β-sitosterol 3-O-β-D-glucopyranoside (Sabha, Hiyasat et al. 2012).

Even though *Allium sativum* and *Allium ursinum* are closely related, the available scientific literature about the latter is significantly modest, although the few *in vitro* and *in vivo* results of comparative studies are promising (Sendl, Elbl et al. 1992; Hiyasat, Sabha et al. 2009). When groups of ten spontaneously hypertensive rats (SHR) were fed diets containing either 1% w/w regular garlic (*Allium sativum*) (AS) or 1% w/w wild garlic (*Allium ursinum*) (AU) for 45 days, the final mean systolic blood pressure (SBP) was reduced significantly compared to controls (C: 189; AS: 175; Au: 173 mmHg) (Preuss, Clouatre et al. 2001). The possible underlying mechanisms include the ability of ramsons to inhibit the activity of angiotensin-converting enzyme (ACE). *In vitro* tests on the water extract from the leaves (at the concentration of 0.3 mg/ml) showed a significantly increased activity on enzyme inhibition when compared to leaves with extract of garlic (58 vs. 30%). *In vivo* experiments on rats fed a standard diet for eight weeks, with 2% of pulverized *A. ursinum* leaves showing a significantly lower plasma ACE activity in the ramsons group as compared to those in the control group (Sendl, Elbl et al. 1992). A study in 2017 highlighted the possible protective effects of *Allium ursinum* lyophilisate supplementation in right-side heart failure caused by pulmonary arterial hypertension (Bombicz, Priksz et al. 2017). The mechanism(s) underlying this effect may involve a direct vasodilating and antiproliferative action of enhanced heme oxygenase 1 (HO-1) derived endogenous carbon monoxide level (Christou, Morita et al. 2000). It was proven several times by investigators that the flavonoidrich plants are potent HO inducers, and this is true for *Allium ursinum* as well (Juhasz, Kertesz et al. 2013; Bombicz, Priksz et al. 2016). Moreover, the free radical scavenging activity of flavonoids may be a probable potent endothel stabilizing mechanism against the pathogenesis of hypertension (Mihaylova, Lante et al. 2014; Bombicz, Priksz et al. 2017). On the other hand, the beneficial effect of wild garlic on the lipid profile in hypercholesterinemia takes a prominent place among the other agents in the treatment of several cardiovascular disorders (Bombicz, Priksz et al. 2016).

4.7.3.2   *Zingiber officinale* (Ginger)

Ginger has been used in traditional medicine for the management and prevention of hypertension and other cardiovascular diseases, therefore it gained the attention of several research teams seeking the underlying antihypertensive mechanism of action. Akinyemi et al. suggest that the possible mechanism through which ginger exerts its BP lowering effect may be through the inhibition of ACE activity and inhibition of sodium nitroprusside-induced lipid-peroxidation. They found that the red ginger showed a greater inhibition of ACE activity than the white one, possibly due to its polyphenol content (Akinyemi, Ademiluyi et al. 2013). To the contrary, another team suggested that the blood pressure-lowering effect of ginger is mediated through the blockade of voltage-dependent calcium channels. They investigated the vasodilatory effect of the crude extract of ginger in rabbit thoracic aorta preparation, and found endothelium-independent blockade of the extract. In more detail, the extract relaxed the phenylephrine-induced vascular contraction at a dose ten times higher than that required against K^+-induced contraction. Ca^{2+}channel-blocking activity was confirmed when ginger shifted the Ca^{2+} dose-response curves to the right, similar to the effect of verapamil (Ghayur and Gilani 2005).

4.7.3.3   *Angelica sinensis* (Danggui)

Radix Angelicae sinensis is often used in Traditional Chinese medicine. More than 70 compounds, including alkyl phthalides, benzenoids, butylphthalide, coumarins, flavones, organic acids, polysaccharides and trepenes, have been isolated from roots of *Angelica sinensis* (Chen, Li et al. 2013). It is used to promote blood flow and has been tested in the treatment of myocardial ischemia-reperfusion injury in animal models (Mak, Chiu et al. 2006; Wu and Hsieh 2011). According to a study, both pre- and post-treament with danggui extract reversed all of the Angiotensine II-induced apoptosis performed on H9c2 cardiomyoblast cells (Huang, Kuo et al. 2014). This result assumed that the herb has Angiotensin II (Ang II) suppressant property. The results of Cao et al. confirm this hypothesis. The 2-kidney, 1-clip (2K, 1C) rat model of hypertension was used to investigate the potential antihypertensive effect of the root of angelicae extract. After ten weeks of treatment, renal Ang II level was significantly lower (along with the mean blood pressure) compared to the control group (Cao, He et al. 2013). The antioxidant properties are assessed by the authors to be in line with the antihypertensive effect.

4.7.4 VASORELAXANT ACTIVITY: DIRECT VASODILATORS

Experimental studies have provided strong evidence that oxidative stress could inactivate nitric oxide, impairing endothelium-dependent vasodilation. Consequently, endothelial dysfunction and subsequent vascular remodelling have a tight relationship with the pathogenesis of hypertension. Any agent that improves overexpression of eNOS and subsequent increase in NO production, might be beneficial in attenuating hypertension.

4.7.4.1 *Allium cepa* (Onion)

The major components of alliums, including onion, are organopolysulfides. The preventive potency of polysulfides in different chronic diseases including cancer, obesity, metabolic syndrome, gastric ulcer and cardiovascular disorders are well-known. The mechanism of polysulfides in cardioprotection are based on its hydrogen sulfide (H_2S) donor property. H_2S, next to nitric oxide (NO) and carbon monoxide (CO), is the third gasotransmitter and synthesized enzymatically from l-cysteine or l-homocysteine. H_2S exerts its endotheliumdependent vasorelaxation by three main mechanisms: 1) activating NO synthase and inhibiting cGMP degradation by phosphodiesterase 5 (PDE5), thus potentiating the effect of NO-cGMP pathway; 2) H_2S-derived polysulfides directly activate protein kinase G; and finally 3) H_2S interacts with NO to form nitroxyl (HNO)-a potent vasorelaxant (Beltowski and Jamroz-Wisniewska 2014). Sakai et al. support this with their *in vivo* experiment on NO synthase inhibitor-induced hypertensive rats and spontaneously hypertensive rats (2003).

4.7.4.2 *Allium sativum* (Garlic)

Garlic has long been used for a variety of cardiovascular conditions, especially hyperlipidemia and hypertension. It is thought to increase nitric oxide production, resulting in smooth muscle relaxation and vasodilatation, the same way as other *Allium* species. Garlic preparations have been found to be superior to placebo, in reducing BP on average by 8–9 mmHg in SBP and 6–7 mmHg in DBP in 20 patients with hypertension, compared to 20 patients with normal BP after a two-month treatment period (Reinhart, Coleman et al. 2008). The results have revealed decreased BP, simultaneously significant reduction of 8hydroxy2deoxyguanosin, levels of nitric oxide and lipid peroxidation, and an increased level of antioxidative vitamin C and E. Several mechanisms of action for the BP-lowering properties of garlic have been postulated, including mediation of intracellular nitric oxide (NO) and hydrogen sulfide (H2S) production, as well as blockage of angiotensin-II production (Beltowski and Jamroz-Wisniewska 2014; Shouk, Abdou et al. 2014). Due to the fact that animal and cell culture experiments were mainly conducted with fresh garlic compounds, containing allicin (S-allyl-cysteine sulfoxide), which has a very low sustained bio-availability in human tissues, the antihypertensive effect of garlic via the proposed ACEI mechanism seems less plausible than its H_2S-stimulating and NO-regulating properties (Lawson and Gardner 2005).

4.7.4.3   *Cinnamomum zeylanicum* (Cinnamon)

Cinnamon is a well-known and worldwide preferred functional food. Many previous studies using the aqueous extract of *C. zeylanicum* have demonstrated the insulin-potentiating properties in the treatment of diabetes and cardiovascular disease (CVD) related to insulin resistance. Naturally occuring type A polyphenols and chromium found in cinnamon have been shown to improve insulin sensitivity (Anderson 2008). Earlier studies suggest that cinnamon prevents the development of insulin resistance, at least in part by enhancing insulin signaling and possibly via the NO pathway in skeletal muscle. An aqueous extract of cinnamon has been shown to improve insulin sensitivity in humans as well (Wang, Anderson et al. 2007). Cinnamon extract also appears to regulate glucose uptake-related genes, such as GLUT-1, GLUT-4, glycogen synthesis 1 and glycogen synthase kinase 3β mRNA expression in adipose tissue (Qin, Polansky et al. 2010). As for its antihypertensive effect, an investigation performed on rat isolated aortic rings showed endothelium, NO, K_{ATP}dependent vasorelaxant by aqueous extract of its stem bark (Nyadjeu, Dongmo et al. 2011).

4.7.5   CALCIUM CHANNEL BLOCKER (CCB)

4.7.5.1   *Piper nigrum* (Black Pepper)

Dried fruits of *Piper nigrum* (black pepper) are commonly used in gastrointestinal disorders. Its principal alkaloid, piperine is responsible for the antispasmodic effect via opioid agonist and Ca^{2+} antagonist activity (Mehmood and Gilani 2010). According to Taqvi, Shah et al. (2008), intravenous administration of piperine caused a dose-dependent decrease in mean arterial preassure in normotensive rats. After that, in a Langendroff's perfused rabbit heart preparation, piperine caused partial inhibition and verapamil caused inhibition of force and rate of ventricular contraction and coronary flow. In addition, piperine inhibited high K^+ precontraction and partially inhibited phenylephrine. Moreover, simirarly to verapamil, piperine caused a rightward shift in calcium-concentration-response curves. Both in rat aorta and in bovine coronary artery preparations, piperine demonstrated K^+ precontraction endothelium-independent vasorelaxation. Based on these data, piperine exerts its blood pressure lowering effect via the blockage of Ca^{2+} channels (Taqvi, Shah et al. 2008).

In conclusion, several different results can be traced in the literature regarding the pathways of how a single herb exerts its antihypertensive effect, probably due to their multiple biologically, and thus medically active compounds. Table 4.4 includes (but is not limited to) the above-mentioned herbs' possible targets in hypertension based on the overviewed scientific literature.

TABLE 4.4
Antihypertensive Herbs and Their Mechanism of Action

Drug name	Effect
Vitis vinifera-resveratrol	Scavenges ROS
	Increase eNOS
	Inhibits Ang II
Lycopersicon esculentum	Scavenges ROS
	Lower LDL
Camellia sinensis	Scavenges ROS
	Anti-inflammatory
	Increased NO
Coenzyme Q10	Scavenges ROS
	Unclear
Taraxacum officinale	Diuretic
	Saluretic
Petroselinum crispus	Inhibition of Na^+-K^+ pump
	Diuretic
Urtica dioica	Saluretic
	NO release
	Ca^{2+} inhibition
Allium ursinum	Scavenges ROS
	Induce HO-1
	ACE inhibition
	Inhibit thrombocyte aggregation
Zingiber officinale	ACE inhibition
	Ca2+ antagonist
Radix Angelicae	Suppress Ang II
	Scavenges ROS
Allium cepa	Increases NO
	Increases H_2S
	Increases eNOS
Allium sativum	Increases NO
	Increases H_2S
	Increases eNOS
	ACE inhibition
Cinnamomum zeylanicum	Increase NO
Piper nigrum	Ca^{2+} antagonist

4.8   INTERACTIONS OF ANTIHYPERTENSIVE HERBS

Nowadays, the use of herbal medication is becoming more frequent, and without medical supervision, it can have the potential to be dangerous or even life-threatening. One of the most important phenomena is the interaction between drugs and herbal medicines. An interaction occurs when the effect of a molecule is changed.

TABLE 4.5
Most Relevant Interactions of Antihypertensive Herbal Agents

Drug Name and PK Effect	Interaction	Effect
Vitis vinifera-(resveratrol) CYP3A4 inhibition, and CYP1A2, CYP2C19 weak inhibition. Clinically significant antiplatelet effects	Anticoagulants or antiplatelet drugs (Aspirin, Clopidogrel, Warfarin)	Reduced platelet aggregation, risk of bleeding.
	NSAID (diclofenac)	Had NO significant effect.
	Food	No interaction found.
	Other herbal medicines	No interaction found.
	Mephenytoin	Resveratrol weakly inhibits the metabolism of (S)-mephenytoin.
	Paclitaxel	Resveratrol moderately inhibits the paclitaxel metabolism.
Lycopersicon esculentum (lycopene)	Colchicine	Colchicine reduces the absorption of lycopene.
	Food (low-fat diet)	Lower lycopene level.
	Food (high-fat monounsaturated fatenriched diet or high carbohydrate low-fat diet)	No significant difference in the serum level of lycopene.
	Herbal medicines (betacarotene)	In a clinical study a single 60 mg dose of betacarotene and 60 mg of lycopene appeared that AUC level of lycopene was dramatically increased. Please, NOTE that 300 mg beta carotene daily for three weeks decreased the level of endogenous lycopene level.
	Lipid lowering drugs: Cholestyramine, Probucol	Reduce the serum level of lycopene.
	Orlistat	Reduce the absorption of lycopene.
	Sucrose polyester ((Olestra) non-absorbable, non-calorific fat ingredient in snack food)	Reduce the serum level of lycopene.
Camellia sinensis Tea contains caffeine, theophhylline, theobromine, falvonoids	Antihypertensives (ACEi, Beta-blockers, Calcium channel blockers, Nitrates)	Moderate increase in blood pressure.
	Antiplatelet drugs: Aspirin, Clopidogrel	Increased absorption of aspirin. Bleeding time was also prolonged. Prothrombin time and thrombin time was not affected.
	Food, Milk	Do not have effect.
	Herbal medicines	No clinical interaction was found.
	Iron compounds	May reduce the absorption of iron.
	Irinotecan	Green tea catechins were unlikely to inhibit the formation of active metabolites of irinotecan.
	Warfarin	Green tea (high level of vitamin-K_1) may reduce the INR

(Continued)

TABLE 4.5 (CONTINUED)
Most Relevant Interactions of Antihypertensive Herbal Agents

Drug Name and PK Effect	Interaction	Effect
Taraxacum officinale Dandelion, Lion's tooth	Ciprofloxacin	Maximum concentration of ciprofloxacin was reduced. The Taraxacum mongolicum extract has high level of magnesium, calcium and iron. These ions have high complex binding capacity with fluoroquinolones (i.e. ciprofloxacin, ofloxacin, norfloxacin, levofloxacin, moxifloxacin, etc.).
Petroselinum crispus Parsley The most important compounds are the natural coumarins and flavonoids	Aminophenazone	In mice study, parsley potentiated and prolonged the analgesic action of aminophanazone (maybe reduced metabolism of aminophenazone by cytochrome P450).
	Lithium	In a case report lithium toxicity occurred after taking the herbal remedy.
	Paracetamol	In mice study, parsley potentiated and prolonged the analgesic action of paracetamol (maybe reduced metabolism of paracetamol by cytochrome P450).
	Pentobarbital	In mice study, parsley extract significantly extended the sleeping time, due to the reduced metabolism of pentobarbital by cytochrome P450.
	Warfarin	In a case report INR was decreased. Important to note that the aforementioned patient took (Nature's Life Greens) which contained a number of plants beside parsley
Urtica dioica Nettle	No clinical interaction.	No relevant pharmacokinetic data found.
Zingiber officinale Ginger, Zingiber, Gan Jiang	The leaves contain flavonoids, therefore, please, see the interactions of flavonoids.	
	Anticoagulants	Many case reports describe markedly raised INRs with coumarin derivates.
	Caffeine	Sho-saiko-to (ginger is one component from 7); In one study the metabolism of caffeine was reduced.
	Nifedipine	Hence nifedipine also has antiplatelet effects, therefore these combinations have synergistic effect.
	Tolbutamide	Sho-saiko-to (ginger is one component from 7); might increase or decrease the absorption of tolbutamide.

(Continued)

TABLE 4.5 (CONTINUED)
Most Relevant Interactions of Antihypertensive Herbal Agents

Drug Name and PK Effect	Interaction	Effect
Radix Angelicae Chinese angelica, Dang Gui, Dong quai. The major components include natural coumarins, and phthalides	Diazepam	In a rat *in vivo* study, the AUC of diazepam was markedly increased, and the muscle relaxant effect of i.v. diazepam was also potentiated. Increased SEDATION!
	Nifedipine	Inhibit the nifedipine oxydase activity, therefore decrease the nifedipine metabolism.
	Oestrogens or Oestrogen antagonists	Hence, Chinese angelica may contain oestrogenic component, therefore it could result additive or oppose effects of oeastrogens. In *in vivo* animal study Chinese extract inhibited the binding of estradiol to the oestrogen receptor and increase uterine growth. Chinese angelica extract could directly stimulate breast cancer growth due to the oppose mechanism of Tamoxifen (oestrogen receptor antagonist). The authors believe that women with estrogen sensitive cancers have to avoid the herbs containing estrogen derivate.
	Tolbutamide	Chinese angelica inhibited the activity of tolbutamide hydroxylase, therefore AUC of tolbutamide was increased.
	Warfarin	Increased prothrombin time and INR.
Allium sativum Ajo, Allium Sulfur-containing compounds, alliin, allicin, flavonoids, kaempferol	ACE inhibitors	In a case report garlic with 15 mg of Lisinopril developed marked hypotension.
	Alcohol	In mice study, garlic inhibited the metabolism of alcohol.
	Antiplatelet drug	Garlic increased the antiplatelet activity. The ajoene compound inhibits the fibrinogen binding to the fibrinogen receptor.
	Benzodiazepine	In clinical studies the garlic tablets do NOT influence the metabolisms of alprazolam and midazolam.
	Chloroxazone	Metabolism of chloroxazone is inhibited by garlic due to the inhibition of P450 CYP2E1.
	Fish oil	Antioxidant and lipidlowering effects are enhanced.

(Continued)

TABLE 4.5 (CONTINUED)
Most Relevant Interactions of Antihypertensive Herbal Agents

Drug Name and PK Effect	Interaction	Effect
	HIV-protease inhibitors	In cell line study allicin decreased the efflux of ritonavir.
		In one human study also reduced the plasma levels of saquiavir.
	Isoniazid	In a study using rabbits, the orally given garlic extract reduced the AUC of isoniazid.
	Paracetamol	Garlic could protect the hepatotoxicity of paracetamol, due to the inhibition of CYP2E1 which responsible for the synthesis of N-acetyl-p-benzoquinoneimine.
	Warfarin	INR increasing. Increased anticoagulant effect.
	Fluindione	INR decreased, decreased anticoagulant effect.
Piper nigrum Pepper	Barbiturates	Study in rats: increased the pentobarbital sleeping time (enzyme inhibition).
Constituents: alkaloids and alkylamines (piperine)	Beta-lactam antibiotics Amoxicillin	Single dose rat study (10/20 mg/kg); Increased AUC level of amoxicillin.
	Cefadroxil	No effect.
	Cefotaxime	Single dose rat study (10/20 mg/kg); Increased AUC level of cefoxatime.
	Carbamazepine	In human study the AUC of carbamazepine was increased.
	Ciclosporin	In an *in vitro* study the transport of ciclosporin by glycoprotein was inhibited by piperine.
	Digoxin	Piperine inhibits the transport of digoxin by PGP.
	Q10	Piperine increased the AUC of Q10 vitamin.
	Rhodiola	Piperine reduce the antidepressant action of rhodiola.
	Turmeric	Increased bioavailability of turmeric.
	Isoniazid	Decreased AUC level of isoniazid.
	Nevirapine	Increased AUC of nevirapine.
	NSAIDs: Diclofenac	In rabbit study: AUC of diclofenac was reduced using Trikatu (ginger, black papper, long pepper).
	Indometacin	Increased plasma maximum level of indomethacin using Trikatu.
	Oxyphenbutazone	In mice and rats: single dose piperine 10 mg/kg increased the AUC and maximum plasma levels of oxyphenbutazone.
	Oxytetracycline	Study in hens: long pepper increased the AUC of oxytetracycline.

(*Continued*)

TABLE 4.5 (CONTINUED)
Most Relevant Interactions of Antihypertensive Herbal Agents

Drug Name and PK Effect	Interaction	Effect
	Phenytoin	Piperine enhance the oral bioavailability of phenytoin. Increased AUC and maximum plasma level.
	Propranolol	Piperine enhanced the oral bioavailability of propranolol. Increased AUC and maximum plasma level.
	Rifampicin	Piperine increased the AUC of rifampicin.
	Theophylline	Piperine dramatically increased the AUC of theophylline.
	Thyroid and antithyroid drugs	2.5 mg/kg piperine lowered the T3 and T4 concentration.
	Verapamil	*In vitro* study: inhibited metabolism of verapamil.
Flavonoids (only related to cardiovascular drugs)	Anticoagulant or antiplatelet drugs	Cocoa flavanols and procyanidins decreased the collagen- and ATP-induced platelet aggregation. Onion soup high in quercetin inhibited collagen stimulated platelet aggregation.
	Calcium channel blockers	In rat and rabbit study: morin, naringin, quercetin increased the AUC of diltiazem nimodipine, and verapamil.
	Digoxin	Pig study: 50 mg/kg of quercetine dramatically increased the plasma level of digoxin. Due to the narrow therapeutic window of digoxin it could be a life-threatening drug interaction.
	Enalapril	In a rat study 2 mg/kg kaempferol increased the AUC level of enalapril.
	Quinine and quinidine	In a rat study: flavonoids increased the AUC level of quinine and quinidine.
	Statins	Kaempferol and naringenin increased the AUC levels of statins. Important to note that this interaction could lead to rhabdomyolysis, and myopathy.

Sources: (European Scientific Cooperative on 2003; Williamson, Driver et al. 2013; Preston and Stockley 2016)

An herbal agent can amplify or negate the mechanisms of action of a drug (tablet, capsule, drops, etc.), leading to dangerous side-effects or complications. A very typical interaction can occur when OTC (overthecounter) medications and herbal medications are used together. Therefore, the authors would like to summarize in Table 4.5 the most important drug interactions of the aforementioned herbal remedies. We would also like to emphasize that the data obtained on drug inter-actions is not always based on clinical evidence and case reports, but on animal experiments or *in vitro* studies. Therefore, it is necessary to consider all the infor-mation, in order to avoid any dangerous effects. Please note, that interaction is a double-edged sword, meaning that they can also be useful, and can be applied, for instance, to increase bioavailability in the case of poorly absorbed or rapidly eliminated molecules.

4.9   CONCLUSIONS

Cardiovascular disease (CVD) remains the leading cause of debility and premature death, and hence is a major public health problem. Hypertension is the main cause of morbidity and mortality associated with CVDs. Even though many drugs, rang-ing from diuretics, renin inhibitors, ACE-inhibitors, ARBs, α-adrenergic blockers, β-adrenergic blockers, vasodilators to calcium channel blockers are used to manage blood pressure levels, only 34% of patients' blood pressure is controlled.

The evidence-based studies strongly indicate the fact that herbs, and other natural nutraceuticals are becoming part of modern medicine in the prevention or treatment of hypertension. The pharmacological action of natural nutraceuticals modulates several molecules in the pathogenesis of blood pressure, including but not limited to: reactive oxygen species (ROS), nitric oxide (NO), hydrogen sulphide (H2S), Angiotensin II and the ReninAngiotensin-Aldosterone (RAA) system. These target sites of pathogenesis show similarity with the mechanism of action of antihyperten-sive drugs. Consequently, herbs were classified in the same manner as drugs are, i.e. diuretics, ACE-inhibitors, direct vasodilators and calcium channel blockers.

The use of complementary and alternative medicine (CAM), including minerals, vitamins and herbal remedies, is widespread and has increased worldwide over the previous years. The rationale of this expanding use of natural sources is the belief that herbs are natural, safe and effective. According to this hypothesis, more than 15 million people in the United States consume high-dose vitamins or herbal remedies. The higher percentage of this population are elderly persons, who also consume mul-tiple medications for comorbid conditions, which increases the risk of adverse herb-drug interactions. Based on these, the authors considered it particularly important to highlight the possible and most harmful adverse interactions of the abovementioned herbs. Mechanism of interaction can easily be characterized when the dynamic or kinetic properties of a certain drug or herb are wellknown. For this reason, this review attempted to collect sufficient information about the herbal remedies from previous scientific studies.

Furthermore, there are also other limitations for the use of herbal therapy for hypertension, such as the lack of enough scientific evidence of safety and efficacy and lack of quality control. Moreover, healthcare professionals have responsibility for careful and thorough communication with patients. Physicians should therefore have good knowledge of herbal remedies, for which, the authors believe, this review will provide valuable assistance.

ACKNOWLEDGMENT

The research was financed by the Higher Education Institutional Excellence Programme (NKFIH-1150-6/2019) of the Ministry of Innovation and Technology in Hungary, within the framework of the Development of Therapeutic Dosies Forms thematic programme of the University of Debrecen. The work/publication is supported by the GINOP-2.3.4-15-2016-00002 project. The project is co-financed by the European Union and the European Regional Development Fund.

ABBREVIATIONS

ABPM:	Ambulatory blood pressure monitoring
ACC:	American College of Cardiology
ACE(I):	Angiotensin-converting enzyme (inhibitor)
ADP:	Adenosine diphosphate
AHA:	American Heart Association
AMPK:	5' AMP-activated protein kinase
AngII:	Angiotensin II
ARB(s):	Angiotensin receptor blocker(s)
ATP:	Adenosine triphosphate
BP:	Blood pressure
BMI:	Body mass index
CCB(s):	Calcium-channel blocker(s)
CKD:	Chronic kidney disease
CO:	Carbon monoxide
CoQ10:	Coenzyme Q10
CT:	Computed tomography
CVD:	Cardiovascular diseases
DASH:	Dietary Approaches to Stop Hypertension (study)
DBP:	Diastolic blood pressure
DNA:	Deoxyribonucleic acid
eNOS:	Endothelial nitric oxide synthase
GLUT:	Glucose transporter
HO-1:	Heme-oxygenase 1 enzyme
LDL:	Low-density lipoprotein
mmHg:	Millimeters of mercury
MUFA:	Monounsaturated fatty acid
NADPH:	Nicotinamide adenine dinucleotide phosphate
NHANES:	National Health and Nutrition Examination Surveys
NO:	Nitric oxide
NSAID(s):	Non-steroidal anti-inflammatory drug(s)
NYHA I–IV stages :	New York Heart Association Functional Classification of heart failure
OTC:	Over-the-counter (= non-prescription)
PDE:	Phosphodiesterase
PUFA:	Poly-unsaturated fatty acid

RAA:	Renin-Angiotensin-Aldosterone system
ROS:	Reactive Oxygen Species
SBP:	Systolic blood pressure
US:	United States (of America)
WHO:	World Health Organization

REFERENCES

(1967). "Effects of treatment on morbidity in hypertension. Results in patients with diastolic blood pressures averaging 115 through 129 mm Hg." *JAMA* 202(11): 1028–1034.

(1977). "Report of the Joint National Committee on detection, evaluation, and treatment of high blood pressure. A cooperative study." *JAMA* 237(3): 255–261.

(1988). "Intersalt: an international study of electrolyte excretion and blood pressure. Results for 24 hour urinary sodium and potassium excretion. Intersalt Cooperative Research Group." *BMJ* 297(6644): 319–328.

(1997). *Dietary Reference Intakes for Calcium, Phosphorus, Magnesium, Vitamin D, and Fluoride*, National Academies Press: Washington (DC).

Aburto, N. J., S. Hanson, et al. (2013). "Effect of increased potassium intake on cardiovascular risk factors and disease: systematic review and meta-analyses." *BMJ* 346: f1378.

Agarwal, A., R. Prasad, et al. (2010). "Effect of green tea extract (catechins) in reducing oxidative stress seen in patients of pulmonary tuberculosis on DOTS Cat I regimen." *Phytomedicine* 17(1): 23–27.

Agarwal, S. and A. V. Rao (1998). "Tomato lycopene and low density lipoprotein oxidation: a human dietary intervention study." *Lipids* 33(10): 981–984.

Aihara, K., O. Kajimoto, et al. (2005). "Effect of powdered fermented milk with Lactobacillus helveticus on subjects with high-normal blood pressure or mild hypertension." *J Am Coll Nutr* 24(4): 257–265.

Akinyemi, A. J., A. O. Ademiluyi, et al. (2013). "Aqueous extracts of two varieties of ginger (Zingiber officinale) inhibit angiotensin I-converting enzyme, iron(II), and sodium nitroprusside-induced lipid peroxidation in the rat heart in vitro." *J Med Food* 16(7): 641–646.

Al Disi, S. S., M. A. Anwar, et al. (2015). "Anti-hypertensive herbs and their mechanisms of action: part I." *Front Pharmacol* 6: 323.

Alderman, M. H., H. Cohen, et al. (1998). "Dietary sodium intake and mortality: the National Health and Nutrition Examination Survey (NHANES I)." *Lancet* 351(9105): 781–785.

Aleixandre, A. and M. Miguel (2016). "Dietary fiber and blood pressure control." *Food Funct* 7(4): 1864–1871.

Ali, A., Y. Ma, et al. (2011). "Chromium effects on glucose tolerance and insulin sensitivity in persons at risk for diabetes mellitus." *Endocr Pract* 17(1): 16–25.

Anderson, R. A. (2008). "Chromium and polyphenols from cinnamon improve insulin sensitivity." *Proc Nutr Soc* 67(1): 48–53.

Ando, K., H. Matsui, et al. (2010). "Protective effect of dietary potassium against cardiovascular damage in salt-sensitive hypertension: possible role of its antioxidant action." *Curr Vasc Pharmacol* 8(1): 59–63.

Appel, L. J., M. W. Brands, et al. (2006). "Dietary approaches to prevent and treat hypertension: a scientific statement from the American Heart Association." *Hypertension* 47(2): 296–308.

Appel, L. J., F. M. Sacks, et al. (2005). "Effects of protein, monounsaturated fat, and carbohydrate intake on blood pressure and serum lipids: results of the OmniHeart randomized trial." *JAMA* 294(19): 2455–2464.

Aybak, M., A. Sermet, et al. (1995). "Effect of oral pyridoxine hydrochloride supplementation on arterial blood pressure in patients with essential hypertension." *Arzneimittelforschung* 45(12): 1271–1273.

Babu, P. V. and D. Liu (2008). "Green tea catechins and cardiovascular health: an update." *Curr Med Chem* 15(18): 1840–1850.

Baggio, E., R. Gandini, et al. (1994). "Italian multicenter study on the safety and efficacy of coenzyme Q10 as adjunctive therapy in heart failure." *Mol Aspects Med* 15: 287–294.

Bailey, R. L., S. R. Akabas, et al. (2017). "Total usual intake of shortfall nutrients varies with poverty among US adults." *J Nutr Educ Behav* 49(8): 639–646 e633.

Balk, E. M., G. P. Adam, et al. (2017). "Global dietary calcium intake among adults: a systematic review." *Osteoporos Int* 28(12): 3315–3324.

Barclay, A. W., P. Petocz, et al. (2008). "Glycemic index, glycemic load, and chronic disease risk—a meta-analysis of observational studies." *Am J Clin Nutr* 87(3): 627–637.

Beltowski, J. and A. Jamroz-Wisniewska (2014). "Hydrogen sulfide and endothelium-dependent vasorelaxation." *Molecules* 19(12): 21183–21199.

Bentley, B. (2006). "A review of methods to measure dietary sodium intake." *J Cardiovasc Nurs* 21(1): 63–67.

Beretta, M. V., F. R. Bernaud, et al. (2018). "Higher fiber intake is associated with lower blood pressure levels in patients with type 1 diabetes." *Arch Endocrinol Metab* 62(1): 47–54.

Bergomi, M., S. Rovesti, et al. (1997). "Zinc and copper status and blood pressure." *J Trace Elem Med Biol* 11(3): 166–169.

Berthold, H. K., B. Michalke, et al. (2012). "Influence of serum selenium concentrations on hypertension: the Lipid Analytic Cologne cross-sectional study." *J Hypertens* 30(7): 1328–1335.

Beveridge, L. A., A. D. Struthers, et al. (2015). "Effect of vitamin D supplementation on blood pressure: a systematic review and meta-analysis incorporating individual patient data." *JAMA Intern Med* 175(5): 745–754.

Bhardwaj, P. and D. Khanna (2013). "Green tea catechins: defensive role in cardiovascular disorders." *Chin J Nat Med* 11(4): 345–353.

Bhatt, S. R., M. F. Lokhandwala, et al. (2011). "Resveratrol prevents endothelial nitric oxide synthase uncoupling and attenuates development of hypertension in spontaneously hypertensive rats." *Eur J Pharmacol* 667(1–3): 258–264.

Billington, E. O., S. M. Bristow, et al. (2017). "Acute effects of calcium supplements on blood pressure: randomised, crossover trial in postmenopausal women." *Osteoporos Int* 28(1): 119–125.

Bisht, S., S. Bhandari, et al. (2012). "Urtica dioica L., an undervalued, economically important plant." *Agric Sci Res J* 2(5): 250–252.

Bombicz, M., D. Priksz, et al. (2016). "Anti-atherogenic properties of *Allium ursinum* liophylisate: impact on lipoprotein homeostasis and cardiac biomarkers in hypercholesterolemic rabbits." *Int J Mol Sci* 17(8): 1284.

Bombicz, M., D. Priksz, et al. (2017). "A novel therapeutic approach in the treatment of pulmonary arterial hypertension: *Allium ursinum* liophylisate alleviates symptoms comparably to sildenafil." *Int J Mol Sci* 18(7): 1436.

Boshtam, M., M. Rafiei, et al. (2002). "Vitamin E can reduce blood pressure in mild hypertensives." *Int J Vitam Nutr Res* 72(5): 309–314.

Cabrera, C., R. Artacho, et al. (2006). "Beneficial effects of green tea—a review." *J Am Coll Nutr* 25(2): 79–99.

Calder, P. C. (2015). "Marine omega-3 fatty acids and inflammatory processes: effects, mechanisms and clinical relevance." *Biochim Biophys Acta* 1851(4): 469–484.

Canatan, H., I. Bakan, et al. (2004). "Relationship among levels of leptin and zinc, copper, and zinc/copper ratio in plasma of patients with essential hypertension and healthy normotensive subjects." *Biol Trace Elem Res* 100(2): 117–123.

Cao, X., T. Luo, et al. (2014). "Resveratrol prevents AngII-induced hypertension via AMPK activation and RhoA/ROCK suppression in mice." *Hypertens Res* 37(9): 803–810.

Cao, Y. J., X. He, et al. (2013). "Effects of imperatorin, the active component from Radix Angelicae (Baizhi), on the blood pressure and oxidative stress in 2K,1C hypertensive rats." *Phytomedicine* 20(12): 1048–1054.

Carnethon, M. R., S. S. Gidding, et al. (2003). "Cardiorespiratory fitness in young adulthood and the development of cardiovascular disease risk factors." *JAMA* 290(23): 3092–3100.

Carpenter, W. E., D. Lam, et al. (2013). "Zinc, copper, and blood pressure: human population studies." *Med Sci Monit* 19: 1–8.

Ceriello, A., E. Motz, et al. (2000). "Hypertension and ascorbic acid." *Lancet* 355(9211): 1272–1273; author reply 1273–1274.

Chagan, L., A. Ioselovich, et al. (2002). "Use of alternative pharmacotherapy in management of cardiovascular diseases." *Am J Manag Care* 8(3): 270–285; quiz 286–278.

Chen, W., J. Guo, et al. (2016). "The beneficial effects of taurine in preventing metabolic syndrome." *Food Funct* 7(4): 1849–1863.

Chen, X. and Y. Wang (2008). "Tracking of blood pressure from childhood to adulthood: a systematic review and meta-regression analysis." *Circulation* 117(25): 3171–3180.

Chen, X. P., W. Li, et al. (2013). "Phytochemical and pharmacological studies on Radix Angelica sinensis." *Chin J Nat Med* 11(6): 577–587.

Chiba, M., S. Saitoh, et al. (2010). "Associations of metabolic factors, especially serum retinol-binding protein 4 (RBP4), with blood pressure in Japanese—the Tanno and Sobetsu study." *Endocr J* 57(9): 811–817.

Chobanian, A. V., G. L. Bakris, et al. (2003). "The seventh report of the joint national committee on prevention, detection, evaluation, and treatment of high blood pressure: the JNC 7 report." *JAMA* 289(19): 2560–2572.

Chrissobolis, S. and F. M. Faraci (2010). "Sex differences in protection against angiotensin II-induced endothelial dysfunction by manganese superoxide dismutase in the cerebral circulation." *Hypertension* 55(4): 905–910.

Christou, H., T. Morita, et al. (2000). "Prevention of hypoxia-induced pulmonary hypertension by enhancement of endogenous heme oxygenase-1 in the rat." *Circ Res* 86(12): 1224–1229.

Cicero, A. F., F. Aubin, et al. (2013). "Do the lactotripeptides isoleucine-proline-proline and valine-proline-proline reduce systolic blood pressure in European subjects? A meta-analysis of randomized controlled trials." *Am J Hypertens* 26(3): 442–449.

Cook, N. R., J. A. Cutler, et al. (2007). "Long term effects of dietary sodium reduction on cardiovascular disease outcomes: observational follow-up of the trials of hypertension prevention (TOHP)." *BMJ* 334(7599): 885–888.

Cook, N. R., E. Obarzanek, et al. (2009). "Joint effects of sodium and potassium intake on subsequent cardiovascular disease: the Trials of Hypertension Prevention follow-up study." *Arch Intern Med* 169(1): 32–40.

Coresh, J., G. L. Wei, et al. (2001). "Prevalence of high blood pressure and elevated serum creatinine level in the United States: findings from the third National Health and Nutrition Examination Survey (1988–1994)." *Arch Intern Med* 161(9): 1207–1216.

Cowan, D. M., W. Zheng, et al. (2009). "Manganese exposure among smelting workers: relationship between blood manganese-iron ratio and early onset neurobehavioral alterations." *Neurotoxicology* 30(6): 1214–1222.

Cremer, A., C. Tambosco, et al. (2018). "Investigating the association of vitamin D with blood pressure and the renin-angiotensin-aldosterone system in hypertensive subjects: a cross-sectional prospective study." *J Hum Hypertens* 32(2): 114–121.

Dakshinamurti, S. and K. Dakshinamurti (2015). "Antihypertensive and neuroprotective actions of pyridoxine and its derivatives." *Can J Physiol Pharmacol* 93(12): 1083–1090.

de Baaij, J. H., J. G. Hoenderop, et al. (2015). "Magnesium in man: implications for health and disease." *Physiol Rev* 95(1): 1–46.

de Jong, F., M. C. Monuteaux, et al. (2012). "Systematic review and meta-analysis of preterm birth and later systolic blood pressure." *Hypertension* 59(2): 226–234.

de Moraes, A. C., L. Gracia-Marco, et al. (2014). "Vitamins and iron blood biomarkers are associated with blood pressure levels in European adolescents. The HELENA study." *Nutrition* 30(11–12): 1294–1300.

DeLuca, H. F. (2004). "Overview of general physiologic features and functions of vitamin D." *Am J Clin Nutr* 80(6 Suppl): 1689S–1696S.

DiNicolantonio, J. J., J. Bhutani, et al. (2015). "Coenzyme Q10 for the treatment of heart failure: a review of the literature." *Open Heart* 2(1): e000326.

Dolinsky, V. W., S. Chakrabarti, et al. (2013). "Resveratrol prevents hypertension and cardiac hypertrophy in hypertensive rats and mice." *Biochim Biophys Acta* 1832(10): 1723–1733.

Doumas, M., C. Tsioufis, et al. (2014). "Non-interventional management of resistant hypertension." *World J Cardiol* 6(10): 1080–1090.

Dr. Duke's Phytochemical and Ethnobotanical Databases. National Agricultural Library, US Department of Agriculture website. Available at: https://phytochem.nal.usda.gov/phytochem/search.

Elliott, P., J. Stamler, et al. (2006). "Association between protein intake and blood pressure: the INTERMAP Study." *Arch Intern Med* 166(1): 79–87.

Engelhard, Y. N., B. Gazer, et al. (2006). "Natural antioxidants from tomato extract reduce blood pressure in patients with grade-1 hypertension: a double-blind, placebo-controlled pilot study." *Am Heart J* 151(1): 100.

European Scientific Cooperative on, P. (2003). *ESCOP Monographs: The Scientific Foundation for Herbal Medicinal Products*, Stuttgart, Thieme.

Evert, A. B., J. L. Boucher, et al. (2013). "Nutrition therapy recommendations for the management of adults with diabetes." *Diabetes Care* 36(11): 3821–3842.

Farzaei, M. H., Z. Abbasabadi, et al. (2013). "Parsley: a review of ethnopharmacology, phytochemistry and biological activities." *J Tradit Chin Med* 33(6): 815–826.

Fels, J., H. Oberleithner, et al. (2010). "Menage a trois: aldosterone, sodium and nitric oxide in vascular endothelium." *Biochim Biophys Acta* 1802(12): 1193–1202.

FitzGerald, R. J., B. A. Murray, et al. (2004). "Hypotensive peptides from milk proteins." *J Nutr* 134(4): 980S–988S.

Forman, J. P., J. B. Scott, et al. (2013). "Effect of vitamin D supplementation on blood pressure in blacks." *Hypertension* 61(4): 779–785.

Forman, J. P., M. J. Stampfer, et al. (2009). "Diet and lifestyle risk factors associated with incident hypertension in women." *JAMA* 302(4): 401–411.

Forouzanfar, M. H., P. Liu, et al. (2017). "Global burden of hypertension and systolic blood pressure of at least 110 to 115 mm Hg, 1990–2015." *JAMA* 317(2): 165–182.

Franklin, S. S., M. G. Larson, et al. (2001). "Does the relation of blood pressure to coronary heart disease risk change with aging? The Framingham Heart Study." *Circulation* 103(9): 1245–1249.

Fujiwara, N., T. Osanai, et al. (2000). "Study on the relationship between plasma nitrite and nitrate level and salt sensitivity in human hypertension: modulation of nitric oxide synthesis by salt intake." *Circulation* 101(8): 856–861.

Funder, J. W., R. M. Carey, et al. (2008). "Case detection, diagnosis, and treatment of patients with primary aldosteronism: an endocrine society clinical practice guideline." *J Clin Endocrinol Metab* 93(9): 3266–3281.

Garrison, R. J., W. B. Kannel, et al. (1987). "Incidence and precursors of hypertension in young adults: the Framingham Offspring Study." *Prev Med* 16(2): 235–251.

Gaschler, M. M. and B. R. Stockwell (2017). "Lipid peroxidation in cell death." *Biochem Biophys Res Commun* 482(3): 419–425.

Getz, G. S. and C. A. Reardon (2007). "Nutrition and cardiovascular disease." *Arterioscler Thromb Vasc Biol* 27(12): 2499–2506.

Ghayour-Mobarhan, M., A. Shapouri-Moghaddam, et al. (2009). "The relationship between established coronary risk factors and serum copper and zinc concentrations in a large Persian cohort." *J Trace Elem Med Biol* 23(3): 167–175.

Ghayur, M. N. and A. H. Gilani (2005). "Ginger lowers blood pressure through blockade of voltage-dependent calcium channels." *J Cardiovasc Pharmacol* 45(1): 74–80.

Gradman, A. H., J. N. Basile, et al. (2011). "Combination therapy in hypertension." *J Clin Hypertens (Greenwich)* 13(3): 146–154.

Grases, F., G. Melero, et al. (1994). "Urolithiasis and phytotherapy." *Int Urol Nephrol* 26(5): 507–511.

Graudal, N. A., T. Hubeck-Graudal, et al. (2017). "Effects of low sodium diet versus high sodium diet on blood pressure, renin, aldosterone, catecholamines, cholesterol, and triglyceride." *Cochrane Database Syst Rev* 4: CD004022.

Grossman, E. and F. H. Messerli (2012). "Drug-induced hypertension: an unappreciated cause of secondary hypertension." *Am J Med* 125(1): 14–22.

Guo, X., X. Zhang, et al. (2013). "Association between pre-hypertension and cardiovascular outcomes: a systematic review and meta-analysis of prospective studies." *Curr Hypertens Rep* 15(6): 703–716.

Haddy, F. J., P. M. Vanhoutte, et al. (2006). "Role of potassium in regulating blood flow and blood pressure." *Am J Physiol Regul Integr Comp Physiol* 290(3): R546–R552.

Hadley, C. W., S. K. Clinton, et al. (2003). "The consumption of processed tomato products enhances plasma lycopene concentrations in association with a reduced lipoprotein sensitivity to oxidative damage." *J Nutr* 133(3): 727–732.

Han, H., X. Fang, et al. (2017). "Dose-response relationship between dietary magnesium intake, serum magnesium concentration and risk of hypertension: a systematic review and meta-analysis of prospective cohort studies." *Nutr J* 16(1): 26.

Hashimoto, M., S. Kim, et al. (2001). "Effect of acute intake of red wine on flow-mediated vasodilatation of the brachial artery." *Am J Cardiol* 88(12): 1457–1460, A1459.

He, J., R. H. Streiffer, et al. (2004). "Effect of dietary fiber intake on blood pressure: a randomized, double-blind, placebo-controlled trial." *J Hypertens* 22(1): 73–80.

Hirsch, A. T., Z. J. Haskal, et al. (2006). "ACC/AHA 2005 Practice Guidelines for the management of patients with peripheral arterial disease (lower extremity, renal, mesenteric, and abdominal aortic): a collaborative report from the American Association for Vascular Surgery/Society for Vascular Surgery, Society for Cardiovascular Angiography and Interventions, Society for Vascular Medicine and Biology, Society of Interventional Radiology, and the ACC/AHA Task Force on Practice Guidelines (Writing Committee to Develop Guidelines for the Management of Patients With Peripheral Arterial Disease): endorsed by the American Association of Cardiovascular and Pulmonary Rehabilitation; National Heart, Lung, and Blood Institute; Society for Vascular Nursing; TransAtlantic Inter-Society Consensus; and Vascular Disease Foundation." *Circulation* 113(11): e463–e654.

Hiyasat, B., D. Sabha, et al. (2009). "Antiplatelet activity of *Allium ursinum* and *Allium sativum*." *Pharmacology* 83(4): 197–204.

Hodgson, J. M., V. Burke, et al. (2006). "Partial substitution of carbohydrate intake with protein intake from lean red meat lowers blood pressure in hypertensive persons." *Am J Clin Nutr* 83(4): 780–787.

Houston, M. (2013). "Nutrition and nutraceutical supplements for the treatment of hypertension: part II." *J Clin Hypertens (Greenwich)* 15(11): 845–851.

Houston, M. (2014). "The role of nutrition and nutraceutical supplements in the treatment of hypertension." *World J Cardiol* 6(2): 38–66.

Houston, M. C. (2007). "Treatment of hypertension with nutraceuticals, vitamins, antioxidants and minerals." *Expert Rev Cardiovasc Ther* 5(4): 681–691.

Houston, M. C. (2011). "The importance of potassium in managing hypertension." *Curr Hypertens Rep* 13(4): 309–317.

Houston, M. C. and K. J. Harper (2008). "Potassium, magnesium, and calcium: their role in both the cause and treatment of hypertension." *J Clin Hypertens (Greenwich)* 10(7 Suppl 2): 3–11.

Huang, C. Y., W. W. Kuo, et al. (2014). "Protective effect of Danggui (Radix Angelicae Sinensis) on angiotensin II-induced apoptosis in H9c2 cardiomyoblast cells." *BMC Complement Altern Med* 14: 358.

Huang, Y., X. Cai, et al. (2014a). "Prehypertension and the risk of stroke: a meta-analysis." *Neurology* 82(13): 1153–1161.

Huang, Y., X. Cai, et al. (2014b). "Prehypertension and Incidence of ESRD: a systematic review and meta-analysis." *Am J Kidney Dis* 63(1): 76–83.

Husemoen, L. L., T. Skaaby, et al. (2016). "Mendelian randomisation study of the associations of vitamin B12 and folate genetic risk scores with blood pressure and fasting serum lipid levels in three Danish population-based studies." *Eur J Clin Nutr* 70(5): 613–619.

J., D. (2005). *Dr. Duke's Phytochemical and Ethnobotanical Databases.*

Jackson, S. L., M. E. Cogswell, et al. (2018). "Association between urinary sodium and potassium excretion and blood pressure among adults in the United States: National Health and Nutrition Examination Survey, 2014." *Circulation* 137(3): 237–246.

Jiang, Y. and W. Zheng (2005). "Cardiovascular toxicities upon manganese exposure." *Cardiovasc Toxicol* 5(4): 345–354.

Juhasz, B., A. Kertesz, et al. (2013). "Cardioprotective effects of sour cherry seed extract (SCSE) on the hypercholesterolemic rabbit heart." *Curr Pharm Des* 19(39): 6896–6905.

Juhola, J., M. Oikonen, et al. (2012). "Childhood physical, environmental, and genetic predictors of adult hypertension: the cardiovascular risk in young Finns study." *Circulation* 126(4): 402–409.

Juonala, M., C. G. Magnussen, et al. (2011). "Childhood adiposity, adult adiposity, and cardiovascular risk factors." *N Engl J Med* 365(20): 1876–1885.

Juraschek, S. P., E. Guallar, et al. (2012). "Effects of vitamin C supplementation on blood pressure: a meta-analysis of randomized controlled trials." *Am J Clin Nutr* 95(5): 1079–1088.

Kanbay, M., Y. Chen, et al. (2011). "Mechanisms and consequences of salt sensitivity and dietary salt intake." *Curr Opin Nephrol Hypertens* 20(1): 37–43.

Katznelson, L., E. R. Laws, Jr., et al. (2014). "Acromegaly: an endocrine society clinical practice guideline." *J Clin Endocrinol Metab* 99(11): 3933–3951.

Kawasaki, T., E. Seki, et al. (2000). "Antihypertensive effect of valyl-tyrosine, a short chain peptide derived from sardine muscle hydrolyzate, on mild hypertensive subjects." *J Hum Hypertens* 14(8): 519–523.

Khan, K., E. Jovanovski, et al. (2018). "The effect of viscous soluble fiber on blood pressure: a systematic review and meta-analysis of randomized controlled trials." *Nutr Metab Cardiovasc Dis* 28(1): 3–13.

Khan, N. and H. Mukhtar (2007). "Tea polyphenols for health promotion." *Life Sci* 81(7): 519–533.

Kim, D. Y., S. H. Kim, et al. (2018). "Association between dietary carbohydrate quality and the prevalence of obesity and hypertension." *J Hum Nutr Diet* 31(5): 587–596.

Kim, H. N., S. H. Kim, et al. (2018). "Effects of zinc, magnesium, and chromium supplementation on cardiometabolic risk in adults with metabolic syndrome: a double-blind, placebo-controlled randomised trial." *J Trace Elem Med Biol* 48: 166–171.

Kim, J. (2013). "Dietary zinc intake is inversely associated with systolic blood pressure in young obese women." *Nutr Res Pract* 7(5): 380–384.

Klatsky, A. L. (2010). "Alcohol and cardiovascular mortality: common sense and scientific truth." *J Am Coll Cardiol* 55(13): 1336–1338.

Kleefstra, N., S. T. Houweling, et al. (2006). "Chromium treatment has no effect in patients with poorly controlled, insulin-treated type 2 diabetes in an obese Western population: a randomized, double-blind, placebo-controlled trial." *Diabetes Care* 29(3): 521–525.

Kleefstra, N., S. T. Houweling, et al. (2007). "Chromium treatment has no effect in patients with type 2 diabetes in a Western population: a randomized, double-blind, placebo-controlled trial." *Diabetes Care* 30(5): 1092–1096.

Kopp, W. (2005). "Pathogenesis and etiology of essential hypertension: role of dietary carbohydrate." *Med Hypotheses* 64(4): 782–787.

Kotchen, T. A. and D. A. McCarron (1998). "Dietary electrolytes and blood pressure: a statement for healthcare professionals from the American Heart Association Nutrition Committee." *Circulation* 98(6): 613–617.

Kraus, B. J., J. L. Sartoretto, et al. (2015). "Novel role for retinol-binding protein 4 in the regulation of blood pressure." *FASEB J* 29(8): 3133–3140.

Kreydiyyeh, S. I. and J. Usta (2002). "Diuretic effect and mechanism of action of parsley." *J Ethnopharmacol* 79(3): 353–357.

Kunutsor, S. K., S. Burgess, et al. (2014). "Vitamin D and high blood pressure: causal association or epiphenomenon?" *Eur J Epidemiol* 29(1): 1–14.

Kuruppu, D., H. C. Hendrie, et al. (2014). "Selenium levels and hypertension: a systematic review of the literature." *Public Health Nutr* 17(6): 1342–1352.

Kusche-Vihrog, K., C. Callies, et al. (2010). "The epithelial sodium channel (ENaC): mediator of the aldosterone response in the vascular endothelium?" *Steroids* 75(8–9): 544–549.

Kusche-Vihrog, K. and H. Oberleithner (2012). "An emerging concept of vascular salt sensitivity." *F1000 Biol Rep* 4: 20.

Laclaustra, M., A. Navas-Acien, et al. (2009). "Serum selenium concentrations and hypertension in the US Population." *Circ Cardiovasc Qual Outcomes* 2(4): 369–376.

Law, M. R., J. K. Morris, et al. (2009). "Use of blood pressure lowering drugs in the prevention of cardiovascular disease: meta-analysis of 147 randomised trials in the context of expectations from prospective epidemiological studies." *BMJ* 338: b1665.

Lawson, L. D. and C. D. Gardner (2005). "Composition, stability, and bioavailability of garlic products used in a clinical trial." *J Agric Food Chem* 53(16): 6254–6261.

Lee, B. K. and Y. Kim (2011). "Relationship between blood manganese and blood pressure in the Korean general population according to KNHANES 2008." *Environ Res* 111(6): 797–803.

Lee, J. H., J. H. O'Keefe, et al. (2008). "Vitamin D deficiency an important, common, and easily treatable cardiovascular risk factor?" *J Am Coll Cardiol* 52(24): 1949–1956.

Lee, Y. K., E. S. Lyu, et al. (2015). "Daily copper and manganese intakes and their relation to blood pressure in normotensive adults." *Clin Nutr Res* 4(4): 259–266.

Lenders, J. W., Q. Y. Duh, et al. (2014). "Pheochromocytoma and paraganglioma: an endocrine society clinical practice guideline." *J Clin Endocrinol Metab* 99(6): 1915–1942.

Lesniak, K. T. and P. M. Dubbert (2001). "Exercise and hypertension." *Curr Opin Cardiol* 16(6): 356–359.

Levy, D., M. G. Larson, et al. (1996). "The progression from hypertension to congestive heart failure." *JAMA* 275(20): 1557–1562.

Lewington, S., R. Clarke, et al. (2002). "Age-specific relevance of usual blood pressure to vascular mortality: a meta-analysis of individual data for one million adults in 61 prospective studies." *Lancet* 360(9349): 1903–1913.

Libien, J. and W. S. Blaner (2007). "Retinol and retinol-binding protein in cerebrospinal fluid: can vitamin A take the 'idiopathic' out of idiopathic intracranial hypertension?" *J Neuroophthalmol* 27(4): 253–257.

Lifton, R. P., A. G. Gharavi, et al. (2001). "Molecular mechanisms of human hypertension." *Cell* 104(4): 545–556.

Lin, Y. L. and J. K. Lin (1997). "(-)-Epigallocatechin-3-gallate blocks the induction of nitric oxide synthase by down-regulating lipopolysaccharide-induced activity of transcription factor nuclear factor-kappaB." *Mol Pharmacol* 52(3): 465–472.

Lind, M. V., A. Larnkjaer, et al. (2017). "Dietary protein intake and quality in early life: impact on growth and obesity." *Curr Opin Clin Nutr Metab Care* 20(1): 71–76.

Liu, R., S. Dang, et al. (2013). "Association between dietary protein intake and the risk of hypertension: a cross-sectional study from rural western China." *Hypertens Res* 36(11): 972–979.

Liu, Z., A. Fang, et al. (2018). "Association of habitually low intake of dietary calcium with blood pressure and hypertension in a population with predominantly plant-based diets." *Nutrients* 10(5): 603.

Livingstone, K. M., D. I. Givens, et al. (2013). "Is fatty acid intake a predictor of arterial stiffness and blood pressure in men? Evidence from the Caerphilly Prospective Study." *Nutr Metab Cardiovasc Dis* 23(11): 1079–1085.

Lonnrot, K., I. Porsti, et al. (1998). "Control of arterial tone after long-term coenzyme Q10 supplementation in senescent rats." *Br J Pharmacol* 124(7): 1500–1506.

Lorell, B. H. and B. A. Carabello (2000). "Left ventricular hypertrophy: pathogenesis, detection, and prognosis." *Circulation* 102(4): 470–479.

Magalhaes, E. I., M. C. Pessoa, et al. (2017). "Dietary calcium intake is inversely associated with blood pressure in Brazilian children." *Int J Food Sci Nutr* 68(3): 331–338.

Mahmood, S. S., D. Levy, et al. (2014). "The Framingham Heart Study and the epidemiology of cardiovascular disease: a historical perspective." *Lancet* 383(9921): 999–1008.

Mak, D. H., P. Y. Chiu, et al. (2006). "Dang-Gui Buxue Tang produces a more potent cardioprotective effect than its component herb extracts and enhances glutathione status in rat heart mitochondria and erythrocytes." *Phytother Res* 20(7): 561–567.

Mancia, G., R. Fagard, et al. (2013). "2013 ESH/ESC guidelines for the management of arterial hypertension: the task force for the management of arterial hypertension of the European Society of Hypertension (ESH) and of the European Society of Cardiology (ESC)." *J Hypertens* 31(7): 1281–1357.

Mattos, C. B., L. V. Viana, et al. (2015). "Increased protein intake is associated with uncontrolled blood pressure by 24-hour ambulatory blood pressure monitoring in patients with type 2 diabetes." *J Am Coll Nutr* 34(3): 232–239.

McDaid, O., B. Stewart-Knox, et al. (2007). "Dietary zinc intake and sex differences in taste acuity in healthy young adults." *J Hum Nutr Diet* 20(2): 103–110.

McDonough, A. A., L. C. Veiras, et al. (2017). "Cardiovascular benefits associated with higher dietary K(+) vs. lower dietary Na(+): evidence from population and mechanistic studies." *Am J Physiol Endocrinol Metab* 312(4): E348–E356.

McRae, M. P. (2016). "Therapeutic benefits of l-arginine: an umbrella review of meta-analyses." *J Chiropr Med* 15(3): 184–189.

Mehmood, M. H. and A. H. Gilani (2010). "Pharmacological basis for the medicinal use of black pepper and piperine in gastrointestinal disorders." *J Med Food* 13(5): 1086–1096.

Mente, A., M. O'Donnell, et al. (2016). "Associations of urinary sodium excretion with cardiovascular events in individuals with and without hypertension: a pooled analysis of data from four studies." *Lancet* 388(10043): 465–475.

Messerli, F. H., R. E. Schmieder, et al. (1997). "Salt. A perpetrator of hypertensive target organ disease?" *Arch Intern Med* 157(21): 2449–2452.

Mihaylova, D. S., A. Lante, et al. (2014). "Study on the antioxidant and antimicrobial activities of *Allium ursinum* L. pressurised-liquid extract." *Nat Prod Res* 28(22): 2000–2005.

Mirhosseini, N., H. Vatanparast, et al. (2017). "The association between serum 25(OH)D status and blood pressure in participants of a community-based program taking vitamin D supplements." *Nutrients* 9(11): 1244.

Mirmiran, P., Z. Bahadoran, et al. (2018). "Dietary sodium to potassium ratio and the incidence of hypertension and cardiovascular disease: a population-based longitudinal study." *Clin Exp Hypertens*: 40(8): 772–779.

Mohammad, Y. M., I. R. Raslan, et al. (2016). "Idiopathic intracranial hypertension induced by topical application of vitamin A." *J Neuroophthalmol* 36(4): 412–413.

Mordukhovich, I., R. O. Wright, et al. (2012). "Associations of toenail arsenic, cadmium, mercury, manganese, and lead with blood pressure in the normative aging study." *Environ Health Perspect* 120(1): 98–104.

Mori, T. A. (2006). "Omega-3 fatty acids and hypertension in humans." *Clin Exp Pharmacol Physiol* 33(9): 842–846.

Morimoto, A., T. Uzu, et al. (1997). "Sodium sensitivity and cardiovascular events in patients with essential hypertension." *Lancet* 350(9093): 1734–1737.

Nagy, B., Z. Nagy-Lorincz, et al. (2017a). "[Hungarian Diet and Nutritional Status Survey - The OTAP2014 study. IV. Microelement intake of the Hungarian population]." *Orv Hetil* 158(21): 803–810.

Nagy, B., Z. Nagy-Lorincz, et al. (2017b). "[Hungarian Diet and Nutritional Status Survey - OTAP2014. III. Macroelement intake of the Hungarian population]." *Orv Hetil* 158(17): 653–661.

Nassiri-Asl, M. and H. Hosseinzadeh (2009). "Review of the pharmacological effects of *Vitis vinifera* (Grape) and its bioactive compounds." *Phytother Res* 23(9): 1197–1204.

Nassiri-Asl, M. and H. Hosseinzadeh (2016). "Review of the pharmacological effects of *Vitis vinifera* (Grape) and its bioactive constituents: an update." *Phytother Res* 30(9): 1392–1403.

Nawrot, T. S., J. A. Staessen, et al. (2007). "Blood pressure and blood selenium: a cross-sectional and longitudinal population study." *Eur Heart J* 28(5): 628–633.

Negishi, H., J. W. Xu, et al. (2004). "Black and green tea polyphenols attenuate blood pressure increases in stroke-prone spontaneously hypertensive rats." *J Nutr* 134(1): 38–42.

Nieman, L. K., B. M. Biller, et al. (2008). "The diagnosis of Cushing's syndrome: an endocrine society clinical practice guideline." *J Clin Endocrinol Metab* 93(5): 1526–1540.

Nkwenkeu, S. F., G. Kennedy, et al. (2002). "Oral manganese intake estimated with dietary records and with direct chemical analysis." *Sci Total Environ* 287(1–2): 147–153.

Noori, N., H. Tabibi, et al. (2013). "Effects of combined lipoic acid and pyridoxine on albuminuria, advanced glycation end-products, and blood pressure in diabetic nephropathy." *Int J Vitam Nutr Res* 83(2): 77–85.

Noreen, E. E. and J. Brandauer (2012). "The effects of supplemental fish oil on blood pressure and morning cortisol in normotensive adults: a pilot study." *J Complement Integr Med* 9.

Nyadjeu, P., A. Dongmo, et al. (2011). "Antihypertensive and vasorelaxant effects of Cinnamomum zeylanicum stem bark aqueous extract in rats." *J Complement Integr Med* 8(1).

O'Donnell, M., A. Mente, et al. (2014). "Urinary sodium and potassium excretion, mortality, and cardiovascular events." *N Engl J Med* 371(7): 612–623.

O'Sullivan, T. A., A. P. Bremner, et al. (2012). "Polyunsaturated fatty acid intake and blood pressure in adolescents." *J Hum Hypertens* 26(3): 178–187.

Oberleithner, H., K. Kusche-Vihrog, et al. (2010). "Endothelial cells as vascular salt sensors." *Kidney Int* 77(6): 490–494.

Oberleithner, H., C. Riethmuller, et al. (2007). "Plasma sodium stiffens vascular endothelium and reduces nitric oxide release." *Proc Natl Acad Sci USA* 104(41): 16281–16286.

Palar, K. and R. Sturm (2009). "Potential societal savings from reduced sodium consumption in the U.S. adult population." *Am J Health Promot* 24(1): 49–57.

Palumbo, G., F. Avanzini, et al. (2000). "Effects of vitamin E on clinic and ambulatory blood pressure in treated hypertensive patients. Collaborative Group of the Primary Prevention Project (PPP)—Hypertension study." *Am J Hypertens* 13(5 Pt 1): 564–567.

Paz Matias, J., D. M. Costa e Silva, et al. (2014). "Effect of zinc supplementation on superoxide dismutase activity in patients with ulcerative rectocolitis." *Nutr Hosp* 31(3): 1434–1437.

Pedrosa, R. P., L. F. Drager, et al. (2011). "Obstructive sleep apnea: the most common secondary cause of hypertension associated with resistant hypertension." *Hypertension* 58(5): 811–817.

Peng, X., R. Zhou, et al. (2014). "Effect of green tea consumption on blood pressure: a meta-analysis of 13 randomized controlled trials." *Sci Rep* 4: 6251.

Perera, P. J., Y. S. Sandamal, et al. (2014). "Benign intracranial hypertension following vitamin A megadose." *Ceylon Med J* 59(1): 31.

Perez, V. and E. T. Chang (2014). "Sodium-to-potassium ratio and blood pressure, hypertension, and related factors." *Adv Nutr* 5(6): 712–741.

Perricone, N. V., D. Bagchi, et al. (2008). "Blood pressure lowering effects of niacin-bound chromium(III) (NBC) in sucrose-fed rats: renin-angiotensin system." *J Inorg Biochem* 102(7): 1541–1548.

Pfister, R., G. Michels, et al. (2014). "Estimated urinary sodium excretion and risk of heart failure in men and women in the EPIC-Norfolk study." *Eur J Heart Fail* 16(4): 394–402.

Pilz, S., M. Gaksch, et al. (2015). "Effects of vitamin D on blood pressure and cardiovascular risk factors: a randomized controlled trial." *Hypertension* 65(6): 1195–1201.

Potenza, M. A., F. L. Marasciulo, et al. (2007). "EGCG, a green tea polyphenol, improves endothelial function and insulin sensitivity, reduces blood pressure, and protects against myocardial I/R injury in SHR." *Am J Physiol Endocrinol Metab* 292(5): E1378–E1387.

Preston, C. L. and I. H. Stockley (2016). *Stockley's Drug Interactions: A Source Book of Interactions, Their Mechanisms, Clinical Importance and Management*, London [etc.], Pharmaceutical Press.

Preuss, H. G., D. Clouatre, et al. (2001). "Wild garlic has a greater effect than regular garlic on blood pressure and blood chemistries of rats." *Int Urol Nephrol* 32(4): 525–530.

Preuss, H. G., B. Echard, et al. (2008). "Comparing metabolic effects of six different commercial trivalent chromium compounds." *J Inorg Biochem* 102(11): 1986–1990.

Preuss, H. G., J. A. Gondal, et al. (1995). "Effects of chromium and guar on sugar-induced hypertension in rats." *Clin Nephrol* 44(3): 170–177.

Preuss, H. G., P. L. Grojec, et al. (1997). "Effects of different chromium compounds on blood pressure and lipid peroxidation in spontaneously hypertensive rats." *Clin Nephrol* 47(5): 325–330.

Preuss, H. G., S. T. Jarrell, et al. (1998). "Comparative effects of chromium, vanadium and gymnema sylvestre on sugar-induced blood pressure elevations in SHR." *J Am Coll Nutr* 17(2): 116–123.

Puglisi, J. P. (2013). "Vitamin D: new implications for mood and blood pressure." *Nurse Pract* 38(12): 47–52.

Qayyum, R., H. M. Qamar, et al. (2016). "Mechanisms underlying the antihypertensive properties of Urtica dioica." *J Transl Med* 14: 254.

Qin, B., M. M. Polansky, et al. (2010). "Cinnamon extract regulates plasma levels of adipose-derived factors and expression of multiple genes related to carbohydrate metabolism and lipogenesis in adipose tissue of fructose-fed rats." *Horm Metab Res* 42(3): 187–193.

Racz-Kotilla, E., G. Racz, et al. (1974). "The action of Taraxacum officinale extracts on the body weight and diuresis of laboratory animals." *Planta Med* 26(3): 212–217.

Rafraf, M., B. Bazyun, et al. (2012). "Impact of vitamin E supplementation on blood pressure and Hs-CRP in type 2 diabetic patients." *Health Promot Perspect* 2(1): 72–79. doi: 10.5681/hpp.2012.009

Rasmussen, C. B., J. K. Glisson, et al. (2012). "Dietary supplements and hypertension: potential benefits and precautions." *J Clin Hypertens (Greenwich)* 14(7): 467–471.

Rebholz, C. M., E. E. Friedman, et al. (2012). "Dietary protein intake and blood pressure: a meta-analysis of randomized controlled trials." *Am J Epidemiol* 176(7 Suppl): S27–S43.

Reboussin, D. M., N. B. Allen, et al. (2018). "Systematic review for the 2017 ACC/AHA/AA PA/ABC/ACPM/AGS/APhA/ASH/ASPC/NMA/PCNA guideline for the prevention, detection, evaluation, and management of high blood pressure in adults: a report of the American College of Cardiology/American Heart Association Task Force on Clinical Practice Guidelines." *J Am Coll Cardiol* 71(19): 2176–2198.

Reinhart, K. M., C. I. Coleman, et al. (2008). "Effects of garlic on blood pressure in patients with and without systolic hypertension: a meta-analysis." *Ann Pharmacother* 42(12): 1766–1771.

Ricci, I., R. Artacho, et al. (2010). "Milk protein peptides with angiotensin I-converting enzyme inhibitory (ACEI) activity." *Crit Rev Food Sci Nutr* 50(5): 390–402.

Richter, C. K., A. C. Skulas-Ray, et al. (2015). "Plant protein and animal proteins: do they differentially affect cardiovascular disease risk?" *Adv Nutr* 6(6): 712–728.

Ried, K., N. Travica, et al. (2016). "The acute effect of high-dose intravenous vitamin C and other nutrients on blood pressure: a cohort study." *Blood Press Monit* 21(3): 160–167.

Rosanoff, A., C. M. Weaver, et al. (2012). "Suboptimal magnesium status in the United States: are the health consequences underestimated?" *Nutr Rev* 70(3): 153–164.

Sabha, D., B. Hiyasat, et al. (2012). "*Allium ursinum* L.: bioassay-guided isolation and identification of a galactolipid and a phytosterol exerting antiaggregatory effects." *Pharmacology* 89(5–6): 260–269.

Sacks, F. M., L. J. Appel, et al. (1999). "A dietary approach to prevent hypertension: a review of the Dietary Approaches to Stop Hypertension (DASH) Study." *Clin Cardiol* 22(7 Suppl): III6–III10.

Sakai, Y., T. Murakami, et al. (2003). "Antihypertensive effects of onion on NO synthase inhibitor-induced hypertensive rats and spontaneously hypertensive rats." *Biosci Biotechnol Biochem* 67(6): 1305–1311.

Saltman, P. (1983). "Trace elements and blood pressure." *Ann Intern Med* 98(5 Pt 2): 823–827.

Saneei, P., A. Salehi-Abargouei, et al. (2014). "Influence of Dietary Approaches to Stop Hypertension (DASH) diet on blood pressure: a systematic review and meta-analysis on randomized controlled trials." *Nutr Metab Cardiovasc Dis* 24(12): 1253–1261.

Santesso, N., E. A. Akl, et al. (2012). "Effects of higher- versus lower-protein diets on health outcomes: a systematic review and meta-analysis." *Eur J Clin Nutr* 66(7): 780–788.

Sargrad, K. R., C. Homko, et al. (2005). "Effect of high protein vs high carbohydrate intake on insulin sensitivity, body weight, hemoglobin A1c, and blood pressure in patients with type 2 diabetes mellitus." *J Am Diet Assoc* 105(4): 573–580.

Saulnier, P. J., E. Gand, et al. (2014). "Sodium and cardiovascular disease." *N Engl J Med* 371(22): 2135–2136.

Savica, V., G. Bellinghieri, et al. (2010). "The effect of nutrition on blood pressure." *Annu Rev Nutr* 30: 365–401.

Schultes, B., A. K. Panknin, et al. (2016). "Glycemic increase induced by intravenous glucose infusion fails to affect hunger, appetite, or satiety following breakfast in healthy men." *Appetite* 105: 562–566.

Schutten, J. C., M. M. Joosten, et al. (2018). "Magnesium and blood pressure: a physiology-based approach." *Adv Chronic Kidney Dis* 25(3): 244–250.

Sendl, A., G. Elbl, et al. (1992). "Comparative pharmacological investigations of *Allium ursinum* and *Allium sativum*." *Planta Med* 58(1): 1–7.

Shouk, R., A. Abdou, et al. (2014). "Mechanisms underlying the antihypertensive effects of garlic bioactives." *Nutr Res* 34(2): 106–115.

Shu, L. and K. Huang (2018). "Effect of vitamin D supplementation on blood pressure parameters in patients with vitamin D deficiency: a systematic review and meta-analysis." *J Am Soc Hypertens* 12(7): 488–496.

Sies, H. and W. Stahl (1995). "Vitamins E and C, beta-carotene, and other carotenoids as antioxidants." *Am J Clin Nutr* 62(6 Suppl): 1315S–1321S.

Skowronska-Jozwiak, E., M. Jaworski, et al. (2017). "Low dairy calcium intake is associated with overweight and elevated blood pressure in Polish adults, notably in premenopausal women." *Public Health Nutr* 20(4): 630–637.

Song, Y. and S. Liu (2012). "Magnesium for cardiovascular health: time for intervention." *Am J Clin Nutr* 95(2): 269–270.

Spagnolo, A., M. Giussani, et al. (2013). "Focus on prevention, diagnosis and treatment of hypertension in children and adolescents." *Ital J Pediatr* 39: 20.

Streppel, M. T., L. R. Arends, et al. (2005). "Dietary fiber and blood pressure: a meta-analysis of randomized placebo-controlled trials." *Arch Intern Med* 165(2): 150–156.

Subasinghe, A. K., S. Arabshahi, et al. (2016). "Association between salt and hypertension in rural and urban populations of low to middle income countries: a systematic review and meta-analysis of population based studies." *Asia Pac J Clin Nutr* 25(2): 402–413.

Sugden, J. A., J. I. Davies, et al. (2008). "Vitamin D improves endothelial function in patients with Type 2 diabetes mellitus and low vitamin D levels." *Diabet Med* 25(3): 320–325.

Sun, Q., B. Wang, et al. (2016). "Taurine supplementation lowers blood pressure and improves vascular function in prehypertension: randomized, double-blind, placebo-controlled study." *Hypertension* 67(3): 541–549.

Tahri, A., S. Yamani, et al. (2000). "Acute diuretic, natriuretic and hypotensive effects of a continuous perfusion of aqueous extract of Urtica dioica in the rat." *J Ethnopharmacol* 73(1–2): 95–100.

Taittonen, L., M. Nuutinen, et al. (1997). "Lack of association between copper, zinc, selenium and blood pressure among healthy children." *J Hum Hypertens* 11(7): 429–433.

Takase, H., T. Sugiura, et al. (2015). "Dietary sodium consumption predicts future blood pressure and incident hypertension in the Japanese normotensive general population." *J Am Heart Assoc* 4(8): e001959.

Tamai, Y., K. Wada, et al. (2011). "Dietary intake of vitamin B12 and folic acid is associated with lower blood pressure in Japanese preschool children." *Am J Hypertens* 24(11): 1215–1221.

Taneja, S. K. and R. Mandal (2007). "Mineral factors controlling essential hypertension—a study in the Chandigarh, India population." *Biol Trace Elem Res* 120(1–3): 61–73.

Tang, W. H., T. Kitai, et al. (2017). "Gut microbiota in cardiovascular health and disease." *Circ Res* 120(7): 1183–1196.

Tangvoraphonkchai, K. and A. Davenport (2018). "Magnesium and cardiovascular disease." *Adv Chronic Kidney Dis* 25(3): 251–260.

Taqvi, S. I., A. J. Shah, et al. (2008). "Blood pressure lowering and vasomodulator effects of piperine." *J Cardiovasc Pharmacol* 52(5): 452–458.

Taylor, B. C., T. J. Wilt, et al. (2011). "Impact of diastolic and systolic blood pressure on mortality: implications for the definition of 'normal'." *J Gen Intern Med* 26(7): 685–690.

Testai, L., S. Chericoni, et al. (2002). "Cardiovascular effects of *Urtica dioica* L. (Urticaceae) roots extracts: in vitro and in vivo pharmacological studies." *J Ethnopharmacol* 81(1): 105–109.

Thor, M. Y., L. Harnack, et al. (2011). "Evaluation of the comprehensiveness and reliability of the chromium composition of foods in the literature." *J Food Compost Anal* 24(8): 1147–1152.

Thrift, A. G., J. J. McNeil, et al. (1996). "Risk factors for cerebral hemorrhage in the era of well-controlled hypertension. Melbourne Risk Factor Study (MERFS) Group." *Stroke* 27(11): 2020–2025.

Tielemans, S. M., W. Altorf-van der Kuil, et al. (2013). "Intake of total protein, plant protein and animal protein in relation to blood pressure: a meta-analysis of observational and intervention studies." *J Hum Hypertens* 27(9): 564–571.

Tielemans, S. M., D. Kromhout, et al. (2014). "Associations of plant and animal protein intake with 5-year changes in blood pressure: the Zutphen Elderly Study." *Nutr Metab Cardiovasc Dis* 24(11): 1228–1233.

Tsai, T. L., C. C. Kuo, et al. (2017). "The decline in kidney function with chromium exposure is exacerbated with co-exposure to lead and cadmium." *Kidney Int* 92(3): 710–720.

Tubek, S. (2007). "Role of zinc in regulation of arterial blood pressure and in the etiopathogenesis of arterial hypertension." *Biol Trace Elem Res* 117(1–3): 39–51.

Ueshima, H., J. Stamler, et al. (2007). "Food omega-3 fatty acid intake of individuals (total, linolenic acid, long-chain) and their blood pressure: INTERMAP study." *Hypertension* 50(2): 313–319.

Umesawa, M., S. Sato, et al. (2009). "Relations between protein intake and blood pressure in Japanese men and women: the Circulatory Risk in Communities Study (CIRCS)." *Am J Clin Nutr* 90(2): 377–384.

van Dijk, R. A., J. A. Rauwerda, et al. (2001). "Long-term homocysteine-lowering treatment with folic acid plus pyridoxine is associated with decreased blood pressure but not with improved brachial artery endothelium-dependent vasodilation or carotid artery stiffness: a 2-year, randomized, placebo-controlled trial." *Arterioscler Thromb Vasc Biol* 21(12): 2072–2079.

Vance, T. M. and O. K. Chun (2015). "Zinc intake is associated with lower cadmium burden in U.S. adults." *J Nutr* 145(12): 2741–2748.

Voortman, T., A. Vitezova, et al. (2015). "Effects of protein intake on blood pressure, insulin sensitivity and blood lipids in children: a systematic review." *Br J Nutr* 113(3): 383–402.

Wagner, H., F. Willer, et al. (1989). "[Biologically active compounds from the aqueous extract of Urtica dioica]." *Planta Med* 55(5): 452–454.

Wang, J. G., R. A. Anderson, et al. (2007). "The effect of cinnamon extract on insulin resistance parameters in polycystic ovary syndrome: a pilot study." *Fertil Steril* 88(1): 240–243.

Wang, Y. F., W. S. Yancy, Jr., et al. (2008). "The relationship between dietary protein intake and blood pressure: results from the PREMIER study." *J Hum Hypertens* 22(11): 745–754.

Wang, Z. Y., Y. Y. Liu, et al. (2018). "l-carnitine and heart disease." *Life Sci* 194: 88–97.

Ward, N. C., J. H. Wu, et al. (2007). "The effect of vitamin E on blood pressure in individuals with type 2 diabetes: a randomized, double-blind, placebo-controlled trial." *J Hypertens* 25(1): 227–234.

Weaver, C. M. (2013). "Potassium and health." *Adv Nutr* 4(3): 368S–377S.

Weber, M., V. Grote, et al. (2014). "Lower protein content in infant formula reduces BMI and obesity risk at school age: follow-up of a randomized trial." *Am J Clin Nutr* 99(5): 1041–1051.

Weinberger, M. H. (1996). "Salt sensitivity of blood pressure in humans." *Hypertension* 27(3 Pt 2): 481–490.

Weiskirchen, S. and R. Weiskirchen (2016). "Resveratrol: how much wine do you have to drink to stay healthy?" *Adv Nutr* 7(4): 706–718.

Wells, E. M., L. R. Goldman, et al. (2012). "Selenium and maternal blood pressure during childbirth." *J Expo Sci Environ Epidemiol* 22(2): 191–197.

Whelton, P. K., R. M. Carey, et al. (2018). "2017 ACC/AHA/AAPA/ABC/ACPM/AGS/APhA /ASH/ASPC/NMA/PCNA guideline for the prevention, detection, evaluation, and management of high blood pressure in adults: a report of the American College of Cardiology/American Heart Association Task Force on Clinical Practice Guidelines." *Hypertension* 71(6): e13–e115.

Whelton, S. P., A. D. Hyre, et al. (2005). "Effect of dietary fiber intake on blood pressure: a meta-analysis of randomized, controlled clinical trials." *J Hypertens* 23(3): 475–481.

Whitworth, J. A. and J. Chalmers (2004). "World health organisation-international society of hypertension (WHO/ISH) hypertension guidelines." *Clin Exp Hypertens* 26(7–8): 747–752.

Widman, L., P. O. Wester, et al. (1993). "The dose-dependent reduction in blood pressure through administration of magnesium. A double blind placebo controlled cross-over study." *Am J Hypertens* 6(1): 41–45.

Williamson, E., S. Driver, et al. (2013). *Stockley's Herbal Medicines Interactions: A Guide to the Interactions of Herbal Medicines*, London, Pharmaceutical Press.

Wilson, C. P., H. McNulty, et al. (2010). "Postgraduate symposium: the MTHFR C677T polymorphism, B-vitamins and blood pressure." *Proc Nutr Soc* 69(1): 156–165.

Wilson, P. W. (1994). "Established risk factors and coronary artery disease: the Framingham Study." *Am J Hypertens* 7(7 Pt 2): 7S–12S.

Wu, C., J. G. Woo, et al. (2017). "Association between urinary manganese and blood pressure: results from National Health and Nutrition Examination Survey (NHANES), 2011–2014." *PLoS One* 12(11): e0188145.

Wu, Y. C. and C. L. Hsieh (2011). "Pharmacological effects of Radix Angelica Sinensis (Danggui) on cerebral infarction." *Chin Med* 6: 32.

Yanagisawa, H., T. Miyazaki, et al. (2014). "Zinc-excess intake causes the deterioration of renal function accompanied by an elevation in systemic blood pressure primarily through superoxide radical-induced oxidative stress." *Int J Toxicol* 33(4): 288–296.

Yang, Q., T. Liu, et al. (2011). "Sodium and potassium intake and mortality among US adults: prospective data from the Third National Health and Nutrition Examination Survey." *Arch Intern Med* 171(13): 1183–1191.

Yin, L., G. Deng, et al. (2018). "Association patterns of urinary sodium, potassium, and their ratio with blood pressure across various levels of salt-diet regions in China." *Sci Rep* 8(1): 6727.

Zhang, J. X., G. P. Zhu, et al. (2017). "Elevated serum retinol-binding protein 4 levels are correlated with blood pressure in prehypertensive Chinese." *J Hum Hypertens* 31(10): 611–615.

Zhang, X., Y. Li, et al. (2016). "Effects of magnesium supplementation on blood pressure: a meta-analysis of randomized double-blind placebo-controlled trials." *Hypertension* 68(2): 324–333.

Zhang, Y. and H. Qiu (2018). "Dietary magnesium intake and hyperuricemia among US adults." *Nutrients* 10(3): 296.

Zhang, Z., M. E. Cogswell, et al. (2013). "Association between usual sodium and potassium intake and blood pressure and hypertension among U.S. adults: NHANES 2005–2010." *PLoS One* 8(10): e75289.

Zhu, X., J. Lin, et al. (2014). "A high-carbohydrate diet lowered blood pressure in healthy Chinese male adolescents." *Biosci Trends* 8(2): 132–137.http://tools.acc.org/ASC VD-Risk-Estimator/.

5 Roles of Daily Diet and Beta-Adrenergic System in the Treatment of Obesity and Diabetes

Ebru Arioglu Inan and Belma Turan

CONTENTS

5.1 INTRODUCTION

Diabetes has been described as a chronic progressive metabolic disorder which affects 422 million people all over the world (WHO 2016). According to estimates by the WHO, diabetes will be the seventh major cause of death by 2030 (Mathers and Loncar 2006). One of the major classes of diabetes, type 1 diabetes, is characterized by insulin deficiency as a result of pancreatic beta cell dysfunction. Type 2 diabetes, on the other hand, could be caused by either impaired insulin secretion or insulin resistance.

Diabetes is one of the leading causes of morbidity and mortality since it further results in such micro- and macrovascular complications as heart failure, stroke, myocardial infarction, nephropathy, neuropathy and retinopathy (Nathan, Cleary et al. 2005; Gilbert 2013). Therefore, the prevention and treatment of diabetes seems to be crucial. Nonpharmacological interventions such as diet and exercise have been advised to prevent progression to type 2 diabetes in the prediabetic stage. Nutrition seems to be a key point since excessive calorie intake first results in increased insulin secretion as a compensatory mechanism, and then leads to gradual beta cell dysfunction.

A diet with a low carbohydrate content and rich in unsaturated fat acids has been suggested as an effective nutrition pattern for patients with type 2 diabetes (Tay, LuscombeMarsh et al. 2015). However, it is more complicated since many other macro- and micronutrients are also recommended for diabetic patients. Thus, a diabetes-specific nutrition formula with balanced macro- and micronutrients has been found to ameliorate postprandial plasma glucose levels (Elia, Ceriello et al. 2005).

Obesity is a metabolic disease characterized by excessive fat accumulation as a result of an imbalance between intake and expenditure of energy. People with a body mass index (BMI) equal or greater than 30 kg/m^2 are defined as obese (WHO 2015). The prevalence of obesity has doubled between 1980 and 2014 and the WHO has reported obesity as a global health threat (Kobyliak, Conte et al. 2016). Obesity has detrimental effects on the cardiovascular system. Calorie restriction accompanied by physical exercise has been recommended to prevent and treat obesity (Lai, Wu et al. 2015).

Obesity and diabetes are related pathologies. In fact, obesity is known as one of the major causes of type 2 diabetes. Beta3-adrenoceptors (β_3-ARs) have an important role both in obesity and diabetes. Stimulating β_3-ARs leads to lipolysis in brown and white adipose tissue (Arch and Kaumann 1993). In addition, improvement in hyperglycemia and hyperinsulinemia was observed after β_3-AR activation in type 2 diabetic mice (Kim, Pennisi et al. 2006).

Novel treatment strategies such as β_3-AR agonists have been the subject of attention recently for treatment of diabetes and obesity. The CL 316,243, a β_3-ARs agonist, has been demonstrated to have antidiabetic effects in rodents (Ghorbani, Shafiee Ardestani et al. 2012). This agonist reversed obesity in fa/fa Zucker rats (Ghorbani and Himms-Hagen 1997). Furthermore, β_3-AR agonist mirabegron has been shown to activate brown adipose tissue in healthy male subjects (Cypess, Weiner et al. 2015).

As nutrition appears to be a key point for prevention and treatment of both type 2 diabetes and obesity, this chapter will focus on the importance of nutrients in the management of diabetes and obesity. Furthermore, the relationship between diet-cardiac health and β_3-ARs will be discussed comprehensively.

5.2 ROLE OF NUTRITION IN THE DEVELOPMENT OF DIABETES AND OBESITY

5.2.1 NUTRITION IN DIABETES

The importance of nutrition in providing glycemic control has been shown in various studies. As a marker of glycemic control, hemoglobin A1c (HbA1c) values were found to be decreased significantly in type 2 diabetic patients after nutrition therapy (Coppell, Kataoka et al. 2010; Andrews, Cooper et al. 2011). In addition to glycemic benefits, nutrition therapy improved most of the metabolic parameters which further lead to diabetic complications (Pastors and Franz 2012).

Nutrition therapy for diabetic patients is based mainly on reduced carbohydrate/fat intake to decrease insulin need. As most type 2 diabetic patients are overweight, a diet with reduced calorie intake has also been useful in terms of weight loss. In addition, weight loss with nutrition therapy has been found to be very effective in preventing prediabetes progress to type 2 diabetes (Youssef 2012).

The glycemic features of foods are relevant in such ways as the amount and type of carbohydrates, and also other food components (ADA 2004). It is emphasized that the total amount of carbohydrates is more important than their source. The ADA has reported that the ratio of carbohydrates and monounsaturated fat in the nutrition should be individualized according to metabolic profile and treatment goals (ADA 2004). Most of the meta-analyses comparing the effectiveness of low carbohydrate diets and high carbohydrate diets in diabetic patients demonstrated that low carbohydrate diets provided better glycemic control; however, some of the meta-analyses claimed no significant difference between the two diets (Nordmann, Nordmann et al. 2006; Kirk, Graves et al. 2008; Ajala, English et al. 2013; Clifton, Condo et al. 2014).

The glycemic index (GI) is an important point for a diabetic diet. Diabetics are recommended to eat foods with low GI which are rich in dietary fiber and wholegrains (Barakatun Nisak, Ruzita et al. 2010). Fiber is useful as it prevents hunger. Taking 50 g of fiber per day has been reported to improve postprandial hyperglycemia (Booth and Cheng 2013; Hamdy, Ganda et al. 2018).

Dietary pattern, defined as food one habitually consumes, is essential in diabetes management. A Mediterranean dietary pattern and other low carbohydrate and high protein patterns were most efficient in terms of glycemic control (ADA 2004). The results of the Dietary Intervention Randomized Controlled Trial (DIRECT) revealed that a Mediterranean diet was significantly effective in decreasing C reactive protein, fasting plasma glucose and insulin resistance (Shai, Schwarzfuchs et al. 2008). The Mediterranean diet pattern is known to include unsaturated fatty acids. Since the type of fatty acids rather than total fat amount is important for a diabetic individual to decrease cardiovascular disease risk (Evert, Boucher et al. 2013), a diet rich in unsaturated fatty acids like the Mediterranean style should be preferred.

Diabetic people should be aware of daily protein intake for normal renal function. Total daily protein amount should not exceed 15–20% of energy intake in these patients (Bantle, Wylie-Rosett et al. 2006).

5.2.1.1 Micronutrients in Diabetes

Micronutrients are as important as macronutrients in a diabetic diet. As most type 2 diabetic patients are overweight, they are usually recommended to lose weight with a low-calorie diet which often causes deficiency in micronutrients such as iron, calcium and vitamin B (Gardner, Kim et al. 2010).

Vitamin D has beneficial effects on glucose homeostasis and low vitamin D levels have been demonstrated as a risk factor for diabetes (Mitri, Muraru et al. 2011). On the other hand, no advantage of Vitamin D on glucose homeostasis has been also reported in clinical trials conducted on persons with normoglycemia, impaired glucose tolerance or type 2 Diabetes (George, Pearson et al. 2012). Similarly, Krul-Poel et al. indicated that Vitamin D did not affect fasting glucose, HbA1C and HOMA-IR values (Krul-Poel, Ter Wee et al. 2017). The relationship between vitamin D levels and diabetes needs to be clarified.

Vitamin K, known for its role in blood coagulation, is thought to have a role in insulin sensitivity and glucose homeostasis. Both insulin sensitivity and two-hour glucose levels after OGTT have been found to be improved due to higher intakes of vitamin K1 (phylloquinone) in a study conducted in men and women (Yoshida, Booth et al. 2008). On the other hand, Kumar et al. showed that fasting glucose, insulin concentration and HOMA-IR values were not changed after 12 months' phylloquinone supplementation (1 mg/day) (Kumar, Binkley et al. 2010). In another study conducted in a prediabetic and premenopausal female population, phylloquinone supplementation for four weeks (1,000 ug/day) resulted in decreased fasting glucose, two-hour post OGTT glucose and insulin levels without improving HOMA-IR values (Rasekhi, Karandish et al. 2015). Choi et al. demonstrated that menatetrenone increases insulin sensitivity in healthy men (Choi, Yu et al. 2011). Furthermore, dietary menaquinone intake was found to be inversely correlated with risk of type 2 diabetes (Beulens, Booth et al. 2013). In STZ diabetic rats, phylloquinone treatment improved pancreas beta cell death and increased insulin secretion, thus providing normoglycemia (Varsha, Thiagarajan et al. 2015). These results were attributed to the beneficial effects of phylloquinone on inflammation and oxidative stress (Manna and Kalita 2016). Similarly, a 12-week menatetrenone treatment has been shown to have beneficial effects on glycemic status in STZ diabetic rats (Iwamoto, Seki et al. 2011).

One of the essential micronutrients in diabetes is zinc. Zinc has an important role in the production, storage and secretion of insulin (Salgueiro, Krebs et al. 2001). Attention was paid to maintaining zinc levels within the normal range in type 2 diabetic patients, since this mineral was found to be related to insulin resistance (Doddigarla, Parwez et al. 2016). Zinc contributes to the signal transduction of the insulin receptor and thus affects glucose levels positively (Haase and Maret 2005). This mineral has been suggested as an antioxidant to protect insulin from free radicals (Robertson 2004). An inverse correlation between HbA1c percentage and serum zinc levels has been demonstrated (Doddigarla, Parwez et al. 2016). In type 2 diabetic patients, fasting glucose values were found to be higher when associated with low zinc levels (Badran, Morsy et al. 2016). Zinc deficiency was reported as a contributing factor for higher levels of glucose, C peptide and HOMA-IR in obese diabetic women (Yerlikaya, Toker et al. 2013). Thus, diabetic patients have been thought to benefit from zinc supplementation. This idea was confirmed by some researchers as they showed that a 30 mg elemental zinc (Parham, Amini et al. 2008) and 50 mg zinc gluconate (Oh and Yoon 2008) treatment was effective in decreasing HcA1c values in type 2 diabetic patients. Zinc supplementation was also useful in managing progression diabetic complications (Parham, Amini et al. 2008).

Another trace element, selenium, has been demonstrated to be essential for normal glucose homeostasis. Serum levels of selenium were found to be attenuated in diabetic patients (Badran, Morsy et al. 2016) which is in line with previous findings (Garg, Gupta et al. 1994). Blood glucose levels were partially improved after selenite treatment in STZ-treated diabetic mice (Zeng, Zhou et al. 2009). However, the findings of the effects of selenium seem to be controversial, since a high selenium intake was also found to be associated with insulin resistance (Wang, Zhang et al. 2014). This effect was suggested to result from both increased reactive oxygen species (ROS) production and decreased ROS levels. Increased selenium levels were correlated with elevated serum insulin levels in mice; however, whether this effect was beneficial or harmful was not elicited (Labunskyy, Lee et al. 2011). It could be suggested that the level of selenium is crucial in terms of its positive or negative effects as both a deficiency or an excessive intake of selenium could be a risk factor for type 2 diabetes (Ogawa-Wong, Berry et al. 2016).

5.2.1.2   Anthocyanins

Anthocyanins are phytochemicals mostly found in fruits and vegetables with red, purple, pink and blue coloring (Castaneda-Ovando, de Lourdes Pacheco-Hernández et al. 2009). Fruits and vegetables rich in anthocyanins are thought to have beneficial effects in type 2 diabetes (Ley, Hamdy et al. 2014). Studies have revealed that anthocyanin consumption decreases the risk of type 2 diabetes; a 7.5 mg/day increase in anthocyanin intake caused a 5% decrease in the risk of type 2 diabetes (Guo, Yang et al. 2016). Anthocyanins extracted from blueberries have been useful for insulin sensitivity in obese patients (Stull, Cash et al. 2010). Furthermore, fasting plasma glucose levels were attenuated as a result of purified anthocyanin treatment in type 2 diabetic patients (Liu, Li et al. 2014). The preventive role of anthocyanins in type 2 diabetes has been explained by their antioxidant capacity, anti-inflammatory capability and their effect on glucose and lipid homeostasis (Guo, Yang et al. 2016).

5.2.1.3   Cocoa Flavanols

Cocoa is rich in monomeric flavanols such as catechin and epicatechin. It also has other phytochemicals such as procyanidins, flavonols and anthocyanins (Sanchez-Rabaneda, Jauregui et al. 2003; Kim, Shim et al. 2014). The beneficial effects of cocoa and its flavanols in type 2 diabetes have been demonstrated in animal studies. Feeding ZDF-rats with a 10% cocoa rich diet for nine weeks resulted in decreased glucose levels and improved insulin sensitivity (Fernandez-Millan, Cordero-Herrera et al. 2015). These findings have been attributed to the effects of cocoa on beta cell mass and function since the cocoa diet prevented oxidative damage and apoptosis in pancreatic beta cells (Fernandez-Millan, Cordero-Herrera et al. 2015). Cocoa also positively affects the liver. A 10% cocoa-rich diet ameliorated insulin resistance in the livers of ZDF-rats through phosphorylated insulin receptor substrate 1 (p(ser)-IRS-1) levels and a glycogen synthase kinase 3 (GSK3)/glycogen synthase pathway (Cordero-Herrera, Martin et al. 2015). Oxidative stress has been found to be attenuated in 10% cocoa-rich diet-fed ZDF-rats (Cordero-Herrera, Martin et al. 2015). Furthermore, epicatechin supplementation for eight weeks (20 mg/kg/day) improved insulin resistance in high fructose-fed rats (Bettaieb, Vazquez Prieto et al. 2014). This improvement was found to be related to the preventive effect of epicatechin on the decreased activity of key components of an insulin signaling pathway (Cordero-Herrera, Martin et al. 2015).

Human studies also support the benefits of cocoa flavanols in diabetes. In a randomized trial, flavanol-rich dark chocolate (500 mg/day polyphenol) significantly decreased the HOMA-IR value and the increased insulin sensitivity index as compared to polyphenol-free white chocolate (Grassi, Lippi et al. 2005). Similarly, in another trial with hypertensive subjects, dark chocolate consumption (88 mg/day flavanol) resulted in enhanced insulin sensitivity and beta cell function (Grassi, Desideri et al. 2008). Furthermore, a 12-week flavanol-high cocoa diet (daily high flavanol cocoa drink with 451 mg flavanol) decreased insulin resistance in overweight and obese adults (Davison, Coates et al. 2008). The effects of cocoa flavanols on insulin resistance have been suggested to be associated with increased nitric oxide bioavailability and decreased oxidative stress (Martin, Goya et al. 2016). Daily flavonoid-enriched chocolate consumption (850 mg flavanols and 100 mg isoflavones) led to significant attenuation in insulin resistance in statin-treated diabetic women (Curtis, Sampson et al. 2012; Curtis, Potter et al. 2013). On the other hand, in diabetic patients, a 30-day flavanol-high treatment did not affect glycemic control (Balzer, Rassaf et al. 2008). Similarly, an eight-week polyphenol-high chocolate (50 mg epicatechin) consumption was ineffective in diabetic patients in regard to insulin resistance or glycemic control (Mellor, Sathyapalan et al. 2010).

5.2.1.4 Resveratrol

Resveratrol is a polyphenolic compound which it has been suggested may be beneficial in diabetes. Cocoa powder, dark chocolate, blueberries, white/red grape juices or seed extracts and wines are major sources of resveratrol (Oyenihi, Oyenihi et al. 2016). The positive effects of resveratrol in diabetes have been related to increased beta cell number (Fiori, Shin et al. 2013) and insulin secretion (Ku, Lee et al. 2012). Resveratrol also stimulated glycogenesis in the liver and skeletal muscle of diabetic rats (Su, Hung et al. 2006). Furthermore, glucose uptake through glucose transporter 4 (GLUT-4) was increased in diabetic rats (Penumathsa, Thirunavukkarasu et al. 2008). Resveratrol could help to improve insulin resistance since it ameliorates IRS-1-mediated insulin signaling in the liver and muscle of diabetic rats (Gonzalez-Rodriguez, Santamaria et al. 2015). The effects of resveratrol on glucose homeostasis has been also reported to be associated with the activation of sirtuin 1 (SIRT-1) and AMP activated protein kinase (AMPK) (Cote, Rasmussen et al. 2015). The antioxidant properties of resveratrol are important for its antidiabetic effect. It decreases the levels of oxidative stress markers and also increases the capacity of antioxidant enzymes (Hamadi, Mansour et al. 2012).

Clinical studies support these positive effects observed in animal experiments. A three-month resveratrol (250 mg/day) administration with a glibenclamide treatment ameliorated glycemic parameters in diabetic patients (Bhatt, Thomas et al. 2012). A fourweek resveratrol (5 mg/twice daily) treatment improved insulin sensitivity in diabetes (Brasnyo, Molnar et al. 2011). Furthermore, a 45-day resveratrol treatment (1 g/day) attenuated levels of fasting plasma glucose, HbA1c and insulin (Movahed, Nabipour et al. 2013). On the other hand, further studies are needed to demonstrate the effects of resveratrol in humans, since recent trials were only conducted on a small group of diabetic patients (Oyenihi, Oyenihi et al. 2016). Another point is determining the optimal dose of resveratrol as its bioavailability is low (Walle, De Legge et al. 2004). Thus, more studies on humans need to be done.

5.2.1.5   Cinnamon

Cinnamon is a spice with several beneficial properties such as antioxidant, antimicrobial and anti-inflammatory effects (Jayaprakasha, Ohnishi-Kameyama et al. 2006; Singh, Maurya et al. 2007; Tung, Chua et al. 2008). Cinnamon has been suggested to have an antidiabetic effect. The mechanisms proposed to support its antidiabetic effects involve stimulation of insulin secretion, increase of glucagon like peptide 1 (GLP-1), delay of gastric emptying and increase of GLUT-4 expression (Bi, Lim et al. 2017). Polyphenolic compounds in cinnamon extract exerted an insulin-like effect in *in vitro* studies (Imparl-Radosevich, Deas et al. 1998). Cinnamon extract ameliorated glucose utilization in fructose-fed rats (Qin, Nagasaki et al. 2004). In diabetic mice, cinnamon attenuated blood glucose levels (Kim, Hyun et al. 2006). The antidiabetic effects of cinnamon were attributed to a compound in it named cinnamaldehyde (Subash Babu, Prabuseenivasan et al. 2007). Cinnamaldehyde treatment in STZ diabetic rats caused significant attenuation in plasma glucose levels (Subash Babu, Prabuseenivasan et al. 2007). In a randomized, double-blind, placebo-controlled clinical trial, a 40-day cinnamon powder administration (1, 3, 6 g/day) reduced fasting plasma glucose in type 2 diabetic male and female patients (Khan, Safdar et al. 2003). The impact of this effect was the same for all doses. This beneficial effect of cinnamon has been attributed to the activation of glycogen synthase, increased glucose uptake and dephosphorylation of the insulin receptor. In another trial, attenuation in fasting plasma glucose and HbA1c was observed due to cinnamon extract treatment (Lu, Sheng et al. 2012). The fact that the patients had different fasting glucose values before the treatment is thought to be a limitation of this trial (Bi, Lim et al. 2017). The glucose-lowering effect of cinnamon should be further investigated.

5.2.1.6   Milk and Dairy Products

Meta analyses revealed that milk and dairy product consumption has no or slight beneficial effect on the risk of diabetes (Tong, Dong et al. 2011; Aune, Norat et al. 2013; Gao, Ning et al. 2013; Maghsoudi, Ghiasvand et al. 2016). It was found that total dairy and yoghurt intake decreased the risk of diabetes (Gijsbers, Ding et al. 2016). This finding has been suggested to be related to their effect on gut microbiota (Astrup 2014). Postprandial glucose levels were decreased due to milk and yoghurt consumption in type 2 diabetic patients (Frid, Nilsson et al. 2005). This effect has been attributed to the whey protein in these products which was shown to stimulate glucose-dependent insulinotropic polypeptide 1 (GIP-1) and further insulin secretion in healthy subjects (Nilsson, Stenberg et al. 2004).

5.2.1.7   Probiotics

Probiotics were found to have a beneficial effect in type 2 diabetes management. In STZ diabetic rats, *Lactobacillus rhamnosus* GG improved glucose tolerance (Tabuchi, Ozaki et al. 2003). Similarly, in db/db mice *Lactobacillus rhamnosus* GG increased insulin sensitivity, which was attributed to decreased ER stress and depressed macrophage activation (Park, Kim et al. 2015). Supplementation of high fat-fed mice with *Lactobacillus* and *Bifidobacterium* bacteria led to decreased insulin and glucose levels (Wang, Tang et al. 2015).

5.2.2   NUTRITION IN OBESITY

Nutrition is an essential point in the prevention and treatment of obesity. Excessive calorie intake, especially of sugar, causes obesity (Te Morenga, Mallard et al. 2012). Dietary natural compounds targeting mechanisms, such as adipocyte function and regulation of food intake, have been reported to have a role in the management of obesity (Spiegelman, Puigserver et al. 2000; Lai, Wu et al. 2017).

5.2.2.1   Probiotics

The role of gut microbiota in energy homeostasis and weight control has been the subject of recent research. Prebiotics and probiotics seem to be important components in the treatment of obesity, as they affect gut microbiota. Probiotic supplementation with *Lactobacillus* bacteria was shown to attenuate adipose tissue weight and body weight gain significantly in high fat high cholesterol-fed mice (Yoo, Kim et al. 2013). Similarly, probiotics with *Lactobacillus* or *Bifidobacterium* bacteria reduced weight gain in high fat-fed mice (Wang, Tang et al. 2015). In obese mice, daily probiotic treatment with *A. muciniphila* for four weeks attenuated body weight (Everard, Belzer et al. 2013). Clinical studies also confirmed the antiobesity properties of probiotics. Consumption of fermented milk containing *Lactobacillus gasseri* SBT2055 for 12 weeks resulted in a markedly decreased area of abdominal visceral and subcutaneous fat (Kadooka, Sato et al. 2010). In this trial, body weight and body mass index (BMI) were also significantly lower in the probiotic treated group.

5.2.2.2   Quercetin

In differentiated OP9 cells, quercetin has been demonstrated to decrease adipogenesis (Seo, Kang et al. 2015). It also resulted in downregulation of expression of transcription factors which have a role in adipogenesis, such as PPARγ, C/EBPα and SREBP-1c. In this study, quercetin increased lipolytic activity in adipocytes and was associated with expression of lipolytic enzymes such as ATGL, HSL and LPL. Another form of quercetin, quercetin 3-O glucoside (Q3G), has also been reported to inhibit lipogenesis (Lee, Seo et al. 2017). It decreased body weight and adipocyte size in epididymal adipose tissue in high fat-fed mice. It suppressed expression levels of some proteins related to lipid metabolism, such as peroxisome proliferator-activated receptor γ (PPARγ), sterol regulatory element-binding transcription factor 1 (SREBP-1c) and fatty acid synthase (FAS). Lee et al. showed the browning effect of quercetin which was attributed to activation of AMPK (Lee, Seo et al. 2017; Lee,

Parks et al. 2017). Furthermore, Arias et al. demonstrated that quercetin combined with resveratrol caused a browning effect in rats fed with an obesogenic diet (Arias, Pico et al. 2017).

In conclusion, these results show, for the first time, that the combination of RSV1Q has a brown-like remodeling effect in rats fed an obesogenic diet. Moreover, iBAT is also a target for the combination of both polyphenols. Whether such outcomes effectively contribute to the antiobesity properties of the combination of these polyphenols remains to be elucidated.

5.2.2.3   Resveratrol

In addition to its antidiabetic effect, resveratrol has also been beneficial in obesity. The mechanisms of its antiobesity action have been suggested to involve downregulation of adipogenic transcription factors, inhibition of lipogenesis, stimulation of lipolysis and enhancement of insulin sensitivity (Aguirre, Fernandez-Quintela et al. 2014). It has been reported that resveratrol improves energy expenditure through increasing mitochondrial biogenesis and function in brown adipose tissue (BAT), liver and skeletal muscle (Aguirre, Fernandez-Quintela et al. 2014; de Oliveira, Nabavi et al. 2016). Resveratrol treatment caused decreased accumulation of lipid droplets in 3T3-L1 and human Simpson-Golabi-Behmel Syndrome (SGBS) preadipocytes (Li, Bouzar et al. 2016). In these cells, resveratrol treatment also attenuated levels of ATPase family AAA domaining protein 3 (ATAD3) which is responsible for mitochondrial mass and lipogenesis in adipocytes. Resveratrol is known to activate AMPK. In 3T3-L1 adipocytes it was demonstrated that resveratrol upregulated AMPK signaling (Mitterberger and Zwerschke 2013; Li, Bouzar et al. 2016). In white adipocytes, resveratrol inhibited lipid accumulation (Li, Bouzar et al. 2016). Resveratrol decreased insulin-mediated glucose conversion to lipids in adipocytes isolated from the epididymal adipose tissue of Wistar rats (Szkudelska, Nogowski et al. 2009). These results indicate that resveratrol could be a novel tool for obesity treatment.

5.2.2.4   Curcumin

Curcumin is an active compound of turmeric and its antiobesity effect has been demonstrated in experimental studies (Alappat and Awad 2010; Jimenez-Osorio, Monroy et al. 2016). Its role in obesity has been suggested to be related to its effects on mitochondrial biogenesis. In 3T3-L1 and primary white adipocytes, curcumin caused AMPK activation and peroxisome proliferator activated receptor γ coactivator 1α (PGC-1α) upregulation. Furthermore, in adipocytes curcumin increased protein expression of carnitin palmitoyltransferase-1 (CPT-1), hormone sensitive lipase (HSL) and phosphorylation of acetyl-coenzyme A carboxylase (ACC), which further led to attenuated lipid accumulation (Lone, Choi et al. 2016). Animal studies confirmed the beneficial effect of curcumin in obesity as browning of the white adipose tissue was stimulated in curcumin fed C57B/6 mice (Wang, Wang et al. 2015). Uncoupling protein 1 (UCP-1) was also upregulated in these mice. These findings suggest that the antiobesity effect of curcumin may be related to mitochondrial biogenesis in white adipose tissue (Lai, Wu et al. 2015).

5.2.2.5 Sulforaphane

Sulforaphane is a compound found in cruciferous vegetables. This compound has been reported to enhance adipocyte mitochondrial biogenesis in 3T3-L1 adipocytes (Zhang, Chen et al. 2016). It increased mitochondrial content and function. This effect further resulted in browning of 3T3-L1 adipocytes. Sulforaphane also increased the expression of UCP-1 protein, lipolysis and fatty acid oxidation in 3T3-L1 adipocytes. Sulforaphane has been also found to inhibit adipogenesis. In C57BL/6 N mice fed with a high-fat diet, sulforaphane reduced body weight, visceral adiposity and adipocyte hypertrophy (Choi, Lee et al. 2014). The effect of this compound on adipogenesis has been related to downregulation of PPARγ and C/EBPα. Furthermore, it also decreased lipogenesis by stimulating the AMPK pathway. Lee et al. indicated that sulforaphane-induced lipolysis resulted from increased HSL gene expression and suggested that it could be related to the AMPK pathway (Lee, Moon et al. 2012).

Roles and effects of nutrients on several parameters of mammalians, commonly and/or separately, in both diabetes and obesity are summarized in Figure 5.1.

5.3 ROLE OF NUTRITION IN THE FUNCTION OF THE CARDIOVASCULAR SYSTEM

The recommendations of American Heart Association (AHA) on nutrition (2015), such as that the fat content in diets has an important role in the pathogenesis of atherosclerosis, particularly pay attention to nutrition-oriented strategies to prevent cardiovascular diseases. In 2015, the AHA reported that the percentage of saturated fatty acids in diets should not exceed 10%, and the type of the fat should consist of polyunsaturated fatty acids (2015). The Mediterranean diet rich in olive oil, vegetables, fruits and low in meat has been suggested to be beneficial for cardiovascular health. A PREDIMED trial, conducted on Spanish adults with high cardiovascular risks, showed that the Mediterranean diet markedly decreased stroke incidence after myocardial infarction (Estruch, Ros et al. 2013). In this trial, the incidence of cardiovascular events was attenuated 30% due to the Mediterranean diet pattern. Similarly, in the Lyon Heart study, the Mediterranean diet significantly prevented cardiac death and nonfatal MI in patients who survived after myocardial infarction (de Lorgeril, Salen et al. 1999). The Mediterranean diet was also demonstrated to effectively decrease incidence of heart failure in men (Tektonidis, Akesson et al. 2015). A dietary approach to stop hypertension (DASH) diet has been also been found beneficial in preventing cardiovascular events. This diet, rich in vegetables, fruits, low fat dairy products and low in saturated or total fat, was successful in preventing coronary artery disease (Fung, Chiuve et al. 2008). The DASH diet was also associated with a decreased risk of mortality in hypertensive patients (Parikh, Lipsitz et al. 2009). Furthermore, it attenuated the incidence of heart failure (Levitan, Wolk et al. 2009).

Salt intake is also an important factor in the management of hypertension. A 5.8 g reduction in salt intake was found to be related to a 3.1 mmHg decrease in systolic blood pressure (Dyer, Elliott et al. 1994). Nuts seem to be another beneficial dietary component for blood pressure control because of their content of high amounts of

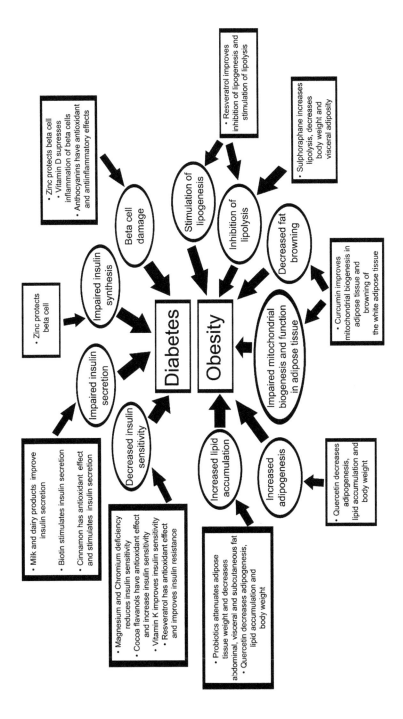

FIGURE 5.1 A summarized representation of the role of nutrition in diabetes and obesity.

magnesium, potassium, calcium, fiber and antioxidants. Wien et al. showed that a diet rich in nuts decreased systolic and/or diastolic blood pressure (Wien, Sabate et al. 2003). People who consume high amounts of nuts have been reported with lower incidences of hypertension (Djousse, Rudich et al. 2009). However, parameters such as weight or salt intake were excluded from this study. Another study, including these parameters and many others, did not find any relation between nut consumption and incidence of hypertension (Martínez-Lapiscina, Pimenta et al. 2010). In fact, most of the clinical studies found that nut consumption did not change blood pressure levels (Sabate, Fraser et al. 1993; Jenkins, Kendall et al. 2002).

Hyperglycemia is known to have unfavorable effects on vascular endothelium. In the hyperglycemic state, factors such as the formation of reactive oxygen species (Roberts and Porter 2013) and advanced glycation end-products (Vistoli, De Maddis et al. 2013) increased vascular permeability or activation of diacylglycerol-protein kinase C pathway (Das Evcimen and King 2007) could lead to negative effects on the vasculature. Thus, the carbohydrate content of a diet seems to be important. It has been revealed that a carbohydrate intake 15% lower than recommended resulted in impaired endothelial function (Merino, Kones et al. 2013). Meta-analysis reported that a low-carbohydrate diet, compared to moderate carbohydrate intake, caused a 1.01% decrease in flow-mediated dilation in overweight or healthy adults without coronary heart disease (Schwingshackl and Hoffmann 2013). This finding is noteworthy since a 1% reduction in flow-mediated dilation is associated with later cardiovascular risk (Inaba, Chen et al. 2010).

The effect of egg consumption on cardiovascular risk has been discussed widely. The results are conflicting. In the Framingham study, no significant relation between egg consumption and total coronary heart disease, myocardial infarction or angina pectoris was reported (Dawber, Nickerson et al. 1982). In this study, it was emphasized that content of the diet, rather than the amount of the eggs consumed, is more important. On the other hand, a positive correlation between high egg intake and coronary heart disease was shown in diabetic patients (Hu, Stampfer et al. 1999). In a physicians' Health Study, consumption of less than seven eggs per week was not related to myocardial infarction or stroke, whereas an intake of seven or higher per week was associated with a higher mortality risk in male physicians (Djousse and Gaziano 2008). Studies on the relation between the risk of diabetes and egg consumption has reported both no risk (Djousse and Gaziano 2008; Zazpe, Beunza et al. 2013) and increased risk (Shi, Yuan et al. 2011; Ericson, Sonestedt et al. 2013; Li, Zhou et al. 2013).

Since flavonoids in fruits have been shown to lower blood pressure, they have been suggested as a preventive for cardiovascular disease (Moline, Bukharovich et al. 2000). Pomegranate juice (50 mL/d) was shown to decrease systolic blood pressure by 5% in a small group of hypertensive patients (Aviram, Dornfeld et al. 2000). On the other hand, a 90-day 240 mL daily pomegranate juice intake had no effect on blood pressure in patients with ischemic coronary artery disease (Gorinstein, Caspi et al. 2006). Proanthocyanin rich extracts from grape seeds attenuated atherosclerosis in cholesterol-fed rabbit aorta (Yamakoshi, Kataoka et al. 1999). Proanthocyanin decreased the number of oxidized LDL positive macrophage-derived foam cells in atherosclerotic lesions. Koga and Meydani suggested that metabolites of flavonoids rather than the intact forms could be effective to reduce cardiovascular risk (Koga and Meydani 2001). They indicated that intragastric administration of plasma metabolites of catechin inhibited U937 cell adhesion and ROS generation in IL-1 beta stimulated cells in human aortic endothelial cells. Plant polyphenols were found to increase endothelium dependent vasorelaxation through nitric oxide (NO)cyclic guanosine monophosphate (cGMP) pathway (Karim, McCormick et al. 2000; Taubert, Berkels et al. 2002; Lorenz, Wessler et al. 2004).

Carotenoids such as lycopene, beta carotene or lutein have been shown to exert preventive effects on cardiovascular disease. Di Mascio et al. showed the ROS scavenging effect of lycopene (Di Mascio, Kaiser et al. 1989). The effectiveness of lycopene to quench singlet oxygen was approximately twice that of beta carotene. The finding that lycopene reduces arterial stiffness and the risk of cardiovascular disease could be partly attributed to decreased oxidative modification of low-density lipoprotein (LDL) (Bohm 2012). This idea was further confirmed by Yeo et al., as they observed a negative correlation between lycopene and oxidized LDL in Korean men (Yeo, Kim et al. 2011). Treatment of male Wistar rats with 1 mL/kg lycopene for 31 days resulted in reduced levels of lipid peroxides and elevated levels of glutathione with increased glutathione peroxidase activity (Bansal, Gupta et al. 2006). Lycopene treatment exerted a cardioprotective effect in male New Zealand white rabbits fed with high cholesterol, since it decreased serum cholesterol levels and increased HDL cholesterol levels (Verghese 2008). MDA levels were significantly attenuated in patients with coronary heart disease after consumption of 200 g of cooked tomatoes for 60 days (Bose and Agrawal 2007). In this case control study, tomato intake also elevated levels of antioxidant enzymes such as SOD, glutathione reductase and glutathione peroxidase. Kim et al. reported the anti-inflammatory and antioxidant effects of lycopene in healthy men (Kim, Paik et al. 2011). Contrarily, no effect of lycopene on endothelial function was shown in non-smoking postmenopausal women who took 46 mg lycopene per day (Stangl, Kuhn et al. 2011).

Probiotics and their metabolites have been suggested in the management of hypertension, since they ameliorate total and LDL cholesterol levels, decrease insulin resistance and also regulate the renin angiotensin system (Khalesi, Sun et al. 2014). Systolic and diastolic blood pressure were reduced significantly in older hypertensive patients who consumed 95 ml of sour milk fermented with *L. helveticus* and *Saccharomyces cerevisiae* per day for eight weeks (Hata, Yamamoto et al. 1996). These beneficial effects of probiotics could be related to their ability to release peptides with angiotensin converting enzyme (ACE) inhibitory properties. *L. helveticus* fermented milk protein casein was shown to release ACE inhibitory tripeptides (Korhonen 2009). The treatment of subjects who have high normal blood pressure with fermented milk tablets containing *L. helveticus* CM4 for four weeks caused an attenuation in diastolic blood pressure in a randomized, placebo-controlled, doubleblind study. On the other hand, in a mild blood pressure group, this treatment decreased systolic blood pressure (Aihara, Kajimoto et al. 2005).

As mentioned in above paragraphs, in the light of literature data, various epidemiologic, experimental and clinical studies have been conducted to help provide dietary recommendations for optimal cardiovascular health. Overall, the role of nutrition in the function of cardiovascular system is summarized in Figure 5.2. As

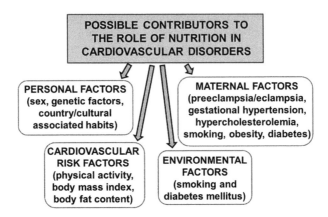

FIGURE 5.2   A summarized representation of the role of nutrition in the function of the mammalian cardiovascular system.

can be seen in this Figure, nutrition can affect the cardiovascular system in the level of heart, vessels or both.

5.4   CARDIAC FUNCTION IN DIABETES AND OBESITY

5.4.1   CARDIAC DYSFUNCTION IN DIABETES

One of the most important complications arising from diabetes is cardiovascular disease. Morbidity and mortality due to cardiac dysfunction is increased in diabetics. The hospitalization ratio for cardiovascular events in diabetic patients was higher compared to nondiabetics (Cavender, Steg et al. 2015). The incidence of heart failure has been found to be increased in diabetics, and diabetes is an independent risk factor for heart failure (MacDonald, Petrie et al. 2008). Russo and Frangogiannis proposed that the increased prevalence of heart failure in diabetic patients may be associated with concomitant pathologies such as hypertension and coronary artery disease in these patients (Russo and Frangogiannis 2016). On the other hand, diabetic cardiomyopathy, independent of any concomitant disease, could result in heart failure.

One of the characteristics of diabetic cardiomyopathy is interstitial and perivascular fibrosis (Russo and Frangogiannis 2016). In diabetic patients, interstitial fibrosis has been found to be related to accumulation of type 1 and 3 collagen (Shimizu, Umeda et al. 1993). STZ-induced diabetes caused diastolic dysfunction accompanied by increased collagen deposition and profibrotic/prohypertrophic gene expression in mice (Huynh, McMullen et al. 2010). Six-week STZ diabetic rats exhibited matrix deposition with increased protein expression of collagen I, collagen IV and fibronectin in the left ventricle (Aragno, Mastrocola et al. 2008). Similarly, STZ-induced diabetic normotensive rats exhibited myocardial fibrosis with increased expression of profibrotic factors, transforming growth factor beta 1 (TGF β1, connective tissue growth factor and matrix proteins. In this study, hypertension did not have additional negative effects on the diabetic heart (Ares-Carrasco, Picatoste et al. 2009). Marked structural changes in STZ-induced diabetic rats are given in Figure 5.3. In these electron micrographs,

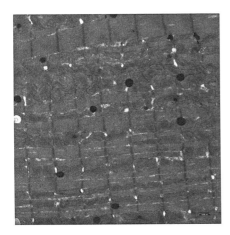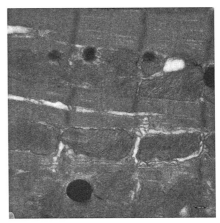

FIGURE 5.3 Representative electron micrographs of STZ-induced diabetic rat heart. Diabetesinduced changes in myofilaments and the Z-lines of myofibers, degeneration of myofibrils, loss of cristae and a granular matrix (left), markedly increased numbers of lipid droplets and signs of fibrosis (right). Magnification: left ×10,000; right ×16,700.

the ultrastructure of diabetic hearts showed marked alterations in myofilaments and Z-lines of myofibers in the mitochondria, including loss of cristae and granular matrix and also increased numbers of lipid droplets and signs of fibrosis.

Additionally, diastolic dysfunction with cardiac fibrosis and hypertrophy was reported in db/db mice (Gonzalez-Quesada, Cavalera et al. 2013). On the other hand, in mice with a genetic model of type 1 diabetes diastolic dysfunction there was no cardiac fibrosis or cardiomyocyte hypertrophy (Basu, Oudit et al. 2009). Cardiac fibrosis in diabetes could be associated with increased renin-angiotensin-aldosterone activity since it was ameliorated by ACE inhibitor (Toblli, Cao et al. 2005). Oxidative stress could be another parameter for cardiac fibrosis. Aragno et al. reported the relationship between oxidative stress and cardiac fibrosis in the diabetic heart as dehydroepiandrosterone (DHEA) treatment of rats ameliorated increased tissue collagen levels and profibrogenic factor expression by improving oxidative stress (Aragno, Mastrocola et al. 2008).

One of the major alterations in the diabetic heart is decreased beta-adrenergic responsiveness. Both noradrenaline and isoprenaline mediated maximum chronotropic responses were found to be decreased in the atria of 14-week diabetic rats (Dincer, Onay et al. 1998). Both mRNA and protein levels of cardiac β_1-ARs was found to be decreased in 14week diabetic rats (Dincer, Bidasee et al. 2001). Similarly, in six-week diabetic rats, reduced levels of cardiac β_1-AR mRNA were shown (Matsuda, Hattori et al. 1999). The mRNA level of β_2-ARs, on the other hand, was markedly increased in 14-week diabetic rat hearts. The protein expression level of β_2-ARs was decreased in this study (Dincer, Bidasee et al. 2001).

5.4.2 CARDIAC FUNCTION IN OBESITY

Hemodynamic changes due to obesity involve increased cardiac output (Alexander 1998) and left ventricular stroke volume (Alpert, Agrawal et al. 2014), decreased peripheral vascular resistance (Alpert, Agrawal et al. 2014) and increased oxygen consumption (Alexander 1998). In obese patients, left ventricular diastolic function was impaired (Herszkowicz, Barbato et al. 2001). Impairment of left ventricular filling has been suggested to be related to left ventricular hypertrophy (Alpert, Lambert et al. 1994; Alpert, Lambert et al. 1995a; Alpert, Lambert et al. 1995b). However, left ventricular systolic dysfunction in obesity is rare (Alpert, Omran et al. 2016).

Left ventricular hypertrophy is a common feature in obesity (Abel, Litwin et al. 2008). Blood pressure contributes to cardiac hypertrophy in obese patients. Both systolic blood pressure and left ventricle end systolic wall stress was shown to correlate positively with the percentage overweight in normotensive obese patients (Alpert, Lambert et al. 1994). The duration of obesity also affects cardiac structural parameters such as left ventricle chamber size, wall thickness and mass (Alpert, Lambert et al. 1995a).

Cardiac hypertrophy was shown in leptin deficient ob/ob mice (Sloan, Tuinei et al. 2011). Left ventricular collagen content was increased in 20-week obese mice (Zaman, Fujii et al. 2004). In this study, both plasma levels and gene expression of plasminogen activator inhibitor 1 (PAI-1) and TGF $\beta 1$ were also elevated. On the other hand, impaired contractility in the absence of myocardial fibrosis in ob/ob mice was demonstrated with decreased PRSW values in the presence of preserved ejection fraction (EF) and dP/dt_{max} values (Van den Bergh, Vanderper et al. 2008). Sarcoplasmic reticulum Ca^{++} ATPase 2a (SERCA2a) affinity for Ca^{2+} was also attenuated in ob/ob mice which was related to regulation and phosphorylation of phospholamban. Thus, impaired contraction and relaxation responses in the heart were attributed to apoptosis and altered Ca^{2+} intake to SERCA2a. Moreover, 16-week high fat feeding of C57/BL6J mice led to cardiac fibrosis and hypertrophy (Calligaris, Lecanda et al. 2013).

Impairment in Ca^{2+} handling and β-ARs has been suggested to have a role in obesity related cardiac dysfunction (Strassheim, Houslay et al. 1992; Relling, Esberg et al. 2006). Lima-Leopoldo et al. reported that altered intracellular Ca^{++} handling through decreased SERCA2a activation results in cardiac dysfunction in obese rats (Lima-Leopoldo, Leopoldo et al. 2011). However, no difference in terms of beta-adrenergic responsiveness to isoprenaline was found between the lean and the obese group in this study.

5.5 DIET AND BETA-ADRENERGIC SYSTEM RELATION IN HEART FUNCTION

Beta-adrenergic responsiveness is one of the major determinants of heart function. Cardiac function has been shown to be deteriorated in subjects who have Western-style eating patterns. On this point, it could be expected that a diet pattern with high fat content would affect the β-adrenergic system in the heart. Carroll et al. revealed that high fat-feeding of New Zealand white rabbits resulted in decreased responsiveness to isoprenaline in the absence of any alteration in β-AR density or affinity (Carroll, Jones et al. 1997; Carroll, Kyser et al. 2002). Isoprenaline induced β-AR mediated contractile response was impaired in high fat-fed C57BL/6 mice despite the preserved cardiac structure (Fu, Hu et al. 2017). Furthermore, the expression levels of proteins such as SERCA2a, phospholamban or troponin I were not changed significantly. The protein expression of β_1-ARs and β_2-ARs and also G protein did not differ between the groups. In contrast, β_2-ARs phosphorylation at PKA site serine 261/262 and GRK site serine 355/356 was increased in the high fat fed group. The contractile response stimulated with isoprenaline in the cardiomyocytes was attenuated in high fat-fed mice. Isoprenaline mediated PKA phosphorylation of phospholamban at serine 16 was also decreased in these cardiomyocytes. Dincer et al. demonstrated that the protein expressions of beta ARs were reduced in obese Ossabaw pigs fed with a high fat diet for 50 weeks (Dincer 2011). This finding has been suggested to be associated with downregulation or desensitization of the receptors as a result of increased activity of the sympathetic nervous system.

However, Lima-Leopoldo et al. reported that feeding with a hypercaloric diet did not cause any impairment in beta-adrenergic responsiveness mediated by isoprenaline in the papillary muscles of obese Wistar rats (Lima-Leopoldo, Leopoldo et al. 2011). Furthermore, in this study, cyclic adenosine monophosphate (cAMP) phosphorylation of proteins associated with Ca^{2+} handling such as SERCA2a and phospholamban was not changed in obesity. Cardiomyocyte contraction amplitude was shown to be decreased in Fisher rats fed with a protein restricted feed after weaning (Penitente, Novaes et al. 2014). Both mean arterial pressure and heart rate were increased significantly in the protein restricted group. This effect was attributed to structural alterations in the cardiovascular system due to reduced amino acid bioavailability. In this study, the fact that the beta AR mediated contractile response mediated by isoprenaline was also depressed has been suggested to arise from reduced sympathetic responsiveness due to low protein intake. Ca^{2+} sparks amplitude was lower and frequency was higher in the protein restricted group which was attributed to altered Ryanodine receptor (RyR2) function. Furthermore, the SERCA2a expression decrease in the restricted group is consistent with inhibited Ca^{2+} intake to SR during cardiomyocyte relaxation, thus contributing to prolongation of relaxation time.

5.6 EFFECT OF β-ARs SYSTEM ON HEART FUNCTION IN DIABETES AND OBESITY

β-ARs are one of seven membrane spanning receptors, GTP binding protein coupled receptors. Three subtypes of β-ARs have been shown to be expressed in cardiac tissue (Wilson and Lincoln 1984; Dincer, Bidasee et al. 2001). Suggested as a fourth subtype, β_4-ARs were later confirmed as a low affinity state of β_1-ARs (Kaumann, Engelhardt et al. 2001). The β_1-ARs are known to be the predominant subtype in the heart. The ratio of β_1-ARs to β_2-ARs in the human heart is 70–80% : 30–20% in the ventricle and 60–70% : 40–30% in the atrium (Brodde 1991). The β_1-ARs are coupled to stimulatory G protein (Gs) whereas β_2-ARs are coupled to both Gs and inhibitory G protein (Gi). Both β_1-ARs and β_2-ARs cause positive inotropic and chronotropic responses through Gs stimulation in the heart. This effect further involves activation of adenylyl cyclase (AC), increased levels of cAMP, activation of protein kinase A (PKA), and phosphorylation of several proteins such as L type Ca^{2+} channels (Zhao, Gutierrez et al. 1994), phospholamban (Simmerman and Jones 1998), troponin I (Sulakhe and Vo 1995) and ryanodine receptors (Marx, Reiken et al. 2000). Phosphorylation of L type Ca^{2+} channels leads to increased Ca^{2+} influx and thus contraction. Phospholamban regulates reuptake of Ca^{2+} into the SR over SERCA2a and has a role in diastolic relaxation. Cardiac β_2-ARs, different from β_1-ARs, also couple to Gi (Kuschel, Zhou et al. 1999a; Kuschel, Zhou et al. 1999b). The β_2-ARs have been shown to exert an antiapoptotic effect in adult mouse cardiomyocytes through Gi coupling (Zhu, Zheng et al. 2001), whereas β_1-ARs are known for apoptotic effects (Zhu, Wang et al. 2003).

The β_3-ARs are also coupled to Gi. This subtype mediates thermogenesis in brown adipocytes and lipolysis in white adipocytes (Lowell and Spiegelman et al. 2000). In the heart, they mediate negative inotropic effects (Gauthier, Tavernier et al. 1996). Different from the other two subtypes, β_3-ARs have introns and lack phosphorylation sites for G protein coupled receptor kinases (GRKs) which provide resistance to catecholamine-induced desensitization (Rozec and Gauthier 2006).

Pott et al. reported that BRL 37344 stimulated positive inotropic effect in the human atrium, which was neutralized in the presence of propranolol (Pott, Brixius et al. 2003). Thus, the effect of BRL 37344 was attributed to β_1-ARs and β_2-ARs (Pott, Brixius et al. 2003). Conversely, in the human atria, β_3-ARs activation resulted in increased contractility and stimulation of L type Ca^{2+} current. It has been confirmed that this effect was dependent on β_3-AR since it was eliminated by β_3-ARs antagonist L 748,337 and not affected by nadolol, an antagonist of β_1-ARs and β_2-ARs (Skeberdis, Gendviliene et al. 2008).

The stimulation of β_3-ARs activates Gi and endothelial nitric oxide synthase (eNOS) (Gauthier, Leblais et al. 1998). Other NOS subtypes, neuronal and inducible NOS (iNOS) are also involved in this pathway (Amour, Loyer et al. 2007). The expression level of β_3-ARs seems to be insignificant in the healthy heart (Dincer, Bidasee et al. 2001). However, they have been shown to be upregulated in several cardiovascular pathologies such as heart failure (Morimoto, Hasegawa et al. 2004) and diabetes (Dincer, Bidasee et al. 2001). This upregulation has been suggested as an adaptive mechanism to prevent the detrimental effects of excessive catecholamine levels in these pathologies, which further becomes maladaptive (Altan, Arioglu et al. 2007).

In the diabetic heart, both responsiveness and expression of β-ARs have been demonstrated to be altered (Dincer, Onay et al. 1998; Dincer, Bidasee et al. 2001). The duration of diabetes seems to be crucial in this alteration since no difference in noradrenaline- or isoprenaline-induced chronotropic responses was reported in eight-week diabetic rat atria, whereas maximum chronotropic effect obtained with noradrenaline was attenuated in 14-week diabetic rat atria (Dincer, Onay et al. 1998). This study revealed that β_1-AR mediated chronotropic responses were reduced in chronic diabetes, although β_2-AR mediated responses were not changed.

Both mRNA and protein expression levels of beta β_1-ARs were markedly reduced in 14-week diabetic rat hearts (Dincer, Bidasee et al. 2001). Similarly, mRNA expression of β_1-ARs was reported to be decreased in six-week diabetic rat hearts (Matsuda, Hattori et al. 1999). The downregulation of β_1-ARs could be related to elevated noradrenaline levels in diabetes since noradrenaline synthesis and release were increased in diabetic cardiomyopathy (Ganguly, Beamish et al. 1987). On the other hand, mRNA levels of β_2-ARs were increased in 14-week diabetic rat hearts, whereas protein levels of this subtype were decreased (Dincer, Bidasee et al. 2001). Reduced protein levels of β_2-ARs did not cause any attenuation in β_2-ARs mediated chronotropic responses (Dincer, Onay et al. 1998). This finding was explained in that a small decrease (17%) in the protein expression of β_2-ARs could be insufficient to alter the chronotropic response (Altan, Arioglu et al. 2007). Thackeray et al. reported that myocardial β-AR binding was reduced 30–40% compared to controls in eight-week high fat-fed STZ-treated hyperglycemic rats (Thackeray, Parsa-Nezhad et al. 2011). In this study, the protein expression level of β_1-ARs was decreased significantly, whereas the density of β_2-ARs was increased in diabetic rats (Thackeray, Parsa-Nezhad et al. 2011).

In obese Zucker rats, both the β-AR-mediated inotropic response by isoprenaline and cAMP mediated inotropic response by forskolin were decreased slightly (Jiang, Carillion et al. 2015). These decreases were more severe in obese diabetic rats compared to obese rats. Furthermore, the density of β_1- and β_2-ARs was decreased markedly in obese rats. Similar to the inotropic response, the decrease in the protein expression level of β_1-ARs was more significant in the obese diabetic rats compared to the obese group. Minhas et al. showed that myocyte contraction stimulated by isoprenaline was attenuated in ob/ob mice which was in line with decreased cell shortening and Ca^{2+} transient (Minhas, Khan et al. 2005). Forskolinmediated AC activation and dibutryl cAMPmediated contractile responses were also depressed in these mice. Furthermore, PKA activation was suppressed in ob/ob mice. No difference in protein expression of β_1-ARs was observed. Dincer et al. found that β_1- and β_2-AR protein expression was significantly decreased in the ventricles of 50-week high fat-fed obese swine, although mRNA expression of these subtypes was not altered (Dincer 2011). Similarly, Vileigas et al. showed that isoprenaline stimulated β-AR contraction response did not change in high fat-fed obese rats (Vileigas, de Deus et al. 2016). They found that the sole component that was changed in the β-adrenergic system of obese rats was the increased protein expression of AC. This increase in AC protein expression did not result in augmented PKA activation. This finding was explained as a reduction in AC activity causing increased protein expression as a compensatory mechanism (Vileigas, de Deus et al. 2016). Ferron et al. did not observe any impairment in the isoprenaline-induced contractile response in the papillary muscles of obese rats (Ferron, Jacobsen et al. 2015). Furthermore, cardiac β_1-AR and β_2-AR and Gsα protein expressions were not changed in these rats. Thus, they attributed the impaired cardiac relaxation response to the changes in Ca^{2+} handling rather than the βadrenergic system.

The difference in the studies in terms of beta-adrenergic responsiveness has been attributed to such factors as duration of the high fat-feeding, percentage of calorie from fat, type of fatty acids or fat source in the diet and animal species or strain (Vileigas, de Deus et al. 2016).

5.7 β_3-ARs IN DIABETES AND OBESITY

As mentioned earlier, β_3-ARs are a member of GPCR. The β_3-ARs have an extracellular Nterminal region and an intracellular C-terminus. The β_3-ARs have been demonstrated to exist in several species such as human (Emorine, Marullo et al. 1989), rat, dog (Strosberg 1997), sheep and goat (Forrest and Hickford 2000). The β_3-ARs differ from β_1-ARs and β_2-ARs as they have introns; additionally, they lack PKA phosphorylation sites, which could explain why they are resistant to short-term agonist-stimulated desensitization in some models (Carpene, Galitzky et al. 1993).

The possible pathways related to β_3-AR activation in diabetes and obesity are given in Figure 5.4. The β_3-ARs are coupled to both Gs and Gi in adipocytes (Soeder, Snedden et al. 1999). Gs coupling results in thermogenesis through activation of AC and PKA, stimulation of p38α and activation of transcription factor 2 (ATF2) (Cao, Medvedev et al. 2001). Gi coupling, on the other hand, leads to lipolysis through the ERK pathway and phosphorylation of hormone sensitive lipase (Robidoux, Kumar et al. 2006). In smooth muscle, the effect of β_3-ARs could be associated with Gs-potassium channel interaction independent of cAMP formation (Ferro 2006). β_3-ARs are coupled to Gi in the heart, and stimulation of this receptor results in activation of NOS and increase in NO and cGMP in ventricles (Gauthier, Leblais et al. 1998). On the other hand, β_3-ARs cause phosphorylation of calcium channels and elevation of intracellular Ca^{2+} levels in human atrial myocytes.

The β_3-AR agonist treatment resulted in weight loss in obese mice without a reduction in food intake (Yoshida, Sakane et al. 1994) which has been attributed to increased thermogenesis in brown adipose tissue (Himms-Hagen, Cui et al. 1994).

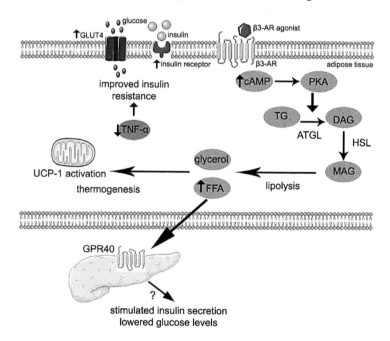

FIGURE 5.4 The possible pathways related to β_3-AR activation in diabetes and obesity. GLUT4, glucose transporter 4; UCP-1, uncoupling protein 1; cAMP, cyclic adenosin monophosphate; PKA, proteinkinase A; TG, triglycerides; DAG, diacylglycerol; MAG, monoacylglycerol; ATGL, adipose triglyceride lipase; HSL, hormone sensitive lipase; GPR40, G protein receptor coupled protein 40; TNF-α, tumor necrosis factor alpha. (Figure drawn using image bank of Servier Medical Art (http://smart.servier.com/) licensed under a Creative Commons Attribution 3.0 Unported License.)

This effect could be related to stimulation of UCP1 (Inokuma, Okamatsu-Ogura et al. 2006). UCP-1 is a mitochondrial carrier protein in brown adipose tissue and has a role in energy consumption as heat. The lipolytic effects of β_3-AR, on the other hand, involve mobilization of fat as free fatty acids from white adipose tissue deposits which are oxidized and used in thermogenesis in brown adipose tissue (Ursino, Vasina et al. 2009). Predisposition to abdominal obesity, it has been suggested, is associated with the presence of the Arg64 allele in the first intracellular loop of β_3-ARs gene which could further lead to insulin resistance (Widen, Lehto et al. 1995). Moreover, a Trp64 Arg mutation of β_3-AR has been found to be related to the early onset of type 2 diabetes in obese Pima North American Indians (Walston, Silver et al. 1995). This mutation is also associated with mild gestational diabetes (Festa, Krugluger et al. 1999).

BRL 26380A, a β_3-AR agonist, was shown to cause weight loss in diet-induced obesity models (Arch and Ainsworth 1983). The β_3-AR agonists have been suggested to decrease visceral fat rather than subcutaneous fat. Actually, increased visceral fat mass was found to be associated with Trp64Arg variant of β_3-AR (Kim-Motoyama, Yasuda et al. 1997). Mirabegron, the first β_3-ARs agonist approved for the treatment of over-active bladder (Leone Roberti Maggiore, Cardozo et al. 2014), was found to increase resting metabolic rate and glucose uptake in brown adipose tissue in humans which was associated with thermogenesis (Cypess, Weiner et al. 2015).

Ghorbani et al. demonstrated the antiobesity effect of β_3-AR stimulation by CL316,243 treatment in obese Zucker rats (Ghorbani, Teimourian et al. 2015). They found that CL316,243 elevated resting metabolic rate and thus increased thermogenesis. Treatment caused a significant atrophy of retroperitoneal fat deposits in obese rats; the white adipocyte was smaller in this group. Furthermore, they reported a reduction in DNA content in retroperitoneal white adipose tissue which points to apoptosis. The β_3-AR activation was shown to induce cellular and metabolic plasticity in white adipose tissue (Granneman, Li et al. 2005). Inflammation, proliferation of stromal cells and multilocular adipocytes, and also stimulation of mitochondrial biogenesis and beta oxidation in multilocular adipocytes are characteristics of this effect of CL316,243 treatment. Inokuma et al. reported that the beneficial effects of CL316,243 stimulated β_3-AR activation in obesity is related to UCP-1 since it was eliminated in UCP-1 knockout mice (Inokuma, Okamatsu-Ogura et al. 2006). These mice, having a genetic deletion of UCP-1, did not exhibit β_3-AR-mediated thermogenesis. Increased oxygen consumption following a CL316,243 injection in wild type mice was also eliminated in UCP-1 knockout mice. Thus, UCP-1 has been suggested to have a role in β_3-AR-mediated brown adipose tissue thermogenesis and energy expenditure. By way of contrast, lack of UCP-1 did not alter lipid mobilization in white adipose tissue. In this study, a two-week CL316,443 treatment was found to be effective in preventing weight gain in mice fed with a cafeteria diet which was eliminated in UCP-1 knockout mice. This difference was attributed to the role of UCP-1 in CL316,243 mediated energy expenditure since energy intake was not changed. CL316,243 decreased lipid droplets in brown adipocytes and adipocytes in white adipose tissue were smaller in UCP-1 knockout mice compared to the wild type.

Another beneficial effect of β_3-AR agonists is their antidiabetic effect. Yoshida et al. demonstrated that chronic β_3-AR agonist treatment ameliorates hyperglycemia, hyperinsulinemia and hyperlipidemia in mice (Yoshida, Sakane et al. 1994). This effect has been suggested to be independent of their antiobesity effect since they don't decrease body weight in low doses (Arch and Wilson 1996). The mechanism of their antidiabetic effect has not been clarified clearly, but has been attributed to increased peripheral insulin sensitivity (Ursino, Vasina et al. 2009).

Acute treatment with CL316,243 did not attenuate blood glucose levels in ZDF-rats. However, long-term treatment improved hyperglycemia and hyperinsulinemia in obese ZDFrats (Liu, Perusse et al. 1998). Increased insulin secretion subsequent to CL316,243 treatment has been suggested to be independent of the direct effect of the agonist, since β_3-AR mRNA was not detected in pancreatic beta cells and the agonist did not stimulate insulin secretion in cultured pancreatic islets (Grujic, Susulic et al. 1997). In obese diabetic KKAy mice, a two-week treatment of CL316,243 led to attenuated serum levels of glucose, insulin, triglyceride, free fatty acid, TNFα expression and an elevated adiponectin level (Fu, Isobe et al. 2008). In the first two days of the treatment, CL316,243 reduced food intake which was normalized after two days. It was reported that this effect could be related to increased free fatty acid levels during the first two days; it was eliminated after day three when the increased levels of fatty acids started to decrease. It was shown that adiponectin receptors 1 and 2, β_3-AR mRNA expression and also adiponectin levels were augmented by the CL316,243 treatment. Finally, the beneficial effects of CL316,243 were attributable to increased mRNA expressions of β_3-ARs, adiponectin receptors 1 and 2, elevated levels of adiponectin and suppressed TNFα. These results are in line with the findings of Ghorbani et al. They showed the relationship between β_3-AR activation by CL316,243 treatment and suppression of TNFα expression in retroperitoneal white adipose tissue which was beneficial to improve insulin resistance (Ghorbani, Shafiee Ardestani et al. 2012). Thus, the authors revealed that stimulation of β_3-AR in adipocytes produce an antidiabetic effect. Similarly, Kim et al. reported that stimulation of β_3-AR by CL316,243 ameliorated hyperglycemia, hyperinsulinemia, lipid profile and metabolism in nonobese diabetic MKR mice (Kim, Pennisi et al. 2006). It also improved insulin sensitivity related to increased hepatic insulin responsiveness and glucose uptake in adipose tissue (Figure 5.4). These results indicate that the antidiabetic effect of β_3-ARs agonists could result from improved insulin action on adipose tissue and liver in MKR mice. In this study, it has been suggested that the beneficial effects on enhanced insulin sensitivity in adipose tissue could be attributed to increased expression levels of UCP-1 in adipose tissue insulin receptors and GLUT4, increased mitochondrial content and impaired adipogenesis (Kato, Ohue et al. 2001).

MacPherson et al. found that CL316,243 stimulated pancreatic insulin secretion and reduced glucose levels in C57BL6 mice (MacPherson, Castellani et al. 2014). This effect was attributed to increased free fatty acid levels stimulated by CL316,243 since it was eliminated when the free fatty acids were reduced with nicotinic acid. It was also shown that these beneficial effects disappeared in mice with a genetic deletion of ATGL, a rate limiting enzyme of lipolysis (Haemmerle, Lass et al. 2006). However, it could not be explained how elevated fatty acids stimulate insulin secretion. It was suggested that the G protein coupled receptor 40 (GPR40), a fatty acid receptor in beta cells (Itoh, Kawamata et al. 2003), could have contributed to this outcome. In GPR40 knockout mice, CL316,243 treatment did not attenuate blood glucose level (Pang, Wu et al. 2010). MacPherson et al. revealed that indirect factors such as interleukin 6 (IL-6) levels could have roles in the CL316,243-mediated decrease in glucose levels, as IL-6 augmented insulin secretion directly or via GLP-1 dependent pathways (Ellingsgaard, Hauselmann et al. 2011). However, their results did not confirm the role of IL-6 in the CL316,243 mediated effect.

5.8   REGULATION OF β_3-ARs IN HEART FROM DIABETIC AND OBESE MAMMALIANS

Gauthier et al. showed the presence of cardiac β_3-ARs in human ventricular endomyocardial biopsies (Gauthier, Tavernier et al. 1996). As mentioned before, distinct from β_1 and β_2-ARs, activation of this subtype has been suggested to exert a negative inotropic effect through the NO-cGMP-protein kinase G (PKG) pathway (Angelone, Filice et al. 2008). Isoprenaline decreased contractile force in the presence of nadolol, the β_1 and β_2-AR antagonist. This effect was thought attributable to β_3-AR. Later, this finding was confirmed by using various β_3-AR agonists such as BRL 37344 and CL316243. Gauthier et al. showed that the negative inotropy obtained with BRL 37344 was not eliminated after metoprolol or nadolol which proved that it was independent of $\beta1$ and β_2-ARs (Gauthier, Tavernier et al. 1996). On the other hand, this effect was inhibited after bupranolol pretreatment, a nonselective β-AR antagonist. The authors also demonstrated the protein expression of β_3-ARs in the endomyocardial biopsies. The signaling pathway of β_3-ARs involves activation of the Gi protein since the effect was not seen when Gi is inhibited with pertussis toxin (Gauthier, Tavernier et al. 1996). The role of NOS in this pathway was also confirmed by using NOS inhibitors. In the failing heart, increased β_3-AR response has been suggested to contribute to cardiac dysfunction (Moniotte and Balligand 2002). The β_3-AR mediated negative inotropy has been thought a preventive mechanism in the case of sympathetic overdrive, but becomes maladaptive in the failing heart since β_3-ARs are not downregulated or desensitized in any way different from β_1 and β_2-ARs (Altan, Arioglu et al. 2007).

There are not enough studies exploring the role of β_3-AR in the diabetic or obese heart. In the diabetic rat heart, β_3-ARs mediated negative inotropic response by BRL 37344 was increased (Kayki-Mutlu, Arioglu-Inan et al. 2014). Nadolol pretreatment did not change the effect. Furthermore, pretreatment with SR 59230A, a β_3-AR antagonist, also did not eliminate negative inotropy. These findings confirmed that the negative inotropic effect resulted from β_3-ARs. Rate of contraction and relaxation mediated by BRL 37344 was increased in the diabetic group. The increase in all these parameters improved after insulin treatment. Supporting this finding, β_3-AR mRNA expression was increased in diabetic hearts which did not decrease to control levels after insulin treatment (Kayki-Mutlu, Arioglu-Inan et al. 2014). The eNOS mRNA expression was also increased in the diabetic group, although this change was not statistically significant. It was also found that BRL37344 stimulated β_3-AR mediated responses were reduced in the left ventricular papillary muscles of diabetic rats, which partially improved after insulin treatment (Arioglu-Inan, Ozakca et al. 2013). In this study, the protein expression level of β_3-ARs increased significantly. Decreased negative inotropic effect, despite upregulation of β_3-ARs, may be related to downstream components of the signaling pathway such as eNOS phosphorylation which could not be investigated in this study.

Moreover, the discrepancy between these two studies could be explained by experimental differences, since a study by Kayki-Mutlu was conducted on a Langendorff perfused whole heart, whereas the other was made on papillary muscle.

Expression levels of both mRNA and the protein of β_3-ARs were increased markedly in 14-week diabetic rat hearts (Dincer, Bidasee et al. 2001). The ratio of β_1, β_2 and β_3-ARs was 62 : 30 : 8 in healthy rat hearts, whereas it altered to 40 : 36 : 23 in diabetes. However, a two-week insulin treatment significantly improved this increase in the protein expression of β_3-ARs. Furthermore, upregulation of β_3-ARs in diabetes was partially improved after a twoweek insulin treatment. Amour et al. demonstrated that β-AR stimulated positive inotropic effect was reduced in diabetic rat hearts and this response was improved to some extent with a β_3-AR antagonist, a nonselective NOS inhibitor or a selective nNOS inhibitor (Amour, Loyer et al. 2007). They also found that the protein expression of β_3-ARs was increased in diabetic hearts. Thus, it was suggested that β_3-AR and β_3-AR signaling through neuronal NOS (nNOS) and NO seem to be important regarding impaired beta-adrenergic responsiveness in diabetic hearts. Increased negative inotropic effect in the heart of STZdiabetic rats was also demonstrated by Okatan et al. (Okatan, Tuncay et al. 2015). The authors found that left ventricular pressure developed with BRL37344 stimulation was significantly higher in the STZ-group compared to the control. The protein expression level of β_3-ARs was also markedly elevated in STZ diabetic rat hearts. Galougahi et al. showed that CL316,243 ameliorated the detrimental effects of hyperglycemia on glutathionylation-induced Na^+-K^+ pump inhibition in cardiomyocytes of New Zealand white rabbit (Karimi Galougahi, Liu et al. 2015). In this study, CL316,243 treatment reduced glutathionylation of Na^+-K^+ pump beta1 subunit in S961-induced hyperglycemia which then resulted in suppression of NADPH oxidase activation and $O_2{}^-$ generation, eNOS recoupling and improved NO mediated Na^+-K^+ pump activity. Thus, they suggested that stimulating β_3-ARs may be an effective target in the prevention and treatment of heart failure seen in diabetes. Furthermore, Turan et al. indicated that increased $O_2{}^-$ generation leads to downregulation of eNOS and inhibition of L type Ca^{2+} channels which contributes to decreased contractility in hyperglycemia (Turan and Tuncay 2014) The role of eNOS in the β_3-ARs mediated negative inotropy has been also shown by other investigators. Furthermore, Amour et al. reported, for the first time, the role of nNOS in this pathway in diabetic cardiomyopathy (Amour, Loyer et al. 2007). They found that expression of nNOS was predominant compared to eNOS in cardiomyocytes. Moreover, it was responsible for generating NO to a great extent compared to eNOS. eNOS, on the other hand, was found in the endothelial cells of hearts. iNOS protein was not detected in these preparations (Amour, Loyer et al. 2007).

The β_3-AR knockout mice myocytes exhibited increased contractile response in the higher concentrations of isoprenaline (Larson, Rainer et al. 2012). Furthermore, sarcomere shortening and Ca^{2+} transients in ob/ob mice were not altered on increasing the doses of BRL 37344 compared to the wild type. However, both sarcomere shortening and Ca^{2+} transients mediated by BRL 37344 were significantly attenuated when ob/ob mice were treated with leptin for 4 weeks. This finding indicates the dependency of β_3-AR mediated effect of leptin in ob/ob myocytes. However, it was confirmed by using β_3-AR knockout myocytes that the attenuation in sarcomere shortening and Ca^{2+} transients is associated with β_3-ARs. In this study, β_3-AR

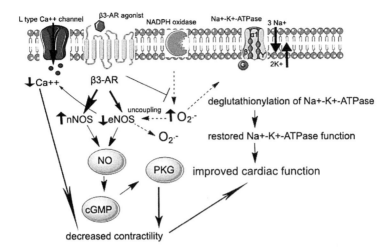

FIGURE 5.5   The possible pathways related to beneficial effects of cardiac β_3-AR stimulation in diabetic heart. nNOS, neuronal nitric oxide synthase; eNOS, endothelial nitric oxide synthase; NO, nitric oxide; NADPH, Nicotinamide adenine dinucleotide phosphate; PKG, protein kinase G; cGMP, cyclic guanosine mono phosphate. (Figure drawn using image bank of Servier Medical Art (http://smart.servier.com/) licensed under a Creative Commons Attribution 3.0 Unported License.)

mRNA expression level was reduced in ob/ob mice, which improved after leptin treatment (Larson, Rainer et al. 2012). Thus, the decreased β_3-AR mediated effect in ob/ob myocytes was suggested to result from the reduced mRNA expression of β_3-ARs. Conversely, Jiang et al. found that the protein expression of β_3-AR was not altered significantly in Zucker obese rats (Jiang, Carillion et al. 2015).

In the light of already published data, the possible pathways related to beneficial effects of cardiac β_3-ARs stimulation in diabetic hearts are summarized in Figure 5.5.

5.9   CONCLUSIONS

Affecting millions of people worldwide, diabetes and obesity are two major health problems. In this regard, the prevention and/or treatment of these diseases have been considered a matter of great importance. Nutrition seems to be the key point to both prevent or treat these pathologies. A balanced diet with macro- and micronutrients could help in the management of diabetes and obesity. Furthermore, foods rich in compounds which have been shown to be beneficial in both glucose and fat metabolism could help to prevent or treat these diseases. β_3-ARs agonists have been demonstrated to produce antidiabetic and antiobesity effects. Future investigations into these features of β-ARs could be an important goal in the management of these pathologies.

ACKNOWLEDGMENT

Figures 5.4 and 5.5 were drawn using image bank of Servier Medical Art (http://smart.servier.com/) licensed under a Creative Commons Attribution 3.0 Unported License.

ABBREVIATIONS

AC:	Adenylyl cyclase
ACC:	Acetyl-coenzyme A carboxylase
ACE:	Angiotensin converting enzyme
AHA:	American Heart Association
AMPK:	Activated protein kinase
ATAD3:	ATPase family AAA domaining protein 3
ATF2:	Activation of transcription factor 2
β-AR:	Beta adrenoceptor
BMI:	Body mass index
cAMP:	Cyclic adenosine monophosphate
cGMP:	Cyclic guanosine mono phosphate
CPT-1:	Carnitine palmitoyltransferase-1
DASH:	Dietary approach to stop hypertension
DHEA:	Dehydroepiandrosterone
EF:	Ejection fraction
eNOS:	Endothelial nitric oxide synthase
FAS:	Fatty acid synthase
GI:	Glycemic index
Gi:	Inhibitory G protein
GIP-1:	Glucose dependent insulinotropic polypeptide 1
GLP-1:	Glucagon like peptide 1
GLUT-4:	Glucose transporter 4
Gs:	Stimulatory G protein
GSK3:	Glycogen synthase kinase 3
HbA1c:	Hemoglobin A1c
HSL:	Hormone sensitive lipase
IL-6:	Interleukin 6
iNOS:	Inducible NOS
LDL:	Low density lipoprotein
nNOS:	Neuronal NOS
NO:	Nitric oxide
p-(ser)-IRS-1:	Phosphorylated insulin receptor substrate 1
PAI-1:	Plasminogen activator inhibitor 1
PGC-1α:	Peroxisome proliferator activated receptor γ coactivator 1α
PKA:	Protein kinase A
PKG:	Protein kinase G
PPARγ:	Peroxisome proliferator-activated receptor γ
ROS:	Reactive oxygen species
RyR2:	Ryanodine receptor 2
SERCA2a:	Sarcoplasmic reticulum Ca^{++} ATPase 2a
SIRT-1:	Sirtuin 1
SREBP-1c:	Sterol regulatory element-binding transcription factor 1
TGF β1:	Transforming growth factor beta 1
TNF-α:	Tumor necrosis factor-α
UCP-1:	Uncoupling protein 1

REFERENCES

(AHA (2015). *"The Facts on Fats. 50 Years of American Heart Association*, Dietary Fats Recommendations" https://www.heart.org/-/media/files/healthy-living/company-collab oration/inap/fats-white-paper-ucm_475005.pdf

Abel, E. D., S. E. Litwin, et al. (2008). "Cardiac remodeling in obesity." *Physiol Rev* 88(2): 389–419.

ADA (2004). "Nutrition principles and recommendations in diabetes." *Diabetes Care* 27(Suppl 1): s36–s36.

Aguirre, L., A. Fernandez-Quintela, et al. (2014). "Resveratrol: anti-obesity mechanisms of action." *Molecules* 19(11): 18632–18655.

Aihara, K., O. Kajimoto, et al. (2005). "Effect of powdered fermented milk with *Lactobacillus helveticus* on subjects with high-normal blood pressure or mild hypertension." *J Am Coll Nutr* 24(4): 257–265.

Ajala, O., P. English, et al. (2013). "Systematic review and meta-analysis of different dietary approaches to the management of type 2 diabetes." *Am J Clin Nutr* 97(3): 505–516.

Alappat, L. and A. B. Awad (2010). "Curcumin and obesity: evidence and mechanisms." *Nutr Rev* 68(12): 729–738.

Alexander, J. K. A. (1998). *Hemodynamic Alterations with Obesity. The Heart and Lung in Obesity*, Armonk, Futura Publishing Co: 45–56.

Alpert, M. A., H. Agrawal, et al. (2014). "Heart failure and obesity in adults: pathophysiology, clinical manifestations and management." *Curr Heart Fail Rep* 11(2): 156–165.

Alpert, M. A., C. R. Lambert, et al. (1994). "Effect of weight loss on left ventricular mass in nonhypertensive morbidly obese patients." *Am J Cardiol* 73(12): 918–921.

Alpert, M. A., C. R. Lambert, et al. (1995a). "Relation of duration of morbid obesity to left ventricular mass, systolic function, and diastolic filling, and effect of weight loss." *Am J Cardiol* 76(16): 1194–1197.

Alpert, M. A., C. R. Lambert, et al. (1995b). "Effect of weight loss on left ventricular diastolic filling in morbid obesity." *Am J Cardiol* 76(16): 1198–1201.

Alpert, M. A., J. Omran, et al. (2016). "Effects of obesity on cardiovascular hemodynamics, cardiac morphology, and ventricular function." *Curr Obes Rep* 5(4): 424–434.

Altan, V. M., E. Arioglu, et al. (2007). "The influence of diabetes on cardiac beta-adrenocep tor subtypes." *Heart Fail Rev* 12(1): 58–65.

Amour, J., X. Loyer, et al. (2007). "Altered contractile response due to increased β3-adrenoceptor stimulation in diabetic cardiomyopathy: the role of nitric oxide synthase 1–derived nitric oxide." *Anesthesiology: J Am Soc Anesthesiol* 107(3): 452–460.

Andrews, R. C., A. R. Cooper, et al. (2011). "Diet or diet plus physical activity versus usual care in patients with newly diagnosed type 2 diabetes: the early ACTID randomised controlled trial." *Lancet* 378(9786): 129–139.

Angelone, T., E. Filice, et al. (2008). "Beta3-adrenoceptors modulate left ventricular relax ation in the rat heart via the NO-cGMP-PKG pathway." *Acta Physiol (Oxf)* 193(3): 229–239.

Aragno, M., R. Mastrocola, et al. (2008). "Oxidative stress triggers cardiac fibrosis in the heart of diabetic rats." *Endocrinology* 149(1): 380–388.

Arch, J. R. and A. T. Ainsworth (1983). "Thermogenic and antiobesity activity of a novel beta-adrenoceptor agonist (BRL 26830A) in mice and rats." *Am J Clin Nutr* 38(4): 549–558.

Arch, J. R. and A. J. Kaumann (1993). "Beta 3 and atypical beta-adrenoceptors." *Med Res Rev* 13(6): 663–729.

Arch, J. R. and S. Wilson (1996). "Prospects for beta 3-adrenoceptor agonists in the treatment of obesity and diabetes." *Int J Obes Relat Metab Disord* 20(3): 191–199.

Ares-Carrasco, S., B. Picatoste, et al. (2009). "Myocardial fibrosis and apoptosis, but not inflammation, are present in long-term experimental diabetes." *Am J Physiol Heart Circ Physiol* 297(6): H2109–H2119.

Arias, N., C. Pico, et al. (2017). "A combination of resveratrol and quercetin induces browning in white adipose tissue of rats fed an obesogenic diet." *Obesity (Silver Spring)* 25(1): 111–121.

Arioglu-Inan, E., I. Ozakca, et al. (2013). "The role of insulin-thyroid hormone interaction on β-adrenoceptor-mediated cardiac responses." *Eur J Pharmacol* 718(1–3): 533–543.

Astrup, A. (2014). "A changing view on saturated fatty acids and dairy: from enemy to friend." *Am J Clin Nutr* 100(6): 1407–1408.

Aune, D., T. Norat, et al. (2013). "Dairy products and the risk of type 2 diabetes: a systematic review and dose-response meta-analysis of cohort studies." *Am J Clin Nutr* 98(4): 1066–1083.

Aviram, M., L. Dornfeld, et al. (2000). "Pomegranate juice consumption reduces oxidative stress, atherogenic modifications to LDL, and platelet aggregation: studies in humans and in atherosclerotic apolipoprotein E-deficient mice." *Am J Clin Nutr* 71(5): 1062–1076.

Badran, M., R. Morsy, et al. (2016). "Assessment of trace elements levels in patients with Type 2 diabetes using multivariate statistical analysis." *J Trace Elem Med Biol* 33: 114–119.

Balzer, J., T. Rassaf, et al. (2008). "Sustained benefits in vascular function through flavanol-containing cocoa in medicated diabetic patients a double-masked, randomized, controlled trial." *J Am Coll Cardiol* 51(22): 2141–2149.

Bansal, P., S. K. Gupta, et al. (2006). "Cardioprotective effect of lycopene in the experimental model of myocardial ischemia-reperfusion injury." *Mol Cell Biochem* 289(1): 1–9.

Bantle, J. P., J. Wylie-Rosett, et al. (2006). "Nutrition recommendations and interventions for diabetes—2006: a position statement of the American Diabetes Association." *Diabetes Care* 29(9): 2140–2157.

Barakatun Nisak, M. Y., A. T. Ruzita, et al. (2010). "Improvement of dietary quality with the aid of a low glycemic index diet in Asian patients with type 2 diabetes mellitus." *J Am Coll Nutr* 29(3): 161–170.

Basu, R., G. Y. Oudit, et al. (2009). "Type 1 diabetic cardiomyopathy in the Akita (Ins2WT/C96Y) mouse model is characterized by lipotoxicity and diastolic dysfunction with preserved systolic function." *Am J Physiol Heart Circ Physiol* 297(6): H2096–H2108.

Bettaieb, A., M. A. Vazquez Prieto, et al. (2014). "(-)-Epicatechin mitigates high-fructose-associated insulin resistance by modulating redox signaling and endoplasmic reticulum stress." *Free Radic Biol Med* 72: 247–256.

Beulens, J. W., S. L. Booth, et al. (2013). "The role of menaquinones (vitamin K(2)) in human health." *Br J Nutr* 110(8): 1357–1368.

Bhatt, J. K., S. Thomas, et al. (2012). "Resveratrol supplementation improves glycemic control in type 2 diabetes mellitus." *Nutr Res* 32(7): 537–541.

Bi, X., J. Lim, et al. (2017). "Spices in the management of diabetes mellitus." *Food Chem* 217: 281–293.

Bohm, V. (2012). "Lycopene and heart health." *Mol Nutr Food Res* 56(2): 296–303.

Booth, G. and A. Y. Cheng (2013). "Canadian Diabetes Association 2013 clinical practice guidelines for the prevention and management of diabetes in Canada. Methods." *Can J Diabetes* 37(Suppl 1): S1–S216.

Bose, K. S. and B. K. Agrawal (2007). "Effect of lycopene from cooked tomatoes on serum antioxidant enzymes, lipid peroxidation rate and lipid profile in coronary heart disease." *Singapore Med J* 48(5): 415–420.

Brasnyo, P., G. A. Molnar, et al. (2011). "Resveratrol improves insulin sensitivity, reduces oxidative stress and activates the Akt pathway in type 2 diabetic patients." *Br J Nutr* 106(3): 383–389.

Brodde, O. E. (1991). "Beta 1- and beta 2-adrenoceptors in the human heart: properties, function, and alterations in chronic heart failure." *Pharmacol Rev* 43(2): 203–242.

Calligaris, S. D., M. Lecanda, et al. (2013). "Mice long-term high-fat diet feeding recapitulates human cardiovascular alterations: an animal model to study the early phases of diabetic cardiomyopathy." *PLoS One* 8(4): e60931.

Cao, W., A. V. Medvedev, et al. (2001). "β-Adrenergic activation of p38 MAP kinase in adipocytes: cAMP induction of the uncoupling protein 1 (UCP1) gene requires p38 MAP kinase." *J Biol Chem* 276(29): 27077–27082.

Carpene, C., J. Galitzky, et al. (1993). "Desensitization of beta-1 and beta-2, but not beta-3, adrenoceptor-mediated lipolytic responses of adipocytes after long-term norepinephrine infusion." *J Pharmacol Exp Ther* 265(1): 237–247.

Carroll, J. F., A. E. Jones, et al. (1997). "Reduced cardiac contractile responsiveness to isoproterenol in obese rabbits." *Hypertension* 30(6): 1376–1381.

Carroll, J. F., C. K. Kyser, et al. (2002). "β-Adrenoceptor density and adenylyl cyclase activity in obese rabbit hearts." *Int J Obes Relat Metab Disord* 26(5): 627–632.

Castaneda-Ovando, A., M. de Lourdes Pacheco-Hernández, et al. (2009). "Chemical studies of anthocyanins: a review." *Food Chem* 113(4): 859–871.

Cavender, M. A., P. G. Steg, et al. (2015). "Impact of diabetes mellitus on hospitalization for heart failure, cardiovascular events, and death: outcomes at 4 years from the Reduction of Atherothrombosis for Continued Health (REACH) registry." *Circulation* 132(10): 923–931.

Choi, H. J., J. Yu, et al. (2011). "Vitamin K2 supplementation improves insulin sensitivity via osteocalcin metabolism: a placebo-controlled trial." *Diabetes Care* 34(9): e147.

Choi, K. M., Y. S. Lee, et al. (2014). "Sulforaphane attenuates obesity by inhibiting adipogenesis and activating the AMPK pathway in obese mice." *J Nutr Biochem* 25(2): 201–207.

Clifton, P. M., D. Condo, et al. (2014). "Long term weight maintenance after advice to consume low carbohydrate, higher protein diets—a systematic review and meta analysis." *Nutr Metab Cardiovasc Dis* 24(3): 224–235.

Coppell, K. J., M. Kataoka, et al. (2010). "Nutritional intervention in patients with type 2 diabetes who are hyperglycaemic despite optimised drug treatment—Lifestyle Over and Above Drugs in Diabetes (LOADD) study: randomised controlled trial." *BMJ* 341: c3337.

Cordero-Herrera, I., M. A. Martin, et al. (2015). "Cocoa-rich diet ameliorates hepatic insulin resistance by modulating insulin signaling and glucose homeostasis in Zucker diabetic fatty rats." *J Nutr Biochem* 26(7): 704–712.

Cote, C. D., B. A. Rasmussen, et al. (2015). "Resveratrol activates duodenal Sirt1 to reverse insulin resistance in rats through a neuronal network." *Nat Med* 21(5): 498–505.

Curtis, P. J., J. Potter, et al. (2013). "Vascular function and atherosclerosis progression after 1 y of flavonoid intake in statin-treated postmenopausal women with type 2 diabetes: a double-blind randomized controlled trial." *Am J Clin Nutr* 97(5): 936–942.

Curtis, P. J., M. Sampson, et al. (2012). "Chronic ingestion of flavan-3-ols and isoflavones improves insulin sensitivity and lipoprotein status and attenuates estimated 10-year CVD risk in medicated postmenopausal women with type 2 diabetes: a 1-year, double-blind, randomized, controlled trial." *Diabetes Care* 35(2): 226–232.

Cypess, A. M., L. S. Weiner, et al. (2015). "Activation of human brown adipose tissue by a β3-adrenergic receptor agonist." *Cell Metab* 21(1): 33–38.

Das Evcimen, N. and G. L. King (2007). "The role of protein kinase C activation and the vascular complications of diabetes." *Pharmacol Res* 55(6): 498–510.

Davison, K., A. M. Coates, et al. (2008). "Effect of cocoa flavanols and exercise on cardiometabolic risk factors in overweight and obese subjects." *Int J Obes (Lond)* 32(8): 1289–1296.

Dawber, T. R., R. J. Nickerson, et al. (1982). "Eggs, serum cholesterol, and coronary heart disease." *Am J Clin Nutr* 36(4): 617–625.

de Lorgeril, M., P. Salen, et al. (1999). "Mediterranean diet, traditional risk factors, and the rate of cardiovascular complications after myocardial infarction: final report of the Lyon Diet Heart Study." *Circulation* 99(6): 779–785.

de Oliveira, M. R., S. F. Nabavi, et al. (2016). "Resveratrol and the mitochondria: from triggering the intrinsic apoptotic pathway to inducing mitochondrial biogenesis, a mechanistic view." *Biochim Biophys Acta* 1860(4): 727–745.

Di Mascio, P., S. Kaiser, et al. (1989). "Lycopene as the most efficient biological carotenoid singlet oxygen quencher." *Arch Biochem Biophys* 274(2): 532–538.

Dincer, U. D. (2011). "Cardiac β-adrenoceptor expression is markedly depressed in Ossabaw swine model of cardiometabolic risk." *Int J Gen Med* 4: 493–499.

Dincer, U. D., K. R. Bidasee, et al. (2001). "The effect of diabetes on expression of beta1-, beta2-, and beta3-adrenoreceptors in rat hearts." *Diabetes* 50(2): 455–461.

Dincer, U. D., A. Onay, et al. (1998). "The effects of diabetes on beta-adrenoceptor mediated responsiveness of human and rat atria." *Diabetes Res Clin Pract* 40(2): 113–122.

Djousse, L. and J. M. Gaziano (2008). "Egg consumption in relation to cardiovascular disease and mortality: the Physicians' Health Study." *Am J Clin Nutr* 87(4): 964–969.

Djousse, L., T. Rudich, et al. (2009). "Nut consumption and risk of hypertension in US male physicians." *Clin Nutr* 28(1): 10–14.

Doddigarla, Z., I. Parwez, et al. (2016). "Correlation of serum chromium, zinc, magnesium and SOD levels with HbA1c in type 2 diabetes: a cross sectional analysis." *Diabetes Metab Syndr* 10(1 Suppl 1): S126–S129.

Dyer, A. R., P. Elliott, et al. (1994). "Urinary electrolyte excretion in 24 hours and blood pressure in the INTERSALT study. II. Estimates of electrolyte-blood pressure associations corrected for regression dilution bias. The INTERSALT Cooperative Research Group." *Am J Epidemiol* 139(9): 940–951.

Elia, M., A. Ceriello, et al. (2005). "Enteral nutritional support and use of diabetes-specific formulas for patients with diabetes: a systematic review and meta-analysis." *Diabetes Care* 28(9): 2267–2279.

Ellingsgaard, H., I. Hauselmann, et al. (2011). "Interleukin-6 enhances insulin secretion by increasing glucagon-like peptide-1 secretion from L cells and alpha cells." *Nat Med* 17(11): 1481–1489.

Emorine, L. J., S. Marullo, et al. (1989). "Molecular characterization of the human beta 3-adrenergic receptor." *Science* 245(4922): 1118–1121.

Ericson, U., E. Sonestedt, et al. (2013). "High intakes of protein and processed meat associate with increased incidence of type 2 diabetes." *Br J Nutr* 109(6): 1143–1153.

Estruch, R., E. Ros, et al. (2013). "Primary prevention of cardiovascular disease with a Mediterranean diet." *N Engl J Med* 368(14): 1279–1290.

Everard, A., C. Belzer, et al. (2013). "Cross-talk between Akkermansia muciniphila and intestinal epithelium controls diet-induced obesity." *Proc Natl Acad Sci USA* 110(22): 9066–9071.

Evert, A. B., J. L. Boucher, et al. (2013). "Nutrition therapy recommendations for the management of adults with diabetes." *Diabetes Care* 36(11): 3821–3842.

Fernandez-Millan, E., I. Cordero-Herrera, et al. (2015). "Cocoa-rich diet attenuates beta cell mass loss and function in young Zucker diabetic fatty rats by preventing oxidative stress and beta cell apoptosis." *Mol Nutr Food Res* 59(4): 820–824.

Ferro, A. (2006). "β-Adrenoceptors and potassium channels." *Naunyn Schmiedebergs Arch Pharmacol* 373(3): 183–185.

Ferron, A. J., B. B. Jacobsen, et al. (2015). "Cardiac dysfunction induced by obesity is not related to β-adrenergic system impairment at the receptor-signalling pathway." *PLoS One* 10(9): e0138605.

Festa, A., W. Krugluger, et al. (1999). "Trp64Arg polymorphism of the beta3-adrenergic receptor gene in pregnancy: association with mild gestational diabetes mellitus." *J Clin Endocrinol Metab* 84(5): 1695–1699.

Fiori, J. L., Y. K. Shin, et al. (2013). "Resveratrol prevents beta-cell dedifferentiation in non-human primates given a high-fat/high-sugar diet." *Diabetes* 62(10): 3500–3513.

Forrest, R. H. and J. G. Hickford (2000). "Rapid communication: nucleotide sequences of the bovine, caprine, and ovine β3-adrenergic receptor genes." *J Anim Sci* 78(5): 1397–1398.

Frid, A. H., M. Nilsson, et al. (2005). "Effect of whey on blood glucose and insulin responses to composite breakfast and lunch meals in type 2 diabetic subjects." *Am J Clin Nutr* 82(1): 69–75.

Fu, L., K. Isobe, et al. (2008). "The effects of beta(3)-adrenoceptor agonist CL-316,243 on adiponectin, adiponectin receptors and tumor necrosis factor-alpha expressions in adipose tissues of obese diabetic KKAy mice." *Eur J Pharmacol* 584(1): 202–206.

Fu, Q., Y. Hu, et al. (2017). "High-fat diet induces protein kinase A and G-protein receptor kinase phosphorylation of β2-adrenergic receptor and impairs cardiac adrenergic reserve in animal hearts." *J Physiol* 595(6): 1973–1986.

Fung, T. T., S. E. Chiuve, et al. (2008). "Adherence to a DASH-style diet and risk of coronary heart disease and stroke in women." *Arch Intern Med* 168(7): 713–720.

Ganguly, P. K., R. E. Beamish, et al. (1987). "Norepinephrine storage, distribution, and release in diabetic cardiomyopathy." *Am J Physiol* 252(6 Pt 1): E734–E739.

Gao, D., N. Ning, et al. (2013). "Dairy products consumption and risk of type 2 diabetes: systematic review and dose-response meta-analysis." *PLoS One* 8(9): e73965.

Gardner, C. D., S. Kim, et al. (2010). "Micronutrient quality of weight-loss diets that focus on macronutrients: results from the A TO Z study." *Am J Clin Nutr* 92(2): 304–312.

Garg, V. K., R. Gupta, et al. (1994). "Hypozincemia in diabetes mellitus." *J Assoc Physicians India* 42(9): 720–721.

Gauthier, C., V. Leblais, et al. (1998). "The negative inotropic effect of beta3-adrenoceptor stimulation is mediated by activation of a nitric oxide synthase pathway in human ventricle." *J Clin Invest* 102(7): 1377–1384.

Gauthier, C., G. Tavernier, et al. (1996). "Functional beta3-adrenoceptor in the human heart." *J Clin Invest* 98(2): 556–562.

George, P. S., E. R. Pearson, et al. (2012). "Effect of vitamin D supplementation on glycaemic control and insulin resistance: a systematic review and meta-analysis." *Diabet Med* 29(8): e142–e150.

Ghorbani, M. and J. Himms-Hagen (1997). "Appearance of brown adipocytes in white adipose tissue during CL 316,243-induced reversal of obesity and diabetes in Zucker fa/fa rats." *Int J Obes Relat Metab Disord* 21(6): 465–475.

Ghorbani, M., M. Shafiee Ardestani, et al. (2012). "Anti diabetic effect of CL 316,243 (a β3-adrenergic agonist) by down regulation of tumour necrosis factor (TNF-α) expression." *PLoS One* 7(10): e45874.

Ghorbani, M., S. Teimourian, et al. (2015). "Apparent histological changes of adipocytes after treatment with CL 316,243, a β-3-adrenergic receptor agonist." *Drug Des Devel Ther* 9: 669–676.

Gijsbers, L., E. L. Ding, et al. (2016). "Consumption of dairy foods and diabetes incidence: a dose-response meta-analysis of observational studies." *Am J Clin Nutr* 103(4): 1111–1124.

Gilbert, R. E. (2013). "Endothelial loss and repair in the vascular complications of diabetes: pathogenetic mechanisms and therapeutic implications." *Circ J* 77(4): 849–856.

Gonzalez-Quesada, C., M. Cavalera, et al. (2013). "Thrombospondin-1 induction in the diabetic myocardium stabilizes the cardiac matrix in addition to promoting vascular rarefaction through angiopoietin-2 upregulation." *Circ Res* 113(12): 1331–1344.

Gonzalez-Rodriguez, A., B. Santamaria, et al. (2015). "Resveratrol treatment restores peripheral insulin sensitivity in diabetic mice in a sirt1-independent manner." *Mol Nutr Food Res* 59(8): 1431–1442.

Gorinstein, S., A. Caspi, et al. (2006). "Red grapefruit positively influences serum triglyceride level in patients suffering from coronary atherosclerosis: studies in vitro and in humans." *J Agric Food Chem* 54(5): 1887–1892.

Granneman, J. G., P. Li, et al. (2005). "Metabolic and cellular plasticity in white adipose tissue I: effects of beta3-adrenergic receptor activation." *Am J Physiol Endocrinol Metab* 289(4): E608–E616.

Grassi, D., G. Desideri, et al. (2008). "Blood pressure is reduced and insulin sensitivity increased in glucose-intolerant, hypertensive subjects after 15 days of consuming high-polyphenol dark chocolate." *J Nutr* 138(9): 1671–1676.

Grassi, D., C. Lippi, et al. (2005). "Short-term administration of dark chocolate is followed by a significant increase in insulin sensitivity and a decrease in blood pressure in healthy persons." *Am J Clin Nutr* 81(3): 611–614.

Grujic, D., V. S. Susulic, et al. (1997). "Beta3-adrenergic receptors on white and brown adipocytes mediate beta3-selective agonist-induced effects on energy expenditure, insulin secretion, and food intake. A study using transgenic and gene knockout mice." *J Biol Chem* 272(28): 17686–17693.

Guo, X., B. Yang, et al. (2016). "Associations of dietary intakes of anthocyanins and berry fruits with risk of type 2 diabetes mellitus: a systematic review and meta-analysis of prospective cohort studies." *Eur J Clin Nutr* 70(12): 1360–1367.

Haase, H. and W. Maret (2005). "Fluctuations of cellular, available zinc modulate insulin signaling via inhibition of protein tyrosine phosphatases." *J Trace Elem Med Biol* 19(1): 37–42.

Haemmerle, G., A. Lass, et al. (2006). "Defective lipolysis and altered energy metabolism in mice lacking adipose triglyceride lipase." *Science* 312(5774): 734–737.

Hamadi, N., A. Mansour, et al. (2012). "Ameliorative effects of resveratrol on liver injury in streptozotocin-induced diabetic rats." *J Biochem Mol Toxicol* 26(10): 384–392.

Hamdy, O., P. O. Ganda, et al. (2018). *Clinical Nutrition Guideline for Overweight and Obese Adults with Type 2 Diabetes (T2D) or Prediabetes, or Those at High Risk for Developing T2D*: SP226–SP231.

Hata, Y., M. Yamamoto, et al. (1996). "A placebo-controlled study of the effect of sour milk on blood pressure in hypertensive subjects." *Am J Clin Nutr* 64(5): 767–771.

Herszkowicz, N., A. Barbato, et al. (2001). "Contribution of Doppler echocardiography to the evaluation of systolic and diastolic function of obese women versus a control group." *Arq Bras Cardiol* 76(3): 189–196.

Himms-Hagen, J., J. Cui, et al. (1994). "Effect of CL-316,243, a thermogenic beta 3-agonist, on energy balance and brown and white adipose tissues in rats." *Am J Physiol* 266(4 Pt 2): R1371–R1382.

Hu, F. B., M. J. Stampfer, et al. (1999). "A prospective study of egg consumption and risk of cardiovascular disease in men and women." *JAMA* 281(15): 1387–1394.

Huynh, K., J. R. McMullen, et al. (2010). "Cardiac-specific IGF-1 receptor transgenic expression protects against cardiac fibrosis and diastolic dysfunction in a mouse model of diabetic cardiomyopathy." *Diabetes* 59(6): 1512–1520.

Imparl-Radosevich, J., S. Deas, et al. (1998). "Regulation of PTP-1 and insulin receptor kinase by fractions from cinnamon: implications for cinnamon regulation of insulin signalling." *Horm Res* 50(3): 177–182.

Inaba, Y., J. A. Chen, et al. (2010). "Prediction of future cardiovascular outcomes by flow-mediated vasodilatation of brachial artery: a meta-analysis." *Int J Cardiovasc Imaging* 26(6): 631–640.

Inokuma, K., Y. Okamatsu-Ogura, et al. (2006). "Indispensable role of mitochondrial UCP1 for antiobesity effect of beta3-adrenergic stimulation." *Am J Physiol Endocrinol Metab* 290(5): E1014–E1021.

Itoh, Y., Y. Kawamata, et al. (2003). "Free fatty acids regulate insulin secretion from pancreatic beta cells through GPR40." *Nature* 422(6928): 173–176.

Iwamoto, J., A. Seki, et al. (2011). "Vitamin K(2) prevents hyperglycemia and cancellous osteopenia in rats with streptozotocin-induced type 1 diabetes." *Calcif Tissue Int* 88(2): 162–168.

Jayaprakasha, G. K., M. Ohnishi-Kameyama, et al. (2006). "Phenolic constituents in the fruits of Cinnamomum zeylanicum and their antioxidant activity." *J Agric Food Chem* 54(5): 1672–1679.

Jenkins, D. J., C. W. Kendall, et al. (2002). "Dose response of almonds on coronary heart disease risk factors: blood lipids, oxidized low-density lipoproteins, lipoprotein(a), homocysteine, and pulmonary nitric oxide: a randomized, controlled, crossover trial." *Circulation* 106(11): 1327–1332.

Jiang, C., A. Carillion, et al. (2015). "Modification of the β-adrenoceptor stimulation pathway in Zucker obese and obese diabetic rat myocardium." *Crit Care Med* 43(7): e241–e249.

Jimenez-Osorio, A. S., A. Monroy, et al. (2016). "Curcumin and insulin resistance—molecular targets and clinical evidences." *Biofactors* 42(6): 561–580.

Kadooka, Y., M. Sato, et al. (2010). "Regulation of abdominal adiposity by probiotics (Lactobacillus gasseri SBT2055) in adults with obese tendencies in a randomized controlled trial." *Eur J Clin Nutr* 64(6): 636–643.

Karim, M., K. McCormick, et al. (2000). "Effects of cocoa extracts on endothelium-dependent relaxation." *J Nutr* 130(Suppl 8S): 2105S–2108S.

Karimi Galougahi, K., C. C. Liu, et al. (2015). "β3-Adrenoceptor activation relieves oxidative inhibition of the cardiac Na+-K+ pump in hyperglycemia induced by insulin receptor blockade." *Am J Physiol Cell Physiol* 309(5): C286–C295.

Kato, H., M. Ohue, et al. (2001). "Mechanism of amelioration of insulin resistance by β3-adrenoceptor agonist AJ-9677 in the KK-Ay/Ta diabetic obese mouse model." *Diabetes* 50(1): 113–122.

Kaumann, A. J., S. Engelhardt, et al. (2001). "Abolition of (-)-CGP 12177-evoked cardiostimulation in double β1/β2-adrenoceptor knockout mice. Obligatory role of β1-adrenoceptors for putative β4-adrenoceptor pharmacology." *Naunyn Schmiedebergs Arch Pharmacol* 363(1): 87–93.

Kayki-Mutlu, G., E. Arioglu-Inan, et al. (2014). "β3-Adrenoceptor-mediated responses in diabetic rat heart." *Gen Physiol Biophys* 33(1): 99–109.

Khalesi, S., J. Sun, et al. (2014). "Effect of probiotics on blood pressure: a systematic review and meta-analysis of randomized, controlled trials." *Hypertension* 64(4): 897–903.

Khan, A., M. Safdar, et al. (2003). "Cinnamon improves glucose and lipids of people with type 2 diabetes." *Diabetes Care* 26(12): 3215–3218.

Kim, H., P. A. Pennisi, et al. (2006). "Effect of adipocyte β3-adrenergic receptor activation on the type 2 diabetic MKR mice." *Am J Physiol Endocrinol Metab* 290(6): E1227–E1236.

Kim, J., J. Shim, et al. (2014). "Cocoa phytochemicals: recent advances in molecular mechanisms on health." *Crit Rev Food Sci Nutr* 54(11): 1458–1472.

Kim, J. Y., J. K. Paik, et al. (2011). "Effects of lycopene supplementation on oxidative stress and markers of endothelial function in healthy men." *Atherosclerosis* 215(1): 189–195.

Kim, S. H., S. H. Hyun, et al. (2006). "Anti-diabetic effect of cinnamon extract on blood glucose in db/db mice." *J Ethnopharmacol* 104(1–2): 119–123.

Kim-Motoyama, H., K. Yasuda, et al. (1997). "A mutation of the beta 3-adrenergic receptor is associated with visceral obesity but decreased serum triglyceride." *Diabetologia* 40(4): 469–472.

Kirk, J. K., D. E. Graves, et al. (2008). "Restricted-carbohydrate diets in patients with type 2 diabetes: a meta-analysis." *J Am Diet Assoc* 108(1): 91–100.

Kobyliak, N., C. Conte, et al. (2016). "Probiotics in prevention and treatment of obesity: a critical view." *Nutr Metab (Lond)* 13: 14.

Koga, T. and M. Meydani (2001). "Effect of plasma metabolites of (+)-catechin and quercetin on monocyte adhesion to human aortic endothelial cells." *Am J Clin Nutr* 73(5): 941–948.

Korhonen, H. (2009). "Milk-derived bioactive peptides: from science to applications." *J Funct Foods 1*: 177–187.

Krul-Poel, Y. H., M. M. Ter Wee, et al. (2017). "Management of endocrine disease: the effect of vitamin D supplementation on glycaemic control in patients with type 2 diabetes mellitus: a systematic review and meta-analysis." *Eur J Endocrinol* 176(1): R1–R14.

Ku, C. R., H. J. Lee, et al. (2012). "Resveratrol prevents streptozotocin-induced diabetes by inhibiting the apoptosis of pancreatic β-cell and the cleavage of poly (ADP-ribose) polymerase." *Endocr J* 59(2): 103–109.

Kumar, R., N. Binkley, et al. (2010). "Effect of phylloquinone supplementation on glucose homeostasis in humans." *Am J Clin Nutr* 92(6): 1528–1532.

Kuschel, M., Y. Y. Zhou, et al. (1999a). "G(i) protein-mediated functional compartmentalization of cardiac beta(2)-adrenergic signaling." *J Biol Chem* 274(31): 22048–22052.

Kuschel, M., Y. Y. Zhou, et al. (1999b). "β2-Adrenergic cAMP signaling is uncoupled from phosphorylation of cytoplasmic proteins in canine heart." *Circulation* 99(18): 2458–2465.

Labunskyy, V. M., B. C. Lee, et al. (2011). "Both maximal expression of selenoproteins and selenoprotein deficiency can promote development of type 2 diabetes-like phenotype in mice." *Antioxid Redox Signal* 14(12): 2327–2336.

Lai, C.-S., J.-C. Wu, et al. (2015). "Molecular mechanism on functional food bioactives for anti-obesity." *Curr Opin Food Sci* 2: 9–13.

Lai, C. S., J. C. Wu, et al. (2017). "Chemoprevention of obesity by dietary natural compounds targeting mitochondrial regulation." *Mol Nutr Food Res* 61(6).

Larson, J. E., P. P. Rainer, et al. (2012). "Dependence of β3-adrenergic signaling on the adipokine leptin in cardiac myocytes." *Int J Obes (Lond)* 36(6): 876–879.

Lee, C. W., J. Y. Seo, et al. (2017). "3-O-Glucosylation of quercetin enhances inhibitory effects on the adipocyte differentiation and lipogenesis." *Biomed Pharmacother* 95: 589–598.

Lee, J. H., M. H. Moon, et al. (2012). "Sulforaphane induced adipolysis via hormone sensitive lipase activation, regulated by AMPK signaling pathway." *Biochem Biophys Res Commun* 426(4): 492–497.

Lee, S. G., J. S. Parks, et al. (2017). "Quercetin, a functional compound of onion peel, remodels white adipocytes to brown-like adipocytes." *J Nutr Biochem* 42: 62–71.

Leone Roberti Maggiore, U., L. Cardozo, et al. (2014). "Mirabegron in the treatment of overactive bladder." *Expert Opin Pharmacother* 15(6): 873–887.

Levitan, E. B., A. Wolk, et al. (2009). "Consistency with the DASH diet and incidence of heart failure." *Arch Intern Med* 169(9): 851–857.

Ley, S. H., O. Hamdy, et al. (2014). "Prevention and management of type 2 diabetes: dietary components and nutritional strategies." *Lancet* 383(9933): 1999–2007.

Li, S., C. Bouzar, et al. (2016). "Resveratrol inhibits lipogenesis of 3T3-L1 and SGBS cells by inhibition of insulin signaling and mitochondrial mass increase." *Biochim Biophys Acta* 1857(6): 643–652.

Li, Y., C. Zhou, et al. (2013). "Egg consumption and risk of cardiovascular diseases and diabetes: a meta-analysis." *Atherosclerosis* 229(2): 524–530.

Lima-Leopoldo, A. P., A. S. Leopoldo, et al. (2011). "Myocardial dysfunction and abnormalities in intracellular calcium handling in obese rats." *Arq Bras Cardiol* 97(3): 232–240.

Liu, X., F. Perusse, et al. (1998). "Mechanisms of the antidiabetic effects of the beta 3-adrenergic agonist CL-316243 in obese Zucker-ZDF rats." *Am J Physiol* 274(5 Pt 2): R1212–R1219.

Liu, Y., D. Li, et al. (2014). "Anthocyanin increases adiponectin secretion and protects against diabetes-related endothelial dysfunction." *Am J Physiol Endocrinol Metab* 306(8): E975–E988.

Lone, J., J. H. Choi, et al. (2016). "Curcumin induces brown fat-like phenotype in 3T3-L1 and primary white adipocytes." *J Nutr Biochem* 27: 193–202.

Lorenz, M., S. Wessler, et al. (2004). "A constituent of green tea, epigallocatechin-3-gallate, activates endothelial nitric oxide synthase by a phosphatidylinositol-3-OH-kinase-, cAMP-dependent protein kinase-, and Akt-dependent pathway and leads to endothelial-dependent vasorelaxation." *J Biol Chem* 279(7): 6190–6195.

Lowell, B. B. and B. M. Spiegelman (2000). "Towards a molecular understanding of adaptive thermogenesis." *Nature* 404(6778): 652–660.

Lu, T., H. Sheng, et al. (2012). "Cinnamon extract improves fasting blood glucose and glycosylated hemoglobin level in Chinese patients with type 2 diabetes." *Nutr Res* 32(6): 408–412.

MacDonald, M. R., M. C. Petrie, et al. (2008). "Impact of diabetes on outcomes in patients with low and preserved ejection fraction heart failure: an analysis of the Candesartan in Heart failure: assessment of Reduction in Mortality and morbidity (CHARM) programme." *Eur Heart J* 29(11): 1377–1385.

MacPherson, R. E., L. Castellani, et al. (2014). "Evidence for fatty acids mediating CL 316,243-induced reductions in blood glucose in mice." *Am J Physiol Endocrinol Metab* 307(7): E563–E570.

Maghsoudi, Z., R. Ghiasvand, et al. (2016). "Empirically derived dietary patterns and incident type 2 diabetes mellitus: a systematic review and meta-analysis on prospective observational studies." *Public Health Nutr* 19(2): 230–241.

Manna, P. and J. Kalita (2016). "Beneficial role of vitamin K supplementation on insulin sensitivity, glucose metabolism, and the reduced risk of type 2 diabetes: a review." *Nutrition* 32(7–8): 732–739.

Martin, M. A., L. Goya, et al. (2016). "Antidiabetic actions of cocoa flavanols." *Mol Nutr Food Res* 60(8): 1756–1769.

Martínez-Lapiscina, E. H., A. M. Pimenta, et al. (2010). "Nut consumption and incidence of hypertension: the SUN prospective cohort." *Nutr Metab Cardiovasc Dis* 20(5): 359–365.

Marx, S. O., S. Reiken, et al. (2000). "PKA phosphorylation dissociates FKBP12.6 from the calcium release channel (ryanodine receptor): defective regulation in failing hearts." *Cell* 101(4): 365–376.

Mathers, C. D. and D. Loncar (2006). "Projections of global mortality and burden of disease from 2002 to 2030." *PLoS Med* 3(11): e442.

Matsuda, N., Y. Hattori, et al. (1999). "Diabetes-induced down-regulation of beta1-adreno-ceptor mRNA expression in rat heart." *Biochem Pharmacol* 58(5): 881–885.

Mellor, D. D., T. Sathyapalan, et al. (2010). "High-cocoa polyphenol-rich chocolate improves HDL cholesterol in Type 2 diabetes patients." *Diabet Med* 27(11): 1318–1321.

Merino, J., R. Kones, et al. (2013). "Negative effect of a low-carbohydrate, high-protein, high-fat diet on small peripheral artery reactivity in patients with increased cardiovascular risk." *Br J Nutr* 109(7): 1241–1247.

Minhas, K. M., S. A. Khan, et al. (2005). "Leptin repletion restores depressed {beta}-adrenergic contractility in ob/ob mice independently of cardiac hypertrophy." *J Physiol* 565(Pt 2): 463–474.

Mitri, J., M. D. Muraru, et al. (2011). "Vitamin D and type 2 diabetes: a systematic review." *Eur J Clin Nutr* 65(9): 1005–1015.

Mitterberger, M. C. and W. Zwerschke (2013). "Mechanisms of resveratrol-induced inhibition of clonal expansion and terminal adipogenic differentiation in 3T3-L1 preadipocytes." *J Gerontol A Biol Sci Med Sci* 68(11): 1356–1376.

Moline, J., I. F. Bukharovich, et al. (2000). "Dietary flavonoids and hypertension: is there a link?" *Med Hypotheses* 55(4): 306–309.

Moniotte, S. and J. L. Balligand (2002). "Potential use of beta(3)-adrenoceptor antagonists in heart failure therapy." *Cardiovasc Drug Rev* 20(1): 19–26.

Morimoto, A., H. Hasegawa, et al. (2004). "Endogenous beta3-adrenoreceptor activation contributes to left ventricular and cardiomyocyte dysfunction in heart failure." *Am J Physiol Heart Circ Physiol* 286(6): H2425–H2433.

Movahed, A., I. Nabipour, et al. (2013). "Antihyperglycemic effects of short term resveratrol supplementation in type 2 diabetic patients." *Evid Based Complement Alternat Med* 2013: 851267.

Nathan, D. M., P. A. Cleary, et al. (2005). "Intensive diabetes treatment and cardiovascular disease in patients with type 1 diabetes." *N Engl J Med* 353(25): 2643–2653.

Nilsson, M., M. Stenberg, et al. (2004). "Glycemia and insulinemia in healthy subjects after lactose-equivalent meals of milk and other food proteins: the role of plasma amino acids and incretins." *Am J Clin Nutr* 80(5): 1246–1253.

Nordmann, A. J., A. Nordmann, et al. (2006). "Effects of low-carbohydrate vs low-fat diets on weight loss and cardiovascular risk factors: a meta-analysis of randomized controlled trials." *Arch Intern Med* 166(3): 285–293.

Ogawa-Wong, A. N., M. J. Berry, et al. (2016). "Selenium and metabolic disorders: an emphasis on type 2 diabetes risk." *Nutrients* 8(2): 80.

Oh, H. M. and J. S. Yoon (2008). "Glycemic control of type 2 diabetic patients after short-term zinc supplementation." *Nutr Res Pract* 2(4): 283–288.

Okatan, E. N., E. Tuncay, et al. (2015). "Profiling of cardiac β-adrenoceptor subtypes in the cardiac left ventricle of rats with metabolic syndrome: comparison with streptozotocin-induced diabetic rats." *Can J Physiol Pharmacol* 93(7): 517–525.

Oyenihi, O. R., A. B. Oyenihi, et al. (2016). "Antidiabetic effects of resveratrol: the way forward in its clinical utility." *J Diabetes Res* 2016: 9737483.

Pang, Z., N. Wu, et al. (2010). "GPR40 is partially required for insulin secretion following activation of beta3-adrenergic receptors." *Mol Cell Endocrinol* 325(1–2): 18–25.

Parham, M., M. Amini, et al. (2008). "Effect of zinc supplementation on microalbuminuria in patients with type 2 diabetes: a double blind, randomized, placebo-controlled, cross-over trial." *Rev Diabet Stud* 5(2): 102–109.

Parikh, A., S. R. Lipsitz, et al. (2009). "Association between a DASH-like diet and mortality in adults with hypertension: findings from a population-based follow-up study." *Am J Hypertens* 22(4): 409–416.

Park, K. Y., B. Kim, et al. (2015). "Lactobacillus rhamnosus GG improves glucose tolerance through alleviating ER stress and suppressing macrophage activation in db/db mice." *J Clin Biochem Nutr* 56(3): 240–246.

Pastors, J. G. and M. J. Franz (2012). "Effectiveness of medical nutrition therapy in diabetes." In: M. J. Franz and A. B. Evert (Eds) *American Diabetes Association Guide to Nutrition Therapy for Diabetes*, Alexandria, VA, USA, American Diabetes Association: 1–18.

Penitente, A. R., R. D. Novaes, et al. (2014). "Basal and β-adrenergic cardiomyocytes contractility dysfunction induced by dietary protein restriction is associated with down-regulation of SERCA2a expression and disturbance of endoplasmic reticulum Ca2+ regulation in rats." *Cell Physiol Biochem* 34(2): 443–454.

Penumathsa, S. V., M. Thirunavukkarasu, et al. (2008). "Resveratrol enhances GLUT-4 translocation to the caveolar lipid raft fractions through AMPK/Akt/eNOS signalling pathway in diabetic myocardium." *J Cell Mol Med* 12(6A): 2350–2361.

Pott, C., K. Brixius, et al. (2003). "The preferential beta3-adrenoceptor agonist BRL 37344 increases force via beta1-/beta2-adrenoceptors and induces endothelial nitric oxide synthase via beta3-adrenoceptors in human atrial myocardium." *Br J Pharmacol* 138(3): 521–529.

Qin, B., M. Nagasaki, et al. (2004). "Cinnamon extract prevents the insulin resistance induced by a high-fructose diet." *Horm Metab Res* 36(2): 119–125.

Rasekhi, H., M. Karandish, et al. (2015). "The effect of vitamin K1 supplementation on sensitivity and insulin resistance via osteocalcin in prediabetic women: a double-blind randomized controlled clinical trial." *Eur J Clin Nutr* 69(8): 891–895.

Relling, D. P., L. B. Esberg, et al. (2006). "High-fat diet-induced juvenile obesity leads to cardiomyocyte dysfunction and upregulation of Foxo3a transcription factor independent of lipotoxicity and apoptosis." *J Hypertens* 24(3): 549–561.

Roberts, A. C. and K. E. Porter (2013). "Cellular and molecular mechanisms of endothelial dysfunction in diabetes." *Diab Vasc Dis Res* 10(6): 472–482.

Robertson, R. P. (2004). "Chronic oxidative stress as a central mechanism for glucose toxicity in pancreatic islet beta cells in diabetes." *J Biol Chem* 279(41): 42351–42354.

Robidoux, J., N. Kumar, et al. (2006). "Maximal beta3-adrenergic regulation of lipolysis involves Src and epidermal growth factor receptor-dependent ERK1/2 activation." *J Biol Chem* 281(49): 37794–37802.

Rozec, B. and C. Gauthier (2006). "beta3-adrenoceptors in the cardiovascular system: putative roles in human pathologies." *Pharmacol Ther* 111(3): 652–673.

Russo, I. and N. G. Frangogiannis (2016). "Diabetes-associated cardiac fibrosis: cellular effectors, molecular mechanisms and therapeutic opportunities." *J Mol Cell Cardiol* 90: 84–93.

Sabate, J., G. E. Fraser, et al. (1993). "Effects of walnuts on serum lipid levels and blood pressure in normal men." *N Engl J Med* 328(9): 603–607.

Salgueiro, M. J., N. Krebs, et al. (2001). "Zinc and diabetes mellitus: is there a need of zinc supplementation in diabetes mellitus patients?" *Biol Trace Elem Res* 81(3): 215–228.

Sanchez-Rabaneda, F., O. Jauregui, et al. (2003). "Liquid chromatographic/electrospray ionization tandem mass spectrometric study of the phenolic composition of cocoa (*Theobroma cacao*)." *J Mass Spectrom* 38(1): 35–42.

Schwingshackl, L. and G. Hoffmann (2013). "Low-carbohydrate diets impair flow-mediated dilatation: evidence from a systematic review and meta-analysis." *Br J Nutr* 110(5): 969–970.

Seo, Y. S., O. H. Kang, et al. (2015). "Quercetin prevents adipogenesis by regulation of transcriptional factors and lipases in OP9 cells." *Int J Mol Med* 35(6): 1779–1785.

Shai, I., D. Schwarzfuchs, et al. (2008). "Weight loss with a low-carbohydrate, Mediterranean, or low-fat diet." *N Engl J Med* 359(3): 229–241.

Shi, Z., B. Yuan, et al. (2011). "Egg consumption and the risk of diabetes in adults, Jiangsu, China." *Nutrition* 27(2): 194–198.

Shimizu, M., K. Umeda, et al. (1993). "Collagen remodelling in myocardia of patients with diabetes." *J Clin Pathol* 46(1): 32–36.

Simmerman, H. K. and L. R. Jones (1998). "Phospholamban: protein structure, mechanism of action, and role in cardiac function." *Physiol Rev* 78(4): 921–947.

Singh, G., S. Maurya, et al. (2007). "A comparison of chemical, antioxidant and antimicrobial studies of cinnamon leaf and bark volatile oils, oleoresins and their constituents." *Food Chem Toxicol* 45(9): 1650–1661.

Skeberdis, V. A., V. Gendviliene, et al. (2008). "β3-adrenergic receptor activation increases human atrial tissue contractility and stimulates the L-type Ca^{2+} current." *J Clin Invest* 118(9): 3219–3227.

Sloan, C., J. Tuinei, et al. (2011). "Central leptin signaling is required to normalize myocardial fatty acid oxidation rates in caloric-restricted ob/ob mice." *Diabetes* 60(5): 1424–1434.

Soeder, K. J., S. K. Snedden, et al. (1999). "The β3-adrenergic receptor activates mitogen-activated protein kinase in adipocytes through a Gi-dependent mechanism." *J Biol Chem* 274(17): 12017–12022.

Spiegelman, B. M., P. Puigserver, et al. (2000). "Regulation of adipogenesis and energy balance by PPARγ and PGC-1." *Int J Obes Relat Metab Disord* 24(Suppl 4): S8–S10.

Stangl, V., C. Kuhn, et al. (2011). "Lack of effects of tomato products on endothelial function in human subjects: results of a randomised, placebo-controlled cross-over study." *Br J Nutr* 105(2): 263–267.

Strassheim, D., M. D. Houslay, et al. (1992). "Regulation of cardiac adenylate cyclase activity in rodent models of obesity." *Biochem J* 283 (Pt 1): 203–208.

Strosberg, A. D. (1997). "Structure and function of the β3-adrenergic receptor." A*nnu Rev Pharmacol Toxicol* 37: 421–450.

Stull, A. J., K. C. Cash, et al. (2010). "Bioactives in blueberries improve insulin sensitivity in obese, insulin-resistant men and women." *J Nutr* 140(10): 1764–1768.

Su, H. C., L. M. Hung, et al. (2006). "Resveratrol, a red wine antioxidant, possesses an insulin-like effect in streptozotocin-induced diabetic rats." *Am J Physiol Endocrinol Metab* 290(6): E1339–E1346.

Subash Babu, P., S. Prabuseenivasan, et al. (2007). "Cinnamaldehyde—a potential antidiabetic agent." *Phytomedicine* 14(1): 15–22.

Sulakhe, P. V. and X. T. Vo (1995). "Regulation of phospholamban and troponin-I phosphorylation in the intact rat cardiomyocytes by adrenergic and cholinergic stimuli: roles of cyclic nucleotides, calcium, protein kinases and phosphatases and depolarization." *Mol Cell Biochem* 149–150: 103–126.

Szkudelska, K., L. Nogowski, et al. (2009). "Resveratrol, a naturally occurring diphenolic compound, affects lipogenesis, lipolysis and the antilipolytic action of insulin in isolated rat adipocytes." *J Steroid Biochem Mol Biol* 113(1–2): 17–24.

Tabuchi, M., M. Ozaki, et al. (2003). "Antidiabetic effect of Lactobacillus GG in streptozotocin-induced diabetic rats." *Biosci Biotechnol Biochem* 67(6): 1421–1424.

Taubert, D., R. Berkels, et al. (2002). "Nitric oxide formation and corresponding relaxation of porcine coronary arteries induced by plant phenols: essential structural features." *J Cardiovasc Pharmacol* 40(5): 701–713.

Tay, J., N. D. Luscombe-Marsh, et al. (2015). "Comparison of low- and high-carbohydrate diets for type 2 diabetes management: a randomized trial." *Am J Clin Nutr* 102(4): 780–790.

Te Morenga, L., S. Mallard, et al. (2012). "Dietary sugars and body weight: systematic review and meta-analyses of randomised controlled trials and cohort studies." *BMJ* 346: e7492.

Tektonidis, T. G., A. Akesson, et al. (2015). "A Mediterranean diet and risk of myocardial infarction, heart failure and stroke: a population-based cohort study." *Atherosclerosis* 243(1): 93–98.

Thackeray, J. T., M. Parsa-Nezhad, et al. (2011). "Reduced CGP12177 binding to cardiac β-adrenoceptors in hyperglycemic high-fat-diet-fed, streptozotocin-induced diabetic rats." *Nucl Med Biol* 38(7): 1059–1066.

Toblli, J. E., G. Cao, et al. (2005). "Reduced cardiac expression of plasminogen activator inhibitor 1 and transforming growth factor β1 in obese Zucker rats by perindopril." *Heart* 91(1): 80–86.

Tong, X., J. Y. Dong, et al. (2011). "Dairy consumption and risk of type 2 diabetes mellitus: a meta-analysis of cohort studies." *Eur J Clin Nutr* 65(9): 1027–1031.

Tung, Y. T., M. T. Chua, et al. (2008). "Anti-inflammation activities of essential oil and its constituents from indigenous cinnamon (Cinnamomum osmophloeum) twigs." *Bioresour Technol* 99(9): 3908–3913.

Turan, B. and E. Tuncay (2014). "Regulation of cardiac β3-adrenergic receptors in hyperglycemia." *Indian J Biochem Biophys* 51(6): 483–492.

Ursino, M. G., V. Vasina, et al. (2009). "The β3-adrenoceptor as a therapeutic target: current perspectives." *Pharmacol Res* 59(4): 221–234.

Van den Bergh, A., A. Vanderper, et al. (2008). "Dyslipidaemia in type II diabetic mice does not aggravate contractile impairment but increases ventricular stiffness." *Cardiovasc Res* 77(2): 371–379.

Varsha, M. K., R. Thiagarajan, et al. (2015). "Vitamin K1 alleviates streptozotocin-induced type 1 diabetes by mitigating free radical stress, as well as inhibiting NF-κB activation and iNOS expression in rat pancreas." *Nutrition* 31(1): 214–222.

Verghese, M., J. E. Richardson, et al. (2008). "Dietary lycopene has a protective effect on cardiovascular disease in New Zealand male rabbits." *J Biol Sci* 8: 268–277.

Vileigas, D. F., A. F. de Deus, et al. (2016). "Saturated high-fat diet-induced obesity increases adenylate cyclase of myocardial beta-adrenergic system and does not compromise cardiac function." *Physiol Rep* 4(17).

Vistoli, G., D. De Maddis, et al. (2013). "Advanced glycoxidation and lipoxidation end products (AGEs and ALEs): an overview of their mechanisms of formation." *Free Radic Res* 47(Suppl 1): 3–27.

Walle, T., M. H. De Legge, et al. (2004). "Resveratrol-[C-14] disposition and metabolism in vivo in humans." *FASEB J.* 18(4):A604

Walston, J., K. Silver, et al. (1995). "Time of onset of non-insulin-dependent diabetes mellitus and genetic variation in the beta 3-adrenergic-receptor gene." *N Engl J Med* 333(6): 343–347.

Wang, J., H. Tang, et al. (2015). "Modulation of gut microbiota during probiotic-mediated attenuation of metabolic syndrome in high fat diet-fed mice." *ISME J* 9(1): 1–15.

Wang, S., X. Wang, et al. (2015). "Curcumin promotes browning of white adipose tissue in a norepinephrine-dependent way." *Biochem Biophys Res Commun* 466(2): 247–253.

Wang, X., W. Zhang, et al. (2014). "High selenium impairs hepatic insulin sensitivity through opposite regulation of ROS." *Toxicol Lett* 224(1): 16–23.

WHO (2015). *Obesity*, World Health Organization: 1–88.

WHO (2016). *Global Report on Diabetes*, World Health Organization: 1–88.

Widen, E., M. Lehto, et al. (1995). "Association of a polymorphism in the beta 3-adrenergic-receptor gene with features of the insulin resistance syndrome in Finns." *N Engl J Med* 333(6): 348–351.

Wien, M. A., J. M. Sabate, et al. (2003). "Almonds vs complex carbohydrates in a weight reduction program." *Int J Obes Relat Metab Disord* 27(11): 1365–1372.

Wilson, C. and C. Lincoln (1984). "Beta-adrenoceptor subtypes in human, rat, guinea pig, and rabbit atria." *J Cardiovasc Pharmacol* 6(6): 1216–1221.

Yamakoshi, J., S. Kataoka, et al. (1999). "Proanthocyanidin-rich extract from grape seeds attenuates the development of aortic atherosclerosis in cholesterol-fed rabbits." *Atherosclerosis* 142(1): 139–149.

Yeo, H. Y., O. Y. Kim, et al. (2011). "Association of serum lycopene and brachial-ankle pulse wave velocity with metabolic syndrome." *Metabolism* 60(4): 537–543.

Yerlikaya, F. H., A. Toker, et al. (2013). "Serum trace elements in obese women with or without diabetes." *Indian J Med Res* 137(2): 339–345.

Yoo, S. R., Y. J. Kim, et al. (2013). "Probiotics *L. plantarum* and *L. curvatus* in combination alter hepatic lipid metabolism and suppress diet-induced obesity." *Obesity (Silver Spring)* 21(12): 2571–2578.

Yoshida, M., S. L. Booth, et al. (2008). "Phylloquinone intake, insulin sensitivity, and glycemic status in men and women." *Am J Clin Nutr* 88(1): 210–215.

Yoshida, T., N. Sakane, et al. (1994). "Anti-obesity and anti-diabetic effects of CL 316,243, a highly specific β3-adrenoceptor agonist, in yellow KK mice." *Life Sci* 54(7): 491–498.

Youssef, G. (2012). "Nutrition therapy and prediabetes." In: M. Franz and A. B. Evert (Eds) *American Diabetes Association Guide to Nutrition Therapy for Diabetes*, Alexandria, VA, USA, American Diabetes Association: 469–500.

Zaman, A. K., S. Fujii, et al. (2004). "Salutary effects of attenuation of angiotensin II on coronary perivascular fibrosis associated with insulin resistance and obesity." *J Mol Cell Cardiol* 37(2): 525–535.

Zazpe, I., J. J. Beunza, et al. (2013). "Egg consumption and risk of type 2 diabetes in a Mediterranean cohort; the sun project." *Nutr Hosp* 28(1): 105–111.

Zeng, J., J. Zhou, et al. (2009). "Effect of selenium on pancreatic proinflammatory cytokines in streptozotocin-induced diabetic mice." *J Nutr Biochem* 20(7): 530–536.

Zhang, H. Q., S. Y. Chen, et al. (2016). "Sulforaphane induces adipocyte browning and promotes glucose and lipid utilization." *Mol Nutr Food Res* 60(10): 2185–2197.

Zhao, X. L., L. M. Gutierrez, et al. (1994). "The α1-subunit of skeletal muscle L-type Ca channels is the key target for regulation by A-kinase and protein phosphatase-1C." *Biochem Biophys Res Commun* 198(1): 166–173.

Zhu, W. Z., S. Q. Wang, et al. (2003). "Linkage of β1-adrenergic stimulation to apoptotic heart cell death through protein kinase A-independent activation of Ca2+/calmodulin kinase II." *J Clin Invest* 111(5): 617–625.

Zhu, W. Z., M. Zheng, et al. (2001). "Dual modulation of cell survival and cell death by beta(2)-adrenergic signaling in adult mouse cardiac myocytes." *Proc Natl Acad Sci USA* 98(4): 1607–1612.

6 High Carbohydrate Diet-Induced Metabolic Syndrome in the Overweight Body
Association between Organ Dysfunction and Insulin Resistance

Belma Turan and Erkan Tuncay

CONTENTS

6.1 INTRODUCTION

The metabolic syndrome (MetS) in humans is a combination of one or more different abnormalities in the body and has multiple metabolic risk factors, including abdominal obesity, glucose intolerance, insulin resistance, family history, physical inactivity and sedentary lifestyle (Grundy 1999; Ford, Giles et al. 2002; Miranda, DeFronzo et al. 2005; Goodman, Daniels et al. 2007). Following a clear definition of *diabetes mellitus* in the literature (Reaven 1991; Brownlee 2001), a consultation group on the definition of diabetes for the World Health Organization (WHO) has defined MetS as a syndrome with a high-risk status for different types of abnormalities in several organs (Scott and Coyne 2014). Metabolic systems are integrated with pathogen-sensing and organ functions. Other signs of MetS include high blood pressure, decreased fasting serum HDL cholesterol, elevated level of fasting serum triglyceride or prediabetes (Davidson 2003). Currently, it is well-known that MetS is closely associated with high risk for cardiovascular diseases (CVD) (Suh and Lee 2014). Although the exact mechanisms of the complex pathways of MetS are under investigation, it is generally accepted that the current food environment contributes to the development of MetS, when our diet is mismatched with our biochemistry (Bremer 2012). Overweightedness in humans is also generally characterized by high intra-abdominal adiposity, as well as others due to convergence with several signaling mechanisms such as obesity, high visceral adiposity, insulin resistance and hyperglycemia. In this regard, the association between high intra-abdominal adiposity, MetS and/or type 2 diabetes (T2D), and between several metabolic abnormality-associated endogenous changes and the increased risk of experiencing serious cardiovascular disorders has been well documented in the literature (Despres and Lemieux 2006; Goodman, Daniels et al. 2007). Parallel to these statements, in recent decades, increasing rates of obesity have driven adipose tissue into the focus of scientific interest (Hotamisligil 2006; Shoelson, Lee et al. 2006; Schenk, Saberi et al. 2008). Promoting the development of insulin resistance and T2D, chronic inflammation was unequivocally shown to be initiated within adipose tissue, thus rendering adipocytes representative of the major interface-connecting metabolism to the immune system (Hotamisligil 2006; Shoelson, Lee et al. 2006; Schenk, Saberi et al. 2008).

As summarized in Figure 6.1, the mechanisms responsible for the development of MetS in the body appear to be multifactorial, and there are great efforts being made by market players spending millions of dollars on developing new therapeutic strategies against MetSassociated activated or inhibited signaling mechanisms and their related complications. The challenge in this area is that the emerging therapeutic agents do not seem very effective in treating obesity or insulin resistance, or reducing further cardio-metabolic risk (Cummings, Henes et al. 2008; Mourad and Le Jeune 2008). All current assessments in this field aim to generate the best treatment modalities, which provide a better health care strategy in a costeffective manner. It seems to be more important to design novel therapeutic approaches focusing on decreasing body weight (adipose tissue mass) and hepatic fat deposition in particular, including *personalized nutrition*, that can positively help treatment and prevention of MetS-associated organ dysfunctions; diet, exercise and lifestyle modification are, however, beyond professional control in most cases, because they depend on the patient's intellectual capacity and their economic situation.

Despite there is important increases in the protection against MetS and/or obesity and/or T2D related complications during recent years, further additional approaches are yet needed for treatment of them with new updated therapeutic guidelines. In these circumstances, the present chapter aims to report on cutting-edge research on an important syndrome CVD developed in the high carbohydrate-fed overweight and insulin resistant MetS body and the role of *personalized nutrition* in therapeutic approaches.

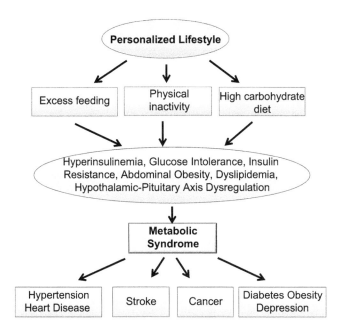

FIGURE 6.1   A proposed mechanism of how a personalized lifestyle plays an important role in the development of metabolic syndrome and thereby several serious organ dysfunctions in the human body.

6.2   CARBOHYDRATE-RICH NUTRITION AND MetS

Carbohydrates, also known as starches and sugars, are the main energy sources of the body and an important amount is included in the content of our daily diet. In the body, excess carbohydrate calories, via excess amounts of sucrose/fructose intake, cause several abnormal processes such as lipogenesis, dyslipidemia, increases in body weight and abdominal size, hyperuremia, ectopic fat deposition, increases in reactive oxygen species (ROS) production, induction of endoplasmic reticulum stress (ER stress) and consequently causing insulin to be released, which in turn stops the fat-burning process thereby causing a condition called insulin resistance. The role of a high carbohydrate diet in humans, mostly due to a high level of sucrose/fructose intake, in inducing metabolic changes leading further to organ dysfunction, is summarized in Figure 6.2. It is also mainly characterized by a high blood glucose level (Volek, Fernandez et al. 2008). Studies have shown that the consumption of sugar-sweetened drinks is associated with excess body weight and increased risk for MetS and/or T2D (Savoca, Evans et al. 2004; Malik, Popkin et al. 2010; Pollock, Bundy et al. 2012). In meta-analysis studies, it was reported that daily consumption of sugar-sweetened beverages was associated with collateral damage giving rise to MetS, thereby underlying the increased risk of development of T2D (Schulze, Manson et al. 2004). Additionally, it should be mentioned, that a high carbohydrate diet leads to MetS posing an increased risk of cardiovascular dysfunction. The effect of a high sucrose intake on systemic metabolism has been exhibited, while myocardial capacity for sucrose uptake and its metabolism have not yet been well described (see

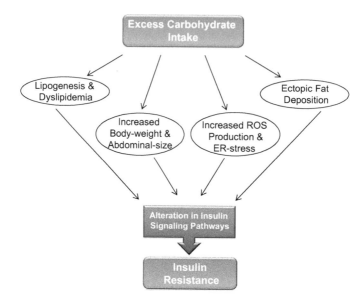

FIGURE 6.2   A pathway demonstrating high carbohydrate intake-related insulin resistance development in humans.

Section 6.3). Associated conditions include hyperuricemia, fatty liver (especially in concurrent obesity) progressing to nonalcoholic fatty liver disease, polycystic ovarian syndrome (in women), erectile dysfunction (in men) and acanthosis nigricans (Figure 6.1 and Figure 6.2).

Long-term MetS leads to multiorgan dysfunction, especially CVD, a major cause of mortality in modern society. MetS is very common in developed countries, and its prevalence is expected to further increase in the near future, in parallel with the rapidly increasing prevalence of obesity and T2D (Eckel, Barouch et al. 2002; Gallagher, Leroith et al. 2010; Dinh, Lankisch et al. 2011). Furthermore, these abnormalities are leading to the loss of cardiomyocytes, CVD and ultimately heart failure in humans (Hansen 1999; Mancia, Bombelli et al. 2010; Kassi, Pervanidou et al. 2011; Matsuzawa, Funahashi et al. 2011). In other words, MetS is the name for a group of factors that increases the risk of heart diseases and other health problems, such as diabetes and stroke. However, it is not very clear yet what is the exact relationship between MetS and development of cardiovascular disorders. Furthermore, MetS can induce increases in the risk for CVD in humans, but there are some contradictory finding in both humans and experimental animals. As an example, although it has been generally documented as a systematic metabolism complication in individuals with high sucrose in their daily diet, its direct cardiac effect cannot be clarified yet. Therefore, we need to know how MetS is affecting the population, as well as why personalized nutrition is applicable in managing it.

Data from the WHO suggests that 65% of the world's population live in countries where being overweight or obese kills more people than being underweight (World Health Organization 2017). The WHO defines 'overweight' as a BMI greater than or equal to 25, and 'obesity' as a BMI greater than or equal to 30. Both overweight and obesity are major risk factors for cardiovascular diseases, specifically heart disease and stroke, and diabetes. The International Diabetes Federation reports that, as of 2011, over 350 million people suffer from diabetes; this number is projected to increase to over half a billion (estimated ~550 million) by 2030. Eighty percent of people with diabetes live in developing countries and in 2011, diabetes caused ~5 million deaths and approximately ~80,000 children were diagnosed with type 1 diabetes. The literature data strongly recommend the importance of dietary carbohydrate restriction as the first approach in diabetes management, although this restriction is not enough to overcome diabetes-induced CVD (Dutta, Podolin et al. 2001; Futh, Dinh et al. 2009). Additionally, it is well accepted that obesity is associated with cardio-metabolic disease, including insulin resistance and diabetes (Kahn and Flier 2000; Qatanani and Lazar 2007). As mentioned in previous paragraphs, insulin resistance is one of the most important risk factors for several pathological conditions such as hypertension (Xi, Yin et al. 2004) and cardiac dysfunction besides the syndromes of T2D (Streja 2004; Lopez-Jaramillo, GomezArbelaez et al. 2014; Fontes-Carvalho, Ladeiras-Lopes et al. 2015). However, it has been demonstrated that high dietary sucrose increases myocardial β-oxidation, reduces glycolytic flux and delays post-ischemic contractile recovery (Gonsolin, Couturier et al. 2007; Jung and Choi 2014).

It is well known that obesity with its accompanying morbidities, both in developed and underdeveloped societies has reached epidemic proportions, generally due to the high carbohydrate content in their daily diet and an unbalanced ratio between energy intake and utilization (Jung and Choi 2014; Fontes-Carvalho, Ladeiras-Lopes et al. 2015). In contrast to a positive energy balance, when energy is needed between meals or during physical exercise, triglycerides stored in adipocytes can be mobilized through lipolysis to release free fatty acids into circulation which are transported to other tissues to be used as an energy source. It is generally accepted that free fatty acids, a product of lipolysis, play a critical role in the development of obesity-related metabolic disturbances, especially insulin resistance (Jung and Choi 2014). In obesity, free fatty acids can directly enter the liver via portal circulation, and increased levels of hepatic free fatty acids induce increased lipid synthesis and gluconeogenesis as well as insulin resistance in the liver (Boden 1997). High levels of circulating free fatty acids can also cause peripheral insulin resistance in both animals and humans (Kelley, Mokan et al. 1993; Boden 1997). Although the exact mechanisms of the complex pathways of MetS are under investigation, it is generally accepted that the current food environment contributes to the development of MetS when our diet is mismatched with our biochemistry (Bremer, Mietus-Snyder et al. 2012).

6.3 MetS AND CARDIOVASCULAR DYSFUNCTION

A high-carbohydrate diet leads to MetS and an increased risk of cardiovascular function, being either insulin resistant dependently or independently of the development of insulin resistance, while the effect of a high carbohydrate diet on systemic metabolism has yet to be established. However, it has been confirmed that abnormal fatty acid metabolism and increased oxidative stress play an intimate role in the pathogenesis of MetS-related cardiovascular diseases. It is already well accepted that MetS includes a cluster of cardiovascular risk factors that includes abdominal obesity, dyslipidemia, hypertension, and impaired glucose tolerance (Grundy 2005; Grundy 2007; Bugger and Abel 2008). Experimental, epidemiological and clinical studies have demonstrated that patients with MetS have significantly elevated cardiovascular morbidity and mortality, not only in developed but also in underdeveloped countries (Grundy 1999; Wong and Malik 2005; Palmieri and Bella 2006; Hotamisligil and Erbay 2008; Zalesin, Franklin et al. 2011; Berwick, Dick et al. 2012; Mandavia, Aroor et al. 2013; Tehrani, Malik et al. 2013; Avila, Osornio-Garduno et al. 2014; Monti, Monti et al. 2014; Xanthakis, Sung et al. 2015). Hypertension and changes in heart rate generally appear early on, with the risk of developing coronary artery disease, arteriosclerosis and heart failure increasing at a later stage (Gallagher, LeRoith et al. 2008).

MetS and morphofunctional characteristics of the left ventricle, as well as left ventricular systolic and diastolic dysfunction (LVSD and LVDD), could develop in patients with MetS (Dursunoglu, Evrengul et al. 2005; Aksoy, Durmus et al. 2014; Erturk, Oner et al. 2015). A clinical study demonstrated the relation between MetS and LV geometry and function, carotid intima-media thickness and arterial stiffness in a community-based cohort of 702 adult subjects. The degree of MetS clustering was found to be strongly correlated with the geometric eccentricity of LV hypertrophy, diastolic dysfunction and arterial changes irrespective of age and blood pressure status, particularly in females. Waist circumference was found to have the most powerful effect on cardiovascular parameters (Ahn, Kim et al. 2010). Furthermore, Nicolini and co-workers (2013) (Nicolini, Martegani et al. 2013) evaluated the influence of gender on LV remodeling in MetS and found that LV remodeling is significantly influenced by gender: the effects of MetS are more pronounced in women, with development of LV concentric hypertrophy/remodeling and preclinical diastolic dysfunction commensurate with those of age-matched men (Nicolini, Martegani et al. 2013). Moreover, some findings support the idea that LVDD can develop in patients with MetS even in the absence of hypertension, though coexistence of hypertension with MetS contributes to further worsening of diastolic functions (Aksoy, Durmus et al. 2014). An early study by Grandi et al. in 2006 (Grandi, Maresca et al. 2006) examined the morphofunctional characteristics of the left ventricle in clinically hypertensive nondiabetic subjects and they explored correlations of MetS with subclinical cardiovascular organ damage, such as increased left ventricular mass and systolic and diastolic dysfunction. Such a focus is relevant because left ventricular structural and functional abnormalities are independent predictors of cardiovascular events including congestive heart failure. Furthermore, their data suggested considering MetS as a distinct condition requiring aggressive treatment strategies because the study showed that MetS was more strongly correlated than its components with subclinical cardiovascular target organ damage.

This multifaceted syndrome is often accompanied by a hyperdynamic circulatory state characterized by increased blood pressure, total blood volume, cardiac output, and metabolic tissue demand. Additionally, it has been widely documented that hypertension generally amplifies the high cardiovascular risk if the disease remains uncontrolled for a long time (Knudson, Dincer et al. 2005; Gallagher, LeRoith et al.

2008; Abdul-Ghani and DeFronzo 2009; Peters 2009; Campos-Pena, Toral-Rios et al. 2017; Gerber and Rutter 2017). In a metaanalysis study by Malik et al. (Malik, Popkin et al. 2010), it was reported that daily consumption of sugar-sweetened beverages was associated with a 15% increased risk of development of T2D. In addition, Schulze et al. (Schulze, Manson et al. 2004) found that those who consumed more than one sugar-sweetened beverage per day had an 83% increased risk of diabetes compared with those who consumed less than one per month. In addition, the cardiovascular effects of a high sugar intake have been reported as a 20% increase in coronary artery disease among men and women (Fung, Malik et al. 2009; de Koning, Malik et al. 2012). Although several studies point to the association between a high carbohydrate diet and MetS, leading to an increased risk of cardiovascular dysfunction, it is not clear yet how high sugar intake alters the synchrony of the cardiovascular system. In addition, albeit that the effects of a high sucrose intake on systemic metabolism has been determined, myocardial capacity for sucrose uptake and its metabolism have not been well characterized yet. Therefore, it is obviously accepted that the MetS components of *diabetes mellitus* and hypertension, as well as isolated MetS, excluding established hypertension or *diabetes mellitus*, are well known to be associated with abnormal cardiac structure and function, particularly LVSD and LVDD, and the elimination of these abnormalities is a goal of cardiovascular disease prevention (Mureddu, Greco et al. 1998; de Simone, Palmieri et al. 2002; Aijaz, Ammar et al. 2008). Experimental, epidemiological, and clinical studies have demonstrated that patients with MetS have significantly elevated cardiovascular morbidity and mortality rates. The primary endpoints of cardio-metabolic risks are coronary and peripheral arterial disease, myocardial infarction, congestive heart failure, arrhythmia and stroke. In the light of already-published data, a summarized block diagram can be presented with a Hodgkin cycle-style positive feedback process to demonstrate the complementary roles of systemic parameters on MetS associated cardiac dysfunction in mammals (Figure 6.3).

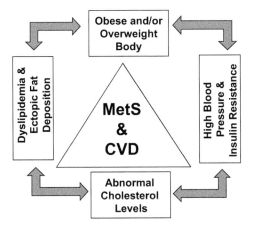

FIGURE 6.3   Inter-relationships between human body status and other factors underlying the development of metabolic syndrome and consequently cardiovascular diseases in humans.

6.4 STRUCTURAL ALTERATIONS AND MOLECULAR MECHANISMS OF CARDIOVASCULAR DYSFUNCTION IN MetS

Personal feeding behavior, fat and glucose metabolism affect the normal metabolic state of the body. In another words, the normal metabolic state of the body is maintained by a daily feeding habit (Pedram and Sun 2014) as well as changes in pro-inflammatory cytokines including interleukin-1 (IL-1), IL-2, IL-6, tumor necrosis factor-alpha (TNF-α), and tumor necrosis factor-beta (TNF-beta) (Aromolaran and Boutjdir 2017). A marked accumulation of adipose tissue is leading to MetS being associated with dyslipidemia (or abnormal levels of serum free fatty acids, FFA), increased secretion of pro-inflammatory cytokines, insulin resistance, hyperglycemia (Sonnenberg, Krakower et al. 2004), fibrosis (Abed, Samuel et al. 2013; Ternacle, Wan et al. 2017), and hyperuricemia (Viazzi, Piscitelli et al. 2017). To date most studies have provided important insights into the impact of individual disorders that contribute to MetS on CVD (Maharani, Kuwabara et al. 2016); however, specific cardiac alterations by MetS remain poorly understood. As presented in Figure 6.3, CVD in MetS is associated with inflammation, decreases in glucose utilization, dyslipidemia, increases in fibrosis, collagen and ectopic fat deposition, abnormal cholesterol levels and insulin resistance. Albarado-Ibanez and co-workers (2013) (Albarado-Ibanez, Avelino-Cruz et al. 2013) examined the structural alterations in the hearts of MetS rats induced in two-month-old male Wistar rats by administering 20% sucrose into the drinking water for eight weeks. Their data demonstrated the sympathetic innervation of the SA node, evidenced by immunofluorescence for tyrosine hydroxylase, and red oil staining showed adipose tissue surrounding nodal cardiomyocytes.

The light microscopy examination of left ventricular tissue of MetS rats fed with 32% sucrose in drinking water for 18 weeks presented evidence that MetS induced important degeneration in the heart tissue such as disorganization of myofibrils, loss of integrity, decrease in myofibril diameter, increase in connective tissue around myofibrils and vessels, intracellular vacuolization, differences in cell organelle composition and increases in lipid accumulation (Okatan, Tuncay et al. 2015). Durak and co-workers obtained further significant irregular, heterogeneous and differentiated appearances in cytoplasm in 22–24 weeks MetS rat heart tissue (Durak, Olgar et al. 2017). Furthermore, they also performed histological investigation of left ventricular heart tissue of 24-week MetS rats by light microscopy in hematoxylin and eosin-stained sections with a normal appearance of myocardial arrangements in longitudinal sections (Figure 6.4A). However, they observed marked loss of eosinophils, increases in the number of lymphatic vessels and vacuolization in MetS rat heart tissue (Figure 6.4B). A transmission electron microscopy examination also demonstrated a markedly irregular and scattered myofibrillar appearance, disturbances and decreases in thickness of Zlines and swollen and deformed mitochondria in the MetS group with a normal appearance in age-matched controls (Figure 6.5A, right and left, respectively). Furthermore, we also observed significant changes in an aortic tissue section in MetS rats such as extensions to lumen from vessel endothelium and pinositotic vesicles with normal structure in age-matched controls (Figure 6.5B, right and left, respectively).

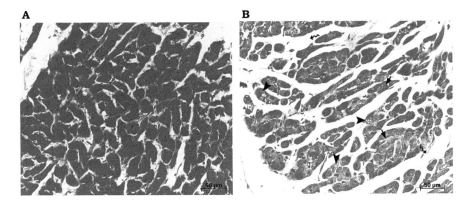

FIGURE 6.4   Light microscopic findings in left ventricle of rats with metabolic syndrome. Micrographs of left ventricular cross-section of control (A) and metabolic syndrome (MetS) groups by using light microscopy (B). Light microscopy examinations represent hematoxylin and eosin-stained transverse heart sections. Normal histological appearance in (A) and loss of eosinophils, increase in lymphatic vessel number and marked vacuolization in (B). Arrow: eosinophil loss; curl arrow: lymphatic vessel; arrowhead: vacuolization. Bar: 100 μm.

The observed alterations in the MetS group, such as marked pale staining and eosinophil loss in some myofibrils and marked heterogeneity in the cytoplasm are linked to intracellular vacuolization and defects of the organelles. Some of the findings in MetS such as intracellular vacuolization were observed previously in streptozotocin-diabetic rat hearts (Ayaz, Can et al. 2002). In addition, the significantly increased intracellular lipid inclusions in MetS group are supporting increased oxidative stress in the heart tissue, as mentioned in other studies (Marfella, Di Filippo et al. 2009). The observed changes in the hemodynamic parameters point out the important cardiac functional changes, being parallel to the structural changes in MetS group. These changes observed in the heart of MetS rats correspond to the changes observed in obese individuals (Kishi, Armstrong et al. 2014).

The underlying mechanisms of MetS associated systemic and organ level syndromes involve mostly the abnormal metabolic state, including insulin resistance and hyperglycemia, activating various adverse systems, such as oxidative stress, eventually leading to cardiac structural and functional alterations, as mentioned previously (Eckel, Barouch et al. 2002; Gallagher, Leroith et al. 2010; Dinh, Lankisch et al. 2011). By connecting the dynamic relationship between the alterations from systemic parameters of the rats and functional parameters of heart, we have strong evidence to make the case that these alterations can be attributed to increased myocardial ROS production and altered Ca^{2+} handling occurring as a consequence of systemic insulin resistance.

Okatan and co-workers also presented that the leptin level was increased while the ghrelin level was decreased in the MetS group (Okatan, Tuncay et al. 2015). It is known that leptin is a mediator of the long-term regulation of energy balance, suppressing food intake and thereby inducing weight loss. On the other hand, ghrelin is a fast-acting hormone, seemingly playing a role in meal initiation. Their data are supporting already-known facts because the body weight of rats from the

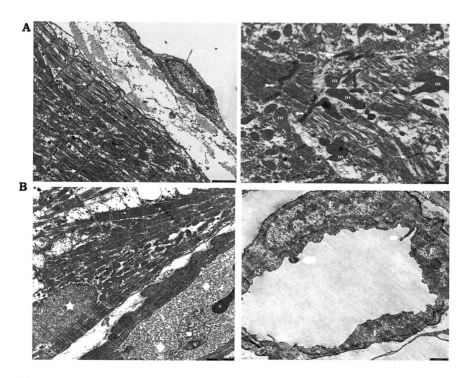

FIGURE 6.5 Electron microscopic examination of heart tissues using transmission electron microscope, TEM. (A) Electron micrograph images in normal rat heart. Normal appearance of myofilaments, Z-lines, and mitochondria (left) as well asenlargement and disruption in myofibril distances, thinning in Z-lines, swollen and partitionation in mitochondria of MetS samples (right). Arrow: Z-lines; star: nucleus; m: mitochondrion; E: endothelial nucleus. Magnification: ×10,000. (B) Vessel section in normal rats by TEM (left). Marked extensions to endothelial lumen and pinocytotic vesicles in MetS rats (right). E: endothelia nucleus; star: muscle tissue nucleus; diamond: vessel lumen; thick arrow: polyploid extensions of endothelial cells. Magnification: ×4,650 (in A) and ×7,750 (in B).

MetS group was significantly higher compared to those of agematched controls. Indeed, in obese subjects the circulating level of the anorexigenic hormone leptin is increased, whereas, surprisingly, the level of the orexigenic hormone ghrelin is decreased (Tschop, Weyer et al. 2001; Hansen, Dall et al. 2002; Faraj, Havel et al. 2003; McLaughlin, Abbasi et al. 2004). It is now established that some obese patients are leptin-resistant (Kimber, Peelman et al. 2008; Roujeau, Jockers et al. 2014). However, the manner in which both the leptin and ghrelin systems contribute to the development or maintenance of obesity is not yet clear. Furthermore, it can be suggested that possible abnormalities in the leptin and ghrelin systems may contribute to the development of cardiovascular dysfunction through obesity (Tritos, Kissinger et al. 2004).

Turan's group and others emphasized that MetS is also characterized with increased oxidative stress, a condition due to an imbalance between production and scavenging of oxidants in the antioxidant system (Vendemiale, Guerrieri et al. 1995;

Bonomini, Rodella et al. 2015; Okatan, Tuncay et al. 2015; Bhatti, Bhadada et al. 2016; Gregorio, De Souza et al. 2016). Increased oxidative stress can play an important role in the pathogenesis of various diseases including MetS (Roberts and Sindhu 2009; Bonomini, Rodella et al. 2015), in part due to its amplification by a concomitant antioxidant deficiency which may favor the propagation of oxidative alterations from intra- to extracellular spaces. Similar to other studies (Galassetti 2012; Korkmaz, Altinoglu et al. 2013; Okatan, Tuncay et al. 2015), our findings, related with significantly increased total oxidative status and decreased total antioxidant status in sera of rats with MetS, support these statements. These changes may further affect the impairment of insulin signaling, which further leads to insulin resistance. Indeed, previous studies using a high sucrose diet have shown a marked glucose intolerance in MetS modeled experimental animals (Pagliassotti, Prach et al. 1996; Brenner, Rimoldi et al. 2003). Additionally, they have shown significantly decreased serum paraoxonase and arylesterase activities in MetS group. Since the antioxidant properties of HDLs are attributable to serum paraoxonase and arylesterase activities (James 2006), these activities seem to be crucial in the relation to increased oxidative stress and organ dysfunction, particularly in terms of cardiovascular pathologies in rats with MetS (Kagota, Maruyama et al. 2013; Eren, Abuhandan et al. 2014).

Studies emphasized the important role of ER stress in the pathogenesis of MetS. More specifically, increased ER stress responses in the liver and adipose tissue of obese mice are known to have a role in the development of systemic insulin resistance and T2D. Furthermore, promotion of ER stress through genetic haploinsufficiency of the *Xbp1* gene triggers obesity and insulin resistance, whereas alleviation of ER stress by treatment with chaperone proteins is protective against metabolic deterioration in obese animals (Ozcan, Cao et al. 2004; Ozcan, Yilmaz et al. 2006). Other studies have also shown that compromising ER function through the manipulation of chaperone proteins can modulate the systemic action of insulin (Hotamisligil 2006). ER stress leads to insulin resistance, at least in part through the serine phosphorylation of insulin receptor substrate 1 (IRS1) by serine/threonine-protein kinase/endoribonuclease (IRE1)-activated c-Jun N-terminal protein kinase 1 (JNK1) (Ozcan, Cao et al. 2004). An interesting recent study also demonstrated a role for ER stress in leptin resistance at the cellular level (Hosoi, Sasaki et al. 2008). These and other findings highlight the importance of the integration of nutrient and inflammatory responses in metabolic homeostasis, and the ways in which dysfunction of the ER could affect this integration and possibly result in chronic metabolic disease (Feng, Zhang et al. 2003; Ozawa, Miyazaki et al. 2005; Myoishi, Hao et al. 2007). Furthermore, Cicek and co-workers (2014) (Cicek, TokcaerKeskin et al. 2014) investigated the acute effects of a dipeptidyl peptidase 4 (DPP-4) inhibitor, sitagliptin, on vascular function in rats with high sucrose diet-induced MetS. In order to elucidate the mechanisms implicated in the effects of DPP-4 inhibition, they tested the involvement of the NO pathway and epigenetic regulation in the MetS. Acute use of sitagliptin protects the vascular function in the rats with MetS in part due to the NO pathway by restoring the depressed aortic relaxation responses mediated by receptors. Application of sitagliptin enhanced the depressed phosphorylation levels of both the endothelial NO synthase and the apoptotic status of protein kinase B, known as Akt, in endothelium-intact thoracic aorta from rats with MetS. A one-hour application of sitagliptin on aortic rings from rats with MetS also induced remarkable histone post-translational modifications such as increased expression of trimethylated histone 3 at lysine residue 27 (H3K27Me3), but not of trimethylated histone 2 at lysine residue 27 (H3K27Me2), resulting in an accumulation of the H3K27Me3. Taking into consideration these data, together with data from Bourgoin and coworkers (2008) (Bourgoin, Bachelard et al. 2008), one can suggest that sitagliptin, aside from glycemic control, may demonstrate its important role in the treatment of patients with MetS.

6.5   ELECTRICAL ALTERATIONS IN THE HEARTS OF MetS INDIVIDUALS

Metabolic syndrome, MetS, includes a group of risk factors, which in turn raise risk factors for CVD and other health problems, such as diabetes and stroke. Additionally, long-term MetS leads to multi-organ dysfunction, especially cardiovascular disease, a major cause of mortality. Although T2D was shown to prolong the QT interval on the ECG and to promote cardiac arrhythmias, this is not so clear for MetS, a precursor state of T2D. The abnormal prolongation of the QT interval is the most prominent electrical remodeling in diabetic hearts; clinically, its prevalence is as high as about 25% in diabetic patients, including Type 1 diabetes (T1D) insulin-dependent *diabetes mellitus* (IDDM) and T2D non-IDDM populations (Rossing, Breum et al. 2001; Veglio, Chinaglia et al. 2004). The QT prolongation is a significant predictor of mortality in IDDM and non-IDDM (Cardoso, Salles et al. 2001; Veglio, Chinaglia et al. 2004), because it is associated with an increased risk of sudden cardiac death in diabetic patients as a consequence of lethal ventricular arrhythmias known as torsades de pointes. Therefore, it seems having sufficient knowledge about the ionic mechanisms underlying diabetic QT potential is essential. Most of the literature data demonstrated that both T1D and T2D have been associated with increases of both the QT interval duration and its dispersion in people of all ages as well as both genders (Marques, George et al. 1997; Landstedt-Hallin, Englund et al. 1999; Veglio, Bruno et al. 2002). Interestingly, QT prolongation has been extensively observed during both hypo- and hyperglycemia (Eckert and Agardh 1998). As the QT interval reflects the dynamics of ventricular repolarization, the activity of repolarizing cardiac potassium channels (I_K) is mainly responsible for QT interval duration. However, Durak and co-workers (2017) (Durak, Olgar et al. 2017) evaluated *in vivo* ECG in rats with MetS and showed that there was significantly decreased QRS amplitude and a shortened QT-interval together with a markedly increased heart rate. Their data observed in rats with MetS are generally mentioned by other investigators in both animal and human studies (Merabet, Fang et al. 2015; Okatan, Tuncay et al. 2015; Korkmaz-Icoz, Lehner et al. 2016). It is accepted that the increased heart rate observed in MetS contributes to the deterioration of LV function by means of impaired LV filling and relaxation and reduced coronary perfusion and cardiac output. These findings nicely correlate with depressed heart work in individuals with MetS.

It is well accepted that MetS is a risk factor for prolonged QT, which may further increase cardiovascular morbidity and mortality in the subjects concerned. A short QTinterval is defined as short QT-syndrome, which is generally a genetic disease of the electrical system of the heart and mutations in the KCNH2, KCNJ2 and KCNQ1 genes. These genetic modifications could lead to short QT-syndrome, which is closely associated with sudden death, generally due to ventricular fibrillation (Gussak, Brugada et al. 2000; Laitinen, Brown et al. 2001; Chen, Xu et al. 2003; Gaita, Giustetto et al. 2003). However, the cause of short QT-syndrome is unclear, while it has been hypothesized the QT shortens when heart rate increases (Maltret, Wiener-Vacher et al. 2014).It has been also proposed that short QTinterval may be acquired (Gaita, Giustetto et al. 2004). For example, it has been shown that some drugs or conditions such as hyperkalemia, acidosis, and hyperthermia could cause induction of short QT-interval even in normal subjects (Patel and Pavri 2009). In this regard, Bai and co-workers (2007) (Bai, Wang et al. 2007) studied the positive effect of phospholipid lysophosphatidylcholine as a metabolic trigger and HERG as an ionic pathway for extracellular K^+ accumulation and 'Short QT Syndrome' in acute myocardial ischemia via blockage of rapid delayed rectifier K^+-current (I_{Kr}) and HERG (the pore-forming subunit of I_{Kr}), due to a hypothesis on extracellular K^+ accumulation and shortening of action potential duration or QT interval, which are pivotal in the genesis of ischemic arrhythmias and sudden cardiac death, although the ionic mechanisms remained obscured. However, Zhang and coworkers (Zhang, Xiao et al. 2006) reported the efficacy of insulin in preventing 'QTprolongation' and associated arrhythmias and mechanisms underlying the effects in a rabbit model of IDDM by demonstrating the restoration effect of insulin on the depressed I_{Kr}/HERG and prevented QT/action potential duration prolongation and the associated arrhythmias, partially due to the antioxidant ability of insulin.

Cardiac ion channel regulation in obesity and the MetS, the relevance to 'long QT' syndrome and atrial fibrillation, and the fact that related mechanisms leading to cardiac electrical remodeling are likely to have a significant medical and economic impact are summarized in a review article by Aromolaran and Boutjdir (2017). The upstroke or initial phase of the cardiac action potential is controlled by the entry of depolarizing Na^+ current (I_{Na}) through voltage-gated Na^+ channels, specifically Nav1.5, which is encoded by the geneSCNA5 (Lieve, Verkerk et al. 2017). Whether and how I_{Na} is directly modulated in MetS/obesity is poorly understood; however, there is indirect evidence for the potential modulation of I_{Na} in MetS/obesity. In this regard, it has been demonstrated that acute 1-h leptin exposure of rabbit atrial myocytes induced increases in peak I_{Na} density (Lin, Chen et al. 2013), while peak I_{Na} density is also increased by pro-inflammatory cytokines (Zhao, Sun et al. 2016) as well as by serum free fatty acids, FFAs (Lin, Chen et al. 2013), in line with the altered functional expression of I_{Na} in MetS/obesity. Obesity is also a key contributor to the expanding prevalence of AF, and according to population-based cohort studies, obese individuals have a 49% increased risk of developing atrial fibrillation, AF, compared to nonobese individuals (Wanahita, Messerli et al. 2008). According to an AHA report, metabolic syndrome affects about 35% of the adult population in the United States, suggesting that MetS may be a risk factor for the onset of atrial fibrillation (AF) in patients. Despite the role of I_{Na} in normal cardiac depolarization, and therefore heart excitability (Luo and Rudy 1991), there is still controversy about its contribution to MetS-associated arrhythmias. O'Connell et al. (O'Connell, Musa et al. 2015) demonstrated that during shortterm exposure of ovine left atrial myocytes to saturated FFA, stearic acid (SA) abbreviated the AP duration, with I_{Na} remaining essentially unchanged; however, elevated FFAs increased I_{Na} density, altered its gating properties and increased cardiac excitability through augmentation of intracellular Ca^{2+} concentration in dog ventricular myocytes (Biet, Morin et al. 2014). On the other hand, hypercholesterolemic rabbits displayed a significantly depressed ventricular peak I_{Na}, leftward shift in the inactivation potential and a slowed time course of recovery when compared to normolipidemic control myocytes (Wu, Su et al. 1997). Considering the implications of altered I_{Na} electrical remodeling, a good understanding of the molecular mechanisms underlying Na^+ channel gating and functional regulation in obese heart is critical for fundamental insights into the prevalent condition of long QT-syndrome, LQTS, and metabolic disease-related arrhythmias in patients. In addition, at molecular level, congenital gain-of-function mutations in the Nav1.5 channel subunits increases I_{Na} density and delays ventricular repolarization leading to QT interval prolongation (Splawski, Shen et al. 2000), while a rat model of diet-induced obesity displayed prolongation of the QRS complex despite unchanged densities of peak I_{Na}, and/or outward K^+ currents measured in ventricular myocytes (Axelsen, Calloe et al. 2015). Furthermore, other observations raise the possibility that I_{Na} functional expression may initially increase and then decrease with progressive weight gain/obesity and/or AF progression (Bosch, Zeng et al. 1999; Sossalla, Kallmeyer et al. 2010).

In the context of metabolic diseases, O'Connell et al. demonstrated that short-term exposure of ovine left atrial myocytes to SA caused a significant reduction of L-type voltageactivated Ca^{2+} channels, I_{CaL} density (O'Connell, Musa et al. 2015), in line with a contribution of FFA-mediated atrial I_{CaL} dysfunction in obesity. The I_{CaL} density was only slightly increased in ventricular myocytes isolated from rabbits fed a cholesterol-rich diet not significantly different from normal chow-fed controls (Luo, Wu et al. 2004). Furthermore, distinct pro-inflammatory cytokines that are involved in obesity have also been shown to alter I_{CaL} density, although these studies have yielded varying results. The effects of obesity on the functional expression of I_{CaL} have also been investigated in animal models with contrasting outcomes. For example, rats fed with a high fat diet for 15 weeks, in order to induce obesity, showed reduced Ca^{2+} influx, while gene expression of CACNAC1 was either decreased (Lima-Leopoldo, Leopoldo et al. 2011) or unchanged (Lima-Leopoldo, Sugizaki et al. 2008). However, in another study, Leopoldo et al. found that mRNA expression of I_{CaL} is increased at 30 weeks (Lima-Leopoldo, Leopoldo et al. Ashrafi et al 2013). (Ashrafi, Yon et al. 2016) also reported increased mRNA levels of ventricular I_{CaL} after eight weeks in high fat diet fed rats, while Leopoldo et al. (Lima-Leopoldo, Leopoldo et al. 2011) found no change in protein expression.

The transient outward K^+ current, I_{to}, one type voltage-gated K^+ channels, is also an important contributor to cardiac action potential (AP), waveform and contributes prominently to the initial and early repolarization phase of atria (Workman, Kane et al. 2001; Virag, Jost et al. 2011) and ventricular AP (Rosati, Pan et al. 2001). In terms of ventricular arrhythmias, decreases in I_{to} would be expected to delay repolarization and prolong APD, making dysregulation of I_{to}, a plausible contributor to LQTS in obese patients (Grandinetti, Chow et al. 2010). The pro-inflammatory cytokine TNF-α has also been shown to depress I_{to} channel function in ventricular myocytes (Grandy and Fiset 2009), while the mRNA and/or protein expression of Kv4.2/Kv4.3

subunits remained essentially unchanged. The impact of TNFαmediated I_{to} reduction on AP-duration (APD) has yielded contrasting results, with one report showing no effect (Fernandez-Velasco, Ruiz-Hurtado et al. 2007), and another prolongation (Grandy and Fiset 2009), suggesting the possibility of reciprocal regulation of other ventricular ion channels. The physiological relevance of ultra-rapid delayed rectifier K^+ current (I_{Kur}) is underscored by data showing that congenital mutations in Kv1.5 channel subunits increase I_{Kur} density and shorten atrial APD, a condition that predisposes one to AF (Christophersen, Olesen et al. 2013). Similarly, Zhang et al. (Zhang, Hartnett et al. 2016) reported a shortened P-R interval and increased atrial Kv1.5 protein expression in mice exposed to a high fat diet for eight weeks, although these data were not correlated with functional Kv1.5 channel data nor was AF induced in these studies. By comparison, in AF patients I_{Kur} current density (Van Wagoner, Pond et al. 1997), mRNA (Lai, Su et al. 1999), and protein expression levels (Brundel, Van Gelder et al. 2001) of Kv1.5 are decreased. Therefore, these observations would suggest that in the mice with diet-induced obesity, the functional expression of Kv1.5 may decrease with the severity of obesity leading to AF induction. Moreover, the selective localization and/or expression of Kv1.5 subunits in the atria and its therapeutic potential demonstrate the need to further elucidate molecular and electrophysiological mechanisms regarding the relative significance and/or contribution of I_{Kur} to the onset and/or progression of AF in obese patients. Since the cardiac delayed rectifier K^+ current, or I_K composed of I_{Kr} and I_{Ks}, is an important regulator of repolarization (Sanguinetti and Jurkiewicz 1990), and while there have been some studies on the electrophysiological effects of obesity-related molecular processes on some cardiac voltagegated channels (O'Connell, Musa et al. 2015; Aromolaran, Colecraft et al. 2016), there is still a lack of studies that have assessed the functional properties of I_K, I_{Kr} and I_{Ks} in obese animal models, most likely due to the lack of expression of these channels in commonly used rodent models (Killeen, Thomas et al. 2008; Aromolaran, Subramanyam et al. 2014). Aromolaran et al. (Aromolaran, Subramanyam et al. 2014) also found that obese atrial myocytes displayed an abbreviated APD and had a significantly larger I_K density compared to the low fat diet controls. The implication of these observations in ventricles is currently unknown; nevertheless, a previous report by Haim et al. (Haim, Wang et al. 2010) has shown that palmitate reduced cardiac contractility, shortened APD and increased the density of voltage-gated K^+ channels in mouse ventricular myocytes.

6.6 CELLULAR ABNORMALITIES IN CA²⁺ HOMEOSTASIS AND INSULIN ACTION IN MAMMALIAN MetS

The mammalian heart muscle, similar to other muscles, acts as a transducer due to its action in energy convergence such as receiving energy in the form of substrates (i.e. chemical energy) and delivering energy in the form of pumped blood to the rest of the body (i.e. mechanical energy). Studies have documented that the important contributing factors of MetS are overweightness/obesity and insulin resistance, *while the fact of the reverse situation of MetSassociated obesity and insulin resistance in humans. The complications further lead to CVD, eventually resulting in the development of more severe diseases, such as T2D cardiomyopathy and congestive heart failure. In particular, insulin resistance-related cardiomyopathy has been described as an early left ventricular diastolic dysfunction that develops before impaired systolic function (Dinh, Lankisch et al. 2011). MetS also increases the probability of supraventricular arrhythmia development such as sick sinus syndrome and the abnormal P-wave. Moreover, the risk of developing atrial fibrillation increases in patients with obesity and MetS (Rosiak, Bolinska et al. 2002). Atrial fibrillation is the most common tachyarrhythmia associated with increased morbidity and mortality (Lin, Chen et al. 2010). Indeed, cardiac arrhythmias, underlain by metabolic disorders, are a pervasive condition that is a rapidly expanding epidemic.

Among observed syndromes, left ventricular systolic and diastolic dysfunction is also a common clinical manifestation of myocardial alterations in patients with MetS and diabetes (Witteles and Fowler 2008; Dinh, Lankisch et al. 2011). Several factors have been proposed as contributors to cardiac dysfunction in MetS/obesity, among them changes in intracellular Ca^{2+} changes and mishandling in cellular Ca^{2+}, which are further precipitating factors in the development of heart failure (Davidoff, Mason et al. 2004; Wold, Dutta et al. 2005; Relling, Esberg et al. 2006; Vasanji, Cantor et al. 2006; Balderas-Villalobos, MolinaMunoz et al. 2013). Nevertheless, it is unclear whether changes in Ca^{2+} handling play a critical role in the development of myocardial dysfunction induced by MetS/obesity. In early stages, metabolic syndrome (in rats with either a fructose-rich diet for six weeks or sucrose-rich drinking water for 16 weeks: Miklos, Kemecsei et al. 2012 and Okatan, Durak et al. 2016, respectively) primarily disturbs sarcoplasmic reticulum (SR) Ca^{2+} ATPase, SERCA2a function in the heart, but consequential hemodynamic dysfunction is prevented by upregulation of the SERCA2a protein level and phosphorylation pathways regulating phospholamban, PLB.

It is known that left ventricular diastolic dysfunction is a common clinical event of myocardial alterations in patients with MetS, similar to diabetes (Schannwell, Schneppenheim et al. 2002; Dinh, Lankisch et al. 2011). These studies also confirmed a depressed SERCA2a function under inhibited Na^+/Ca^{2+}-exchanger, an NCX condition in MetS-group. Coincident with the alteration of cardiac function, there are a variety of metabolic and biochemical abnormalities that include changes in contractile apparatus and activity of SERCA2a, besides other actors associated with intracellular Ca^{2+}-homeostasis (Malhotra, Penpargkul et al. 1981; Ganguly, Pierce et al. 1983; Bouchard and Bose 1991; Choi, Zhong et al. 2002). Since myocardial contractility is primarily controlled by Ca^{2+}-cycling into/out of cytoplasm and cardiac SR Ca^{2+} release channels, RyR2 and SERCA2a, which play important roles in its removal following muscular contraction, studies showed the possible contribution of altered RyR2 and SERCA2a to defective regulation of Ca^{2+}-cycling in cardiomyocyte from MetS rats (Allo, Lincoln et al. 1991; Vasanji, Cantor et al. 2006; Okatan, Durak et al. 2016), although the presence of some contradictions on this subject in the literature should be noted (Eckel, Gerlach-Eskuchen et al. 1991; Noda, Hayashi et al. 1992). Moreover, it has been shown that MetS/obesity-related altered functional expression of RyR2 or inositol triphosphate receptors (IP3R), involved in regulating the intracellular Ca^{2+}concentration, are also likely to contribute to the pathogenesis of arrhythmias, assessing the contribution of RyR/IP3R signaling pathways to Ca^{2+} mishandling in obesity and metabolic disorders (Splawski, Shen et al. 2000;

Fauconnier, Lacampagne et al. 2005; Lima-Leopoldo, Sugizaki et al. 2008; Lima-Leopoldo, Leopoldo et al. 2011).

Insulin resistance is one of the important risk factors for several pathological conditions, including hypertension (Xi, Yin et al. 2004; Guo, Gong et al. 2005), MetS and T2D (Streja 2004; Lopez-Jaramillo, Lopez-Lopez et al. 2014; Fontes-Carvalho, LadeirasLopes et al. 2015; Lee and McDonald 2015), while it even predicts coronary heart disease in either adult or elderly nondiabetic subjects (Lempiainen, Mykkanen et al. 1999; Kim, Ju et al. 2013). Furthermore, it has been demonstrated that high dietary sucrose triggers hyperinsulinemia, increases myocardial β-oxidation, reduces glycolytic flux and delays postischemic contractile recovery. Interestingly, it has been shown that insulin resistance is a greater risk for coronary heart disease than diabetes, indicating how nondiabetics can have a higher risk of coronary heart disease than diabetics when insulin resistance is elevated (Kim, Ju et al. 2013). Indeed, this essential hormone, insulin, regulates various physiological and pathophysiological functions in the heart. These include myocardial energy metabolism, cardiomyocyte contractility, protein production, hypertrophy and cardiomyopathy in patients with T1D and T2D, and also ion-transport mechanisms (Brownsey, Boone et al. 1997), while chronic heart failure can be associated with insulin resistance (Paolisso, De Riu et al. 1991). In normal mammalian hearts (Lucchesi, Medina et al. 1972) and isolated cardiac muscle preparations (Lucchesi, Medina et al. 1972; von Lewinski, Bruns et al. 2005), insulin exerts an increase in contractile force (positive inotropic effects), while its sub-cellular mechanism remains unclear. Since the well-known action of insulin is promoting entry of glucose into the cell, and subsequent stimulation of the carbohydrate metabolic pathway, it is very logical to interpret this as a positive otropic action of insulin via an association between insulin and enhancement of glycolytic flux in the myocardium (Lucchesi, Medina et al. 1972). However, this concept was not substantiated by previous studies (Lucchesi, Medina et al. 1972; von Lewinski, Bruns et al. 2005), which provided evidence that the increase in contractile force produced by insulin is independent of its action on glucose transport in the myocardium. However, the most of the experimental data related with insulin action seems to be contradictory. Because, it is not clear yet whether the observed improvements in the hemodynamic parameters are due to insulin associated a direct inotropic or peripheral vasodilatory effect.

6.7   CONCLUSIONS AND PREVENTION APPROACHES

MetS is becoming a worldwide matter with severe risk factors for diseases such as diabetes and prediabetes, abdominal obesity, high cholesterol, and high blood pressure that mostly create a platform for several CVDs. In the present times of plenty, not only in developed societies but also in underdeveloped countries, obesity with its accompanying morbidities has reached epidemic proportions, mostly due to the high carbohydrate content in their daily diet with an unbalanced ratio between energy intake and utilization (Hossain, Kawar et al. 2007; Phillips and Prins 2008; Sharma, Okere et al. 2008). Therefore, in view of the increasing prevalence of MetS throughout the world (Sharma, Okere et al. 2008), it is clearly accepted that more can be done to provide affordable treatment options for this condition. These options include local well-controlled natural products with an appropriate safety profile, as they may have been available in foods for many generations. In this respect, flavonoids are accepted to be potential options for treatment of MetS, as they are ubiquitous as secondary metabolites in plants and reduce the risk of coronary heart disease and diabetes (Tripoli, Giammanco et al. 2005; Alam, Kauter et al. 2013). However, some studies stated these potentially beneficial measures are effective in only a minority of people, primarily due to a lack of compliance with lifestyle and diet changes (Katzmarzyk, Leon et al. 2003). Furthermore, the International Obesity Taskforce states that interventions on a sociopolitical level are required to reduce development of the metabolic syndrome in populations (James, Rigby et al. 2004). Coping with all these strategies requires good patient monitoring with conventional and/or new therapeutic agents. However, it has been demonstrated that weight loss of as little as 4–5 kg, with its pleiotropic benefits and optimal safety profile, seems to be one of the most effective way to handle the MetS and/or T2D associated complications (Pories, Swanson et al. 1995; Harris 2000; Pi-Sunyer, Schweizer et al. 2007); in most T2D patients, present clinical praxis fails to attain sustained weight loss and glycemic control (Nathan, Buse et al. 2009). However, current treatment options do not correct the current CVD.

Literature data has also emphasized the contribution of these new parameters of left ventricular diastolic function, evidently deteriorated due to the increasing number of MetS factors (Rossi, Ruiz de Azua et al. 2015; Francisqueti, Minatel et al. 2017; Haberka, Lelek et al. 2017). In summary, both experimental and clinical investigations have shown that the structure of the left ventricle is significantly impaired in patients with MetS, together with left ventricular dysfunctions, both systolic and diastolic. Thus, it can be concluded that the impaired global left ventricular function is actually the result of impairment of several factors in MetS individuals. The degree of structural and functional damage increased with the number of risk factors for MetS. Further studies are necessary for complementing the influence of MetS on the structure and function of the heart. However, there are available current data providing essential information about the importance and need of further additional data to clarify the molecular mechanisms underlying MetS-induced cardiac dysfunction.

ACKNOWLEDGMENTS

The data presented here is supported through a grant by TUBITAK SBAG-214S254.

ABBREVIATIONS

AF:	Atrial fibrillation
Akt:	Protein kinase B
AP:	Action potential
APD:	AP-duration
BMI:	Body mass index
CVD:	Cardiovascular diseases
DPP-4:	Dipeptidyl peptidase 4
ER stress:	Endoplasmic reticulum stress

FFA:	Free fatty acids
H3K27Me2/H3K27Me3:	Trimethylated histone 2/3 at lysine residue 27
IDDM:	Insulin-dependent *diabetes mellitus* (IDDM)
IL:	Interleukin
IP3R:	Inositol triphosphate receptors
IRE1:	Serine/Threonine-Protein Kinase/Endoribonuclease
IRS1:	Insulin receptor substrate 1
JNK1:	c-Jun N-terminal protein kinase 1
LQTS:	Long QT-syndrome
LVDD:	Left ventricular diastolic dysfunction
LVSD:	Left ventricular systolic dysfunction
MetS:	Metabolic syndrome
NCX:	Na^+/Ca^{2+}-exchanger
NO:	Nitric oxide
ROS:	Reactive oxygen species
RyR2:	Cardiac SR Ca^{2+} release channels
SERCA2a:	Cardiac SR Ca^{2+} ATPase
SR:	Sarcoplasmic reticulum
T1D:	Type 1 diabetes
T2D:	Type 2 diabetes
TNF-β:	Tumor necrosis factor-beta
TNF-α:	Tumor necrosis factor-alpha
WHO:	World Health Organization

REFERENCES

Abdul-Ghani, M. A. and R. A. DeFronzo (2009). "Plasma glucose concentration and prediction of future risk of type 2 diabetes." *Diabetes Care* 32(Suppl 2): S194–S198.

Abed, H. S., C. S. Samuel, et al. (2013). "Obesity results in progressive atrial structural and electrical remodeling: implications for atrial fibrillation." *Heart Rhythm* 10(1): 90–100.

Ahn, M. S., J. Y. Kim, et al. (2010). "Cardiovascular parameters correlated with metabolic syndrome in a rural community cohort of Korea: the ARIRANG study." *J Korean Med Sci* 25(7): 1045–1052.

Aijaz, B., K. A. Ammar, et al. (2008). "Abnormal cardiac structure and function in the metabolic syndrome: a population-based study." *Mayo Clin Proc* 83(12): 1350–1357.

Aksoy, S., G. Durmus, et al. (2014). "Is left ventricular diastolic dysfunction independent from presence of hypertension in metabolic syndrome? An echocardiographic study." *J Cardiol* 64(3): 194–198.

Alam, M. A., K. Kauter, et al. (2013). "Naringin improves diet-induced cardiovascular dysfunction and obesity in high carbohydrate, high fat diet-fed rats." *Nutrients* 5(3): 637–650.

Albarado-Ibanez, A., J. E. Avelino-Cruz, et al. (2013). "Metabolic syndrome remodels electrical activity of the sinoatrial node and produces arrhythmias in rats." *PLoS One* 8(11): c76534.

Allo, S. N., T. M. Lincoln, et al. (1991). "Non-insulin-dependent diabetes-induced defects in cardiac cellular calcium regulation." *Am J Physiol* 260(6 Pt 1): C1165–C1171.

American Heart Association (2009). "FACTS bridging the gap CVD health disparities". https ://www.heart.org/idc/groups/heart-public/@wcm/@hcm.

Aromolaran, A. S. and M. Boutjdir (2017). "Cardiac ion channel regulation in obesity and the metabolic syndrome: relevance to long QT syndrome and atrial fibrillation." *Front Physiol* 8: 431.

Aromolaran, A. S., H. M. Colecraft, et al. (2016). "High-fat diet-dependent modulation of the delayed rectifier K(+) current in adult guinea pig atrial myocytes." *Biochem Biophys Res Commun* 474(3): 554–559.

Aromolaran, A. S., P. Subramanyam, et al. (2014). "LQT1 mutations in KCNQ1 C-terminus assembly domain suppress IKs using different mechanisms." *Cardiovasc Res* 104(3): 501–511.

Ashrafi, R., M. Yon, et al. (2016). "Altered left ventricular ion channel transcriptome in a high-fat-fed rat model of obesity: insight into obesity-induced arrhythmogenesis." *J Obes* 2016: 7127898.

Avila, G., D. S. Osornio-Garduno, et al. (2014). "Functional and structural impact of pirfenidone on the alterations of cardiac disease and diabetes mellitus." *Cell Calcium* 56(5): 428–435.

Axelsen, L. N., K. Calloe, et al. (2015). "Diet-induced pre-diabetes slows cardiac conductance and promotes arrhythmogenesis." *Cardiovasc Diabetol* 14: 87.

Ayaz, M., B. Can, et al. (2002). "Protective effect of selenium treatment on diabetes-induced myocardial structural alterations." *Biol Trace Elem Res* 89(3): 215–226.

Bai, Y., J. Wang, et al. (2007). "Phospholipid lysophosphatidylcholine as a metabolic trigger and HERG as an ionic pathway for extracellular K accumulation and 'short QT syndrome' in acute myocardial ischemia." *Cell Physiol Biochem* 20(5): 417–428.

Balderas-Villalobos, J., T. Molina-Munoz, et al. (2013). "Oxidative stress in cardiomyocytes contributes to decreased SERCA2a activity in rats with metabolic syndrome." *Am J Physiol Heart Circ Physiol* 305(9): H1344–H1353.

Berwick, Z. C., G. M. Dick, et al. (2012). "Heart of the matter: coronary dysfunction in metabolic syndrome." *J Mol Cell Cardiol* 52(4): 848–856.

Bhatti, G. K., S. K. Bhadada, et al. (2016). "Metabolic syndrome and risk of major coronary events among the urban diabetic patients: North Indian Diabetes and Cardiovascular Disease Study-NIDCVD-2." *J Diabetes Complications* 30(1): 72–78.

Biet, M., N. Morin, et al. (2014). "Lasting alterations of the sodium current by short-term hyperlipidemia as a mechanism for initiation of cardiac remodeling." *Am J Physiol Heart Circ Physiol* 306(2): H291–H297.

Boden, G. (1997). "Role of fatty acids in the pathogenesis of insulin resistance and NIDDM." *Diabetes* 46(1): 3–10.

Bonomini, F., L. F. Rodella, et al. (2015). "Metabolic syndrome, aging and involvement of oxidative stress." *Aging and Disease* 6(2): 109–120.

Bosch, R. F., X. Zeng, et al. (1999). "Ionic mechanisms of electrical remodeling in human atrial fibrillation." *Cardiovasc Res* 44(1): 121–131.

Bouchard, R. A. and D. Bose (1991). "Influence of experimental diabetes on sarcoplasmic reticulum function in rat ventricular muscle." *Am J Physiol* 260(2 Pt 2): H341–H354.

Bourgoin, F., H. Bachelard, et al. (2008). "Endothelial and vascular dysfunctions and insulin resistance in rats fed a high-fat, high-sucrose diet." *Am J Physiol Heart Circ Physiol* 295(3): H1044–H1055.

Bremer, A. A. (2012). "Insulin resistance in pediatric disease." *Pediatr Ann* 41(2): e1–e7.

Bremer, A. A., M. Mietus-Snyder, et al. (2012). "Toward a unifying hypothesis of metabolic syndrome." *Pediatrics* 129(3): 557–570.

Brenner, R. R., O. J. Rimoldi, et al. (2003). "Desaturase activities in rat model of insulin resistance induced by a sucrose-rich diet." *Lipids* 38(7): 733–742.

Brownlee, M. (2001). "Biochemistry and molecular cell biology of diabetic complications." *Nature* 414(6865): 813–820.

Brownsey, R. W., A. N. Boone, et al. (1997). "Actions of insulin on the mammalian heart: metabolism, pathology and biochemical mechanisms." *Cardiovasc Res* 34(1): 3–24.

Brundel, B. J., I. C. Van Gelder, et al. (2001). "Alterations in potassium channel gene expression in atria of patients with persistent and paroxysmal atrial fibrillation: differential regulation of protein and mRNA levels for K+ channels." *J Am Coll Cardiol* 37(3): 926–932.

Bugger, H. and E. D. Abel (2008). "Molecular mechanisms for myocardial mitochondrial dysfunction in the metabolic syndrome." *Clin Sci (Lond)* 114(3): 195–210.

Campos-Pena, V., D. Toral-Rios, et al. (2017). "Metabolic syndrome as a risk factor for alzheimer's disease: is Aβ a crucial factor in both pathologies?" *Antioxid Redox Signal* 26(10): 542–560.

Cardoso, C., G. Salles, et al. (2001). "Clinical determinants of increased QT dispersion in patients with diabetes mellitus." *Int J Cardiol* 79(2–3): 253–262.

Chen, Y. H., S. J. Xu, et al. (2003). "KCNQ1 gain-of-function mutation in familial atrial fibrillation." *Science* 299(5604): 251–254.

Choi, K. M., Y. Zhong, et al. (2002). "Defective intracellular Ca(2+) signaling contributes to cardiomyopathy in Type 1 diabetic rats." *Am J Physiol Heart Circ Physiol* 283(4): H1398–H1408.

Christophersen, I. E., M. S. Olesen, et al. (2013). "Genetic variation in KCNA5: impact on the atrial-specific potassium current IKur in patients with lone atrial fibrillation." *Eur Heart J* 34(20): 1517–1525.

Cicek, F. A., Z. Tokcaer-Keskin, et al. (2014). "Di-peptidyl peptidase-4 inhibitor sitagliptin protects vascular function in metabolic syndrome: possible role of epigenetic regulation." *Mol Biol Rep* 41(8): 4853–4863.

Cummings, D. M., S. Henes, et al. (2008). "Insulin resistance status: predicting weight response in overweight children." *Arch Pediatr Adolesc Med* 162(8): 764–768.

Davidoff, A. J., M. M. Mason, et al. (2004). "Sucrose-induced cardiomyocyte dysfunction is both preventable and reversible with clinically relevant treatments." *Am J Physiol Endocrinol Metab* 286(5): E718–E724.

Davidson, M. B. (2003). "Metabolic syndrome/insulin resistance syndrome/pre-diabetes: new section in diabetes care." *Diabetes Care* 26(11): 3179.

de Koning, L., V. S. Malik, et al. (2012). "Sweetened beverage consumption, incident coronary heart disease, and biomarkers of risk in men." *Circulation* 125(14): 1735–1741, S1731.

de Simone, G., V. Palmieri, et al. (2002). "Association of left ventricular hypertrophy with metabolic risk factors: the HyperGEN study." *J Hypertens* 20(2): 323–331.

Despres, J. P. and I. Lemieux (2006). "Abdominal obesity and metabolic syndrome." *Nature* 444(7121): 881–887.

Dinh, W., M. Lankisch, et al. (2011). "Metabolic syndrome with or without diabetes contributes to left ventricular diastolic dysfunction." *Acta Cardiol* 66(2): 167–174.

Durak, A., Y. Olgar, et al. (2017). "Onset of decreased heart work is correlated with increased heart rate and shortened QT interval in high-carbohydrate fed overweight rats." *Can J Physiol Pharmacol* 95(11): 1335–1342.

Dursunoglu, D., H. Evrengul, et al. (2005). "Do female patients with metabolic syndrome have masked left ventricular dysfunction?" *Anadolu Kardiyol Derg* 5(4): 283–288.

Dutta, K., D. A. Podolin, et al. (2001). "Cardiomyocyte dysfunction in sucrose-fed rats is associated with insulin resistance." *Diabetes* 50(5): 1186–1192.

Eckel, J., E. Gerlach-Eskuchen, et al. (1991). "Alpha-adrenoceptor-mediated increase in cytosolic free calcium in isolated cardiac myocytes." *J Mol Cell Cardiol* 23(5): 617–625.

Eckel, R. H., W. W. Barouch, et al. (2002). "Report of the National Heart, Lung, and Blood Institute-National Institute of Diabetes and Digestive and Kidney Diseases Working Group on the pathophysiology of obesity-associated cardiovascular disease." *Circulation* 105(24): 2923–2928.

Eckert, B. and C. D. Agardh (1998). "Hypoglycaemia leads to an increased QT interval in normal men." *Clin Physiol* 18(6): 570–575.

Eren, E., M. Abuhandan, et al. (2014). "Serum paraoxonase/arylesterase activity and oxidative stress status in children with metabolic syndrome." *J Clin Res Pediatr Endocrinol* 6(3): 163–168.

Erturk, M., E. Oner, et al. (2015). "The role of isovolumic acceleration in predicting subclinical right and left ventricular systolic dysfunction in patient with metabolic syndrome." *Anatol J Cardiol* 15(1): 42–49.

Faraj, M., P. J. Havel, et al. (2003). "Plasma acylation-stimulating protein, adiponectin, leptin, and ghrelin before and after weight loss induced by gastric bypass surgery in morbidly obese subjects." *J Clin Endocrinol Metab* 88(4): 1594–1602.

Fauconnier, J., A. Lacampagne, et al. (2005). "Ca2+-dependent reduction of IK1 in rat ventricular cells: a novel paradigm for arrhythmia in heart failure?" *Cardiovasc Res* 68(2): 204–212.

Feng, B., D. Zhang, et al. (2003). "Niemann-Pick C heterozygosity confers resistance to lesional necrosis and macrophage apoptosis in murine atherosclerosis." *Proc Natl Acad Sci USA* 100(18): 10423–10428.

Fernandez-Velasco, M., G. Ruiz-Hurtado, et al. (2007). "TNF-alpha downregulates transient outward potassium current in rat ventricular myocytes through iNOS overexpression and oxidant species generation." *Am J Physiol Heart Circ Physiol* 293(1): H238–H245.

Fontes-Carvalho, R., R. Ladeiras-Lopes, et al. (2015). "Diastolic dysfunction in the diabetic continuum: association with insulin resistance, metabolic syndrome and type 2 diabetes." *Cardiovasc Diabetol* 14: 4.

Ford, E. S., W. H. Giles, et al. (2002). "Prevalence of the metabolic syndrome among US adults: findings from the third National Health and Nutrition Examination Survey." *JAMA* 287(3): 356–359.

Francisqueti, F. V., I. O. Minatel, et al. (2017). "Effect of gamma-oryzanol as therapeutic agent to prevent cardiorenal metabolic syndrome in animals submitted to high sugar-fat diet." *Nutrients* 9(12).

Fung, T. T., V. Malik, et al. (2009). "Sweetened beverage consumption and risk of coronary heart disease in women." *Am J Clin Nutr* 89(4): 1037–1042.

Futh, R., W. Dinh, et al. (2009). "Soluble P-selectin and matrix metalloproteinase 2 levels are elevated in patients with diastolic dysfunction independent of glucose metabolism disorder or coronary artery disease." *Exp Clin Cardiol* 14(3): e76–e79.

Gaita, F., C. Giustetto, et al. (2003). "Short QT syndrome: a familial cause of sudden death." *Circulation* 108(8): 965–970.

Gaita, F., C. Giustetto, et al. (2004). "Short QT syndrome: pharmacological treatment." *J Am Coll Cardiol* 43(8): 1494–1499.

Galassetti, P. (2012). "Inflammation and oxidative stress in obesity, metabolic syndrome, and diabetes." *Exp Diabetes Res* 2012: 943706.

Gallagher, E. J., D. LeRoith, et al. (2008). "The metabolic syndrome—from insulin resistance to obesity and diabetes." *Endocrinol Metab Clin North Am* 37(3): 559–579, vii.

Gallagher, E. J., D. Leroith, et al. (2010). "Insulin resistance in obesity as the underlying cause for the metabolic syndrome." *Mt Sinai J Med* 77(5): 511–523.

Ganguly, P. K., G. N. Pierce, et al. (1983). "Defective sarcoplasmic reticular calcium transport in diabetic cardiomyopathy." *Am J Physiol* 244(6): E528–E535.

Gerber, P. A. and G. A. Rutter (2017). "The role of oxidative stress and hypoxia in pancreatic beta-cell dysfunction in diabetes mellitus." *Antioxid Redox Signal* 26(10): 501–518.

Gonsolin, D., K. Couturier, et al. (2007). "High dietary sucrose triggers hyperinsulinemia, increases myocardial beta-oxidation, reduces glycolytic flux and delays post-ischemic contractile recovery." *Mol Cell Biochem* 295(1–2): 217–228.

Goodman, E., S. R. Daniels, et al. (2007). "Definition of metabolic syndrome." *J Pediatr* 150(4): e36; author reply e36–e37.

Grandi, A. M., A. M. Maresca, et al. (2006). "Metabolic syndrome and morphofunctional characteristics of the left ventricle in clinically hypertensive nondiabetic subjects." *Am J Hypertens* 19(2): 199–205.

Grandinetti, A., D. C. Chow, et al. (2010). "Association of increased QTc interval with the cardiometabolic syndrome." *J Clin Hypertens (Greenwich)* 12(4): 315–320.

Grandy, S. A. and C. Fiset (2009). "Ventricular K+ currents are reduced in mice with elevated levels of serum TNFalpha." *J Mol Cell Cardiol* 47(2): 238–246.

Gregorio, B. M., D. B. De Souza, et al. (2016). "The potential role of antioxidants in metabolic syndrome." *Curr Pharmaceut Des* 22(7): 859–869.

Grundy, S. M. (1999). "Hypertriglyceridemia, insulin resistance, and the metabolic syndrome." *Am J Cardiol* 83(9b): 25f–29f.

Grundy, S. M. (2005). "Metabolic syndrome: therapeutic considerations." *Handb Exp Pharmacol* (170): 107–133.

Grundy, S. M. (2007). "Metabolic syndrome: a multiplex cardiovascular risk factor." *J Clin Endocrinol Metab* 92(2): 399–404.

Guo, J. Z., Y. C. Gong, et al. (2005). "[A clinical intervention study among 463 essential hypertensive patients with metabolic syndrome]." *Zhonghua Xin Xue Guan Bing Za Zhi* 33(2): 132–136.

Gussak, I., P. Brugada, et al. (2000). "Idiopathic short QT interval: a new clinical syndrome?" *Cardiology* 94(2): 99–102.

Haberka, M., M. Lelek, et al. (2017). "Novel combined index of cardiometabolic risk related to periarterial fat improves the clinical prediction for coronary artery disease complexity." *Atherosclerosis* 268: 76–83.

Haim, T. E., W. Wang, et al. (2010). "Palmitate attenuates myocardial contractility through augmentation of repolarizing Kv currents." *J Mol Cell Cardiol* 48(2): 395–405.

Hansen, B. C. (1999). "The metabolic syndrome X." *Ann NY Acad Sci* 892: 1–24.

Hansen, T. K., R. Dall, et al. (2002). "Weight loss increases circulating levels of ghrelin in human obesity." *Clin Endocrinol (Oxf)* 56(2): 203–206.

Harris, M. I. (2000). "Health care and health status and outcomes for patients with type 2 diabetes." *Diabetes Care* 23(6): 754–758.

Hosoi, T., M. Sasaki, et al. (2008). "Endoplasmic reticulum stress induces leptin resistance." *Mol Pharmacol* 74(6): 1610–1619.

Hossain, P., B. Kawar, et al. (2007). "Obesity and diabetes in the developing world—a growing challenge." *N Engl J Med* 356(3): 213–215.

Hotamisligil, G. S. (2006). "Inflammation and metabolic disorders." *Nature* 444(7121): 860–867.

Hotamisligil, G. S. and E. Erbay (2008). "Nutrient sensing and inflammation in metabolic diseases." *Nat Rev Immunol* 8(12): 923–934.

International Diabetes Federation (2011). *IDF DIABETES ATLAS Fifth Edition.* http://www.idf.org/diabetesatlas/5e/the-global-burden

James, P. T., N. Rigby, et al. (2004). "The obesity epidemic, metabolic syndrome and future prevention strategies." *Eur J Cardiovasc Prev Rehabil* 11(1): 3–8.

James, R. W. (2006). "A long and winding road: defining the biological role and clinical importance of paraoxonases." *Clin Chem Lab Med* 44(9): 1052–1059.

Jung, U. J. and M. S. Choi (2014). "Obesity and its metabolic complications: the role of adipokines and the relationship between obesity, inflammation, insulin resistance, dyslipidemia and nonalcoholic fatty liver disease." *Int J Mol Sci* 15(4): 6184–6223.

Kagota, S., K. Maruyama, et al. (2013). "Chronic oxidative-nitrosative stress impairs coronary vasodilation in metabolic syndrome model rats." *Microvasc Res* 88: 70–78.

Kahn, B. B. and J. S. Flier (2000). "Obesity and insulin resistance." *J Clin Invest* 106(4): 473–481.

Kassi, E., P. Pervanidou, et al. (2011). "Metabolic syndrome: definitions and controversies." *BMC Med* 9: 48.

Katzmarzyk, P. T., A. S. Leon, et al. (2003). "Targeting the metabolic syndrome with exercise: evidence from the HERITAGE Family Study." *Med Sci Sports Exerc* 35(10): 1703–1709.

Kelley, D. E., M. Mokan, et al. (1993). "Interaction between glucose and free fatty acid metabolism in human skeletal muscle." *J Clin Invest* 92(1): 91–98.

Killeen, M. J., G. Thomas, et al. (2008). "Effects of potassium channel openers in the isolated perfused hypokalaemic murine heart." *Acta Physiol (Oxf)* 193(1): 25–36.

Kim, J. K., Y. S. Ju, et al. (2013). "High pulse pressure and metabolic syndrome are associated with proteinuria in young adult women." *BMC Nephrol* 14: 45.

Kimber, W., F. Peelman, et al. (2008). "Functional characterization of naturally occurring pathogenic mutations in the human leptin receptor." *Endocrinology* 149(12): 6043–6052.

Kishi, S., A. C. Armstrong, et al. (2014). "Association of obesity in early adulthood and middle age with incipient left ventricular dysfunction and structural remodeling: the CARDIA study (Coronary Artery Risk Development in Young Adults)." *JACC Heart Fail* 2(5): 500–508.

Knudson, J. D., U. D. Dincer, et al. (2005). "Leptin resistance extends to the coronary vasculature in prediabetic dogs and provides a protective adaptation against endothelial dysfunction." *Am J Physiol Heart Circ Physiol* 289(3): H1038–H1046.

Korkmaz, G. G., E. Altinoglu, et al. (2013). "The association of oxidative stress markers with conventional risk factors in the metabolic syndrome." *Metab Clin Exp* 62(6): 828–835.

Korkmaz-Icoz, S., A. Lehner, et al. (2016). "Left ventricular pressure-volume measurements and myocardial gene expression profile in type 2 diabetic Goto-Kakizaki rats." *Am J Physiol Heart Circ Physiol* 311(4): H958–H971.

Lai, L. P., M. J. Su, et al. (1999). "Changes in the mRNA levels of delayed rectifier potassium channels in human atrial fibrillation." *Cardiology* 92(4): 248–255.

Laitinen, P. J., K. M. Brown, et al. (2001). "Mutations of the cardiac ryanodine receptor (RyR2) gene in familial polymorphic ventricular tachycardia." *Circulation* 103(4): 485–490.

Landstedt-Hallin, L., A. Englund, et al. (1999). "Increased QT dispersion during hypoglycaemia in patients with type 2 diabetes mellitus." *J Intern Med* 246(3): 299–307.

Lee, T. C. and E. G. McDonald (2015). "Clinician understanding of cholesterol treatment guidelines." *JAMA* 313(23): 2381–2382.

Lempiainen, P., L. Mykkanen, et al. (1999). "Insulin resistance syndrome predicts coronary heart disease events in elderly nondiabetic men." *Circulation* 100(2): 123–128.

Lieve, K. V., A. O. Verkerk, et al. (2017). "Gain-of-function mutation in SCN5A causes ventricular arrhythmias and early onset atrial fibrillation." *Int J Cardiol* 236: 187–193.

Lima-Leopoldo, A. P., A. S. Leopoldo, et al. (2011). "Myocardial dysfunction and abnormalities in intracellular calcium handling in obese rats." *Arq Bras Cardiol* 97(3): 232–240.

Lima-Leopoldo, A. P., A. S. Leopoldo, et al. (2013). "Influence of long-term obesity on myocardial gene expression." *Arq Bras Cardiol* 100(3): 229–237.

Lima-Leopoldo, A. P., M. M. Sugizaki, et al. (2008). "Obesity induces upregulation of genes involved in myocardial Ca2+ handling." *Braz J Med Biol Res* 41(7): 615–620.

Lin, Y. K., Y. J. Chen, et al. (2010). "Potential atrial arrhythmogenicity of adipocytes: implications for the genesis of atrial fibrillation." *Med Hypotheses* 74(6): 1026–1029.

Lin, Y. K., Y. C. Chen, et al. (2013). "Leptin modulates electrophysiological characteristics and isoproterenol-induced arrhythmogenesis in atrial myocytes." *J Biomed Sci* 20: 94.

Lopez-Jaramillo, P., D. Gomez-Arbelaez, et al. (2014). "The role of leptin/adiponectin ratio in metabolic syndrome and diabetes." *Horm Mol Biol Clin Investig* 18(1): 37–45.

Lopez-Jaramillo, P., J. Lopez-Lopez, et al. (2014). "The goal of blood pressure in the hypertensive patient with diabetes is defined: now the challenge is go from recommendations to practice." *Diabetol Metab Syndr* 6(1): 31.

Lucchesi, B. R., M. Medina, et al. (1972). "The positive inotropic action of insulin in the canine heart." *Eur J Pharmacol* 18(1): 107–115.

Luo, C. H. and Y. Rudy (1991). "A model of the ventricular cardiac action potential. Depolarization, repolarization, and their interaction." *Circ Res* 68(6): 1501–1526.

Luo, T. Y., C. C. Wu, et al. (2004). "Dietary cholesterol affects sympathetic nerve function in rabbit hearts." *J Biomed Sci* 11(3): 339–345.

Maharani, N., M. Kuwabara, et al. (2016). "Hyperuricemia and atrial fibrillation." *Int Heart J* 57(4): 395–399.

Malhotra, A., S. Penpargkul, et al. (1981). "The effect of streptozotocin-induced diabetes in rats on cardiac contractile proteins." *Circ Res* 49(6): 1243–1250.

Malik, V. S., B. M. Popkin, et al. (2010). "Sugar-sweetened beverages and risk of metabolic syndrome and type 2 diabetes: a meta-analysis." *Diabetes Care* 33(11): 2477–2483.

Maltret, A., S. Wiener-Vacher, et al. (2014). "Type 2 short QT syndrome and vestibular dysfunction: mirror of the Jervell and Lange-Nielsen syndrome?" *Int J Cardiol* 171(2): 291–293.

Mancia, G., M. Bombelli, et al. (2010). "Impact of different definitions of the metabolic syndrome on the prevalence of organ damage, cardiometabolic risk and cardiovascular events." *J Hypertens* 28(5): 999–1006.

Mandavia, C. H., A. R. Aroor, et al. (2013). "Molecular and metabolic mechanisms of cardiac dysfunction in diabetes." *Life Sci* 92(11): 601–608.

Marfella, R., C. Di Filippo, et al. (2009). "Myocardial lipid accumulation in patients with pressure-overloaded heart and metabolic syndrome." *J Lipid Res* 50(11): 2314–2323.

Marques, J. L., E. George, et al. (1997). "Altered ventricular repolarization during hypoglycaemia in patients with diabetes." *Diabet Med* 14(8): 648–654.

Matsuzawa, Y., T. Funahashi, et al. (2011). "The concept of metabolic syndrome: contribution of visceral fat accumulation and its molecular mechanism." *J Atheroscler Thromb* 18(8): 629–639.

McLaughlin, T., F. Abbasi, et al. (2004). "Plasma ghrelin concentrations are decreased in insulin-resistant obese adults relative to equally obese insulin-sensitive controls." *J Clin Endocrinol Metab* 89(4): 1630–1635.

Merabet, N., Y. H. Fang, et al. (2015). "Selective heart rate reduction improves metabolic syndrome-related left ventricular diastolic dysfunction." *J Cardiovasc Pharmacol* 66(4): 399–408.

Miklos, Z., P. Kemecsei, et al. (2012). "Early cardiac dysfunction is rescued by upregulation of SERCA2a pump activity in a rat model of metabolic syndrome." *Acta Physiol (Oxf)* 205(3): 381–393.

Miranda, P. J., R. A. DeFronzo, et al. (2005). "Metabolic syndrome: definition, pathophysiology, and mechanisms." *Am Heart J* 149(1): 33–45.

Monti, M., A. Monti, et al. (2014). "Correlation between epicardial fat and cigarette smoking: CT imaging in patients with metabolic syndrome." *Scand Cardiovasc J* 48(5): 317–322.

Mourad, J. J. and S. Le Jeune (2008). "Blood pressure control, risk factors and cardiovascular prognosis in patients with diabetes: 30 years of progress." *J Hypertens Suppl* 26(3): S7–S13.

Mureddu, G. F., R. Greco, et al. (1998). "Relation of insulin resistance to left ventricular hypertrophy and diastolic dysfunction in obesity." *Int J Obes Relat Metab Disord* 22(4): 363–368.

Myoishi, M., H. Hao, et al. (2007). "Increased endoplasmic reticulum stress in atherosclerotic plaques associated with acute coronary syndrome." *Circulation* 116(11): 1226–1233.

Nathan, D. M., J. B. Buse, et al. (2009). "Medical management of hyperglycemia in type 2 diabetes: a consensus algorithm for the initiation and adjustment of therapy: a consensus statement of the American Diabetes Association and the European Association for the Study of Diabetes." *Diabetes Care* 32(1): 193–203.

Nicolini, E., G. Martegani, et al. (2013). "Left ventricular remodeling in patients with metabolic syndrome: influence of gender." *Nutr Metab Cardiovasc Dis* 23(8): 771–775.

Noda, N., H. Hayashi, et al. (1992). "Cytosolic Ca2+ concentration and pH of diabetic rat myocytes during metabolic inhibition." *J Mol Cell Cardiol* 24(4): 435–446.

O'Connell, R. P., H. Musa, et al. (2015). "Free fatty acid effects on the atrial myocardium: membrane ionic currents are remodeled by the disruption of T-tubular architecture." *PLoS One* 10(8): e0133052.

Okatan, E. N., A. T. Durak, et al. (2016). "Electrophysiological basis of metabolic-syndrome-induced cardiac dysfunction." *Can J Physiol Pharmacol* 94(10): 1064–1073.

Okatan, E. N., E. Tuncay, et al. (2015). "Profiling of cardiac beta-adrenoceptor subtypes in the cardiac left ventricle of rats with metabolic syndrome: comparison with streptozotocin-induced diabetic rats." *Can J Physiol Pharmacol* 93(7): 517–525.

Ozawa, K., M. Miyazaki, et al. (2005). "The endoplasmic reticulum chaperone improves insulin resistance in type 2 diabetes." *Diabetes* 54(3): 657–663.

Ozcan, U., Q. Cao, et al. (2004). "Endoplasmic reticulum stress links obesity, insulin action, and type 2 diabetes." *Science* 306(5695): 457–461.

Ozcan, U., E. Yilmaz, et al. (2006). "Chemical chaperones reduce ER stress and restore glucose homeostasis in a mouse model of type 2 diabetes." *Science* 313(5790): 1137–1140.

Pagliassotti, M. J., P. A. Prach, et al. (1996). "Changes in insulin action, triglycerides, and lipid composition during sucrose feeding in rats." *Am J Physiol Regul Integr Comp Physiol* 271(5): R1319–R1326.

Palmieri, V. and J. N. Bella (2006). "Metabolic syndrome and left ventricular structure and functional abnormalities." *Am J Hypertens* 19(2): 206–207.

Paolisso, G., S. De Riu, et al. (1991). "Insulin resistance and hyperinsulinemia in patients with chronic congestive heart failure." *Metab Clin Exp* 40(9): 972–977.

Patel, U. and B. B. Pavri (2009). "Short QT syndrome: a review." *Cardiol Rev* 17(6): 300–303.

Pedram, P. and G. Sun (2014). "Hormonal and dietary characteristics in obese human subjects with and without food addiction." *Nutrients* 7(1): 223–238.

Peters, A. L. (2009). "Patient and treatment perspectives: revisiting the link between type 2 diabetes, weight gain, and cardiovascular risk." *Cleve Clin J Med* 76(Suppl 5): S20–S27.

Phillips, L. K. and J. B. Prins (2008). "The link between abdominal obesity and the metabolic syndrome." *Curr Hypertens Rep* 10(2): 156–164.

Pi-Sunyer, F. X., A. Schweizer, et al. (2007). "Efficacy and tolerability of vildagliptin monotherapy in drug-naive patients with type 2 diabetes." *Diabetes Res Clin Pract* 76(1): 132–138.

Pollock, N. K., V. Bundy, et al. (2012). "Greater fructose consumption is associated with cardiometabolic risk markers and visceral adiposity in adolescents." *J Nutr* 142(2): 251–257.

Pories, W. J., M. S. Swanson, et al. (1995). "Who would have thought it? An operation proves to be the most effective therapy for adult-onset diabetes mellitus." *Ann Surg* 222(3): 339–350; discussion 350–332.

Qatanani, M. and M. A. Lazar (2007). "Mechanisms of obesity-associated insulin resistance: many choices on the menu." *Genes Dev* 21(12): 1443–1455.

Reaven, G. M. (1991). "Insulin resistance and compensatory hyperinsulinemia: role in hypertension, dyslipidemia, and coronary heart disease." *Am Heart J* 121(4 Pt 2): 1283–1288.

Relling, D. P., L. B. Esberg, et al. (2006). "High-fat diet-induced juvenile obesity leads to cardiomyocyte dysfunction and upregulation of Foxo3a transcription factor independent of lipotoxicity and apoptosis." *J Hypertens* 24(3): 549–561.

Roberts, C. K. and K. K. Sindhu (2009). "Oxidative stress and metabolic syndrome." *Life Sci* 84(21–22): 705–712.

Rosati, B., Z. Pan, et al. (2001). "Regulation of KChIP2 potassium channel beta subunit gene expression underlies the gradient of transient outward current in canine and human ventricle." *J Physiol* 533(Pt 1): 119–125.

Rosiak, M., H. Bolinska, et al. (2002). "P wave dispersion and P wave duration on SAECG in predicting atrial fibrillation in patients with acute myocardial infarction." *Ann Noninvasive Electrocardiol* 7(4): 363–368.

Rossi, M., I. Ruiz de Azua, et al. (2015). "CK2 acts as a potent negative regulator of receptor-mediated insulin release in vitro and in vivo." *Proc Natl Acad Sci USA* 112(49): E6818–E6824.

Rossing, P., L. Breum, et al. (2001). "Prolonged QTc interval predicts mortality in patients with Type 1 diabetes mellitus." *Diabet Med* 18(3): 199–205.

Roujeau, C., R. Jockers, et al. (2014). "New pharmacological perspectives for the leptin receptor in the treatment of obesity." *Front Endocrinol (Lausanne)* 5: 167.

Sanguinetti, M. C. and N. K. Jurkiewicz (1990). "Two components of cardiac delayed rectifier K+ current. Differential sensitivity to block by class III antiarrhythmic agents." *J Gen Physiol* 96(1): 195–215.

Savoca, M. R., C. D. Evans, et al. (2004). "The association of caffeinated beverages with blood pressure in adolescents." *Arch Pediatr Adolesc Med* 158(5): 473–477.

Schannwell, C. M., M. Schneppenheim, et al. (2002). "Left ventricular diastolic dysfunction as an early manifestation of diabetic cardiomyopathy." *Cardiology* 98(1–2): 33–39.

Schenk, S., M. Saberi, et al. (2008). "Insulin sensitivity: modulation by nutrients and inflammation." *J Clin Invest* 118(9): 2992–3002.

Schulze, M. B., J. E. Manson, et al. (2004). "Sugar-sweetened beverages, weight gain, and incidence of type 2 diabetes in young and middle-aged women." *JAMA* 292(8): 927–934.

Scott, M. G. and D. W. Coyne (2014). "Should we sweat the small (micro) things?" *Clin Chem* 60(3): 435–437.

Sharma, N., I. C. Okere, et al. (2008). "High-sugar diets increase cardiac dysfunction and mortality in hypertension compared to low-carbohydrate or high-starch diets." *J Hypertens* 26(7): 1402–1410.

Shoelson, S. E., J. Lee, et al. (2006). "Inflammation and insulin resistance." *J Clin Invest* 116(7): 1793–1801.

Sonnenberg, G. E., G. R. Krakower, et al. (2004). "A novel pathway to the manifestations of metabolic syndrome." *Obes Res* 12(2): 180–186.

Sossalla, S., B. Kallmeyer, et al. (2010). "Altered Na(+) currents in atrial fibrillation effects of ranolazine on arrhythmias and contractility in human atrial myocardium." *J Am Coll Cardiol* 55(21): 2330–2342.

Splawski, I., J. Shen, et al. (2000). "Spectrum of mutations in long-QT syndrome genes. KVLQT1, HERG, SCN5A, KCNE1, and KCNE2." *Circulation* 102(10): 1178–1185.

Streja, D. (2004). "Metabolic syndrome and other factors associated with increased risk of diabetes." *Clin Cornerstone* 6(Suppl 3): S14–S29.

Suh, S. and M. K. Lee (2014). "Metabolic syndrome and cardiovascular diseases in Korea." *J Atheroscler Thromb* 21(Suppl 1): S31–S35.

Tehrani, D. M., S. Malik, et al. (2013). "Coronary artery calcium screening in persons with metabolic syndrome and diabetes: implications for prevention." *Metab Syndr Relat Disord* 11(3): 143–148.

Ternacle, J., F. Wan, et al. (2017). "Short-term high-fat diet compromises myocardial function: a radial strain rate imaging study." *Eur Heart J Cardiovasc Imaging* 18(11): 1283–1291.

Tripoli, E., M. Giammanco, et al. (2005). "The phenolic compounds of olive oil: structure, biological activity and beneficial effects on human health." *Nutr Res Rev* 18(1): 98–112.

Tritos, N. A., K. V. Kissinger, et al. (2004). "Association between ghrelin and cardiovascular indexes in healthy obese and lean men." *Clin Endocrinol (Oxf)* 60(1): 60–66.

Tschop, M., C. Weyer, et al. (2001). "Circulating ghrelin levels are decreased in human obesity." *Diabetes* 50(4): 707–709.

Van Wagoner, D. R., A. L. Pond, et al. (1997). "Outward K+ current densities and Kv1.5 expression are reduced in chronic human atrial fibrillation." *Circ Res* 80(6): 772–781.

Vasanji, Z., E. J. Cantor, et al. (2006). "Alterations in cardiac contractile performance and sarcoplasmic reticulum function in sucrose-fed rats is associated with insulin resistance." *Am J Physiol Cell Physiol* 291(4): C772–C780.

Veglio, M., G. Bruno, et al. (2002). "Prevalence of increased QT interval duration and dispersion in type 2 diabetic patients and its relationship with coronary heart disease: a population-based cohort." *J Intern Med* 251(4): 317–324.

Veglio, M., A. Chinaglia, et al. (2004). "QT interval, cardiovascular risk factors and risk of death in diabetes." *J Endocrinol Invest* 27(2): 175–181.

Vendemiale, G., F. Guerrieri, et al. (1995). "Mitochondrial oxidative-phosphorylation and intracellular glutathione compartmentation during rat-liver regeneration." *Hepatology* 21(5): 1450–1454.

Viazzi, F., P. Piscitelli, et al. (2017). "Metabolic syndrome, serum uric acid and renal risk in patients with T2D." *PLoS One* 12(4): e0176058.

Virag, L., N. Jost, et al. (2011). "Analysis of the contribution of I(to) to repolarization in canine ventricular myocardium." *Br J Pharmacol* 164(1): 93–105.

Volek, J. S., M. L. Fernandez, et al. (2008). "Dietary carbohydrate restriction induces a unique metabolic state positively affecting atherogenic dyslipidemia, fatty acid partitioning, and metabolic syndrome." *Prog Lipid Res* 47(5): 307–318.

von Lewinski, D., S. Bruns, et al. (2005). "Insulin causes [Ca2+]i-dependent and [Ca2+]i-independent positive inotropic effects in failing human myocardium." *Circulation* 111(20): 2588–2595.

Wanahita, N., F. H. Messerli, et al. (2008). "Atrial fibrillation and obesity—results of a meta-analysis." *Am Heart J* 155(2): 310–315.

Witteles, R. M. and M. B. Fowler (2008). "Insulin-resistant cardiomyopathy clinical evidence, mechanisms, and treatment options." *J Am Coll Cardiol* 51(2): 93–102.

Wold, L. E., K. Dutta, et al. (2005). "Impaired SERCA function contributes to cardiomyocyte dysfunction in insulin resistant rats." *J Mol Cell Cardiol* 39(2): 297–307.

Wong, N. D. and S. Malik (2005). "C-reactive protein for cardiovascular risk assessment in the metabolic syndrome: response to Kholeif et al." *Diabetes Care* 28(10): 2598–2599.

Workman, A. J., K. A. Kane, et al. (2001). "The contribution of ionic currents to changes in refractoriness of human atrial myocytes associated with chronic atrial fibrillation." *Cardiovasc Res* 52(2): 226–235.

World Health Organization (2017). "Obesity and overweight." Retrieved December, 18 2017 from http://www.who.int/mediacentre/factsheets/fs311/en/.

Wu, C. C., M. J. Su, et al. (1997). "Comparison of aging and hypercholesterolemic effects on the sodium inward currents in cardiac myocytes." *Life Sci* 61(16): 1539–1551.

Xanthakis, V., J. H. Sung, et al. (2015). "Relations between subclinical disease markers and type 2 diabetes, metabolic syndrome, and incident cardiovascular disease: the Jackson Heart Study." *Diabetes Care* 38(6): 1082–1088.

Xi, S., W. Yin, et al. (2004). "A minipig model of high-fat/high-sucrose diet-induced diabetes and atherosclerosis." *Int J Exp Pathol* 85(4): 223–231.

Zalesin, K. C., B. A. Franklin, et al. (2011). "Impact of obesity on cardiovascular disease." *Med Clin North Am* 95(5): 919–937.

Zhang, F., S. Hartnett, et al. (2016). "High fat diet induced alterations of atrial electrical activities in mice." *Am J Cardiovasc Dis* 6(1): 1–9.

Zhang, Y., J. Xiao, et al. (2006). "Restoring depressed HERG K+ channel function as a mechanism for insulin treatment of abnormal QT prolongation and associated arrhythmias in diabetic rabbits." *Am J Physiol Heart Circ Physiol* 291(3): H1446–H1455.

Zhao, Y., Q. Sun, et al. (2016). "Regulation of SCN3B/scn3b by Interleukin 2 (IL-2): IL-2 modulates SCN3B/scn3b transcript expression and increases sodium current in myocardial cells." *BMC Cardiovasc Disord* 16: 1.

7 Caloric Restriction in Obesity and Diabetic Heart Disease

Edith Hochhauser, Maayan Waldman and Michael Arad

CONTENTS

7.1   INTRODUCTION

7.1.1   CARDIOVASCULAR CONSEQUENCES OF DIABETES

The imbalance between energy intake and energy expenditure has led to the increasing prevalence of obesity. Obesity is related to and leads to a series of metabolic derangements such as dyslipidemia, hypertension, insulin resistance and diabetes. These metabolic perturbations tend to be associated with each other and all increase the risk of cardiovascular diseases (CVD). Insulin resistance is one of the most important consequences of obesity and is defined as a reduced response of target tissues to insulin (DeFronzo and Tripathy 2009). The expansion of white adipose tissue (WAT) occurring in obese individuals leads to chronic (Hotamisligil 2006) adipokines secretion and lipids accumulation, all characterized by increased serum levels of proinflammatory cytokines TNFα, IL-6, IL-1β and monocyte chemoattractant protein-1 (MCP1). The reduction in the intrinsic adiponectin signaling is involved in the development of metabolic dysfunction and diabetes (Ouchi, Parker et al. 2011).

Large-scale epidemiologic studies have generally demonstrated that all-cause mortality increases in linear fashion as overweight and obesity increases (Thom and Lean 2017). On average, median survival is reduced by 2–4 years in those maintaining a BMI of 30–35 kg/m^2, and by 8–10 years at a BMI of 40–45 kg/m^2 (Lean, Powrie et al. 1990). This reduced life expectancy is largely caused by cardiovascular disease and some cancers and is further reduced when type II *diabetes mellitus* (T2DM) is present (Lean, Powrie et al. 1990).

Approximately one-third of adults and 20% of teenagers in the US are obese (Fang, Dong et al. 2008; Flegal, Carroll et al. 2010; Ogden, Carroll et al. 2010; Vandanmagsar, Youm et al. 2011), and the estimated global number of adults affected with diabetes is expected to increase from 135 million in 1995 to 300 million by 2025 (King, Aubert et al. 1998). Obesity causes adverse alterations in adipose tissue that predisposes individuals to metabolic dysregulation, insulin resistance and metabolic dysfunction (Sam and Mazzone 2014). Diabetes is a risk factor for atherosclerosis in large arteries (carotids, aorta and femoral arteries) as well as coronary atherosclerosis, increasing the risk of myocardial infarction (MI), stroke, limb loss and renal failure (Iltis, Kober et al. 2005). CVD is considered the leading cause of mortality in individuals with T2DM. Population-based studies have shown that the risk of heart failure is increased 2–3fold by diabetes, especially in individuals with MI (Ansley and Wang 2013), hypertension (Tocci, Sciarretta et al. 2008) and atrial fibrillation (Du, Ninomiya et al. 2009). Diabetes in heart failure (HF) patients predicts poor prognosis independently of coronary artery disease and the left ventricular ejection fraction (MacDonald, Petrie et al. 2008).

The relationship between diabetes and HF is bi-directional. Not only are patients with diabetes at increased risk of HF, but patients with HF having a greater likelihood of developing diabetes. HF predicted the development of T2DM independently of age, gender, family history of diabetes, body mass index (BMI), waist/hip ratio, systolic and diastolic blood pressure and therapy for HF (Amato, Paolisso et al. 1997; Dei Cas, Spigoni et al. 2013). A possible pathophysiologic explanation is that HF increases the propensity to insulin resistance (IR), a condition associated with an increased risk of developing T2DM and CV disease (CVD). HF is associated with marked IR, characterized by both fasting and stimulated hyperinsulinemia (Swan, Anker et al. 1997). It is therefore apparent that diabetes and HF are inter-related and need to be treated comprehensively.

Diabetes promotes adverse myocardial remodeling which is aggravated in the presence of hypertension (Paul 2003; Patel and Mehta 2012). In patients with aortic stenosis, diabetes was associated with increased myocardial fibrosis, increased LV mass and reduced systolic function despite similar aortic value gradients and independent of coexisting coronary artery diseases (Falcao-Pires, Hamdani et al. 2011; Lindman, Arnold et al. 2011). Diabetic cardiomyopathy is defined as a defect in ventricular contractile function occurring independently of CAD and hypertension (Hayat, Patel et al. 2004). The term now includes diastolic dysfunction that is now considered as an early feature of diabetic cardiomyopathy, usually preceding the development of systolic dysfunction (Schannwell, Schneppenheim et al. 2002;

Varga, Giricz et al. 2015). The etiology of diabetic cardiomyopathy includes the two main defects in diabetes such as insulin resistance and hyperglycemia triggering a cascade of specific myocyte abnormalities. There is intrinsic myocyte stiffening, characterized by increased diastolic tension. This is followed by interstitial fibrosis, collagen cross-linking by advanced glycosylation products and eventual systolic dysfunction (Paul 2003). Thus, diabetic myocardial remodeling appears to have significant impact on the evolution of HF, irrespective of the etiology (Eguchi, Kario et al. 2005).

Figure 7.1 illustrates the different pathways contributing to the development of diabetic cardiomyopathy. The known mechanisms contributing to diabetic cardiomyopathy are impaired calcium homeostasis and glucose intolerance, increased activation of the renin-angiotensin system (RAS), elevation of protein kinase C (PKC) signaling, changes in metabolism, mitochondrial dysfunction, fibrosis, oxidative stress, adiposity and increased inflammation (Boudina and Abel 2007; Schilling and Mann 2012). It has been reported that the myocardial defensive effects of preconditioning are exhausted in the presence of chronic *diabetes mellitus* as a result of chronic over-activation of survival pathways such as protein kinase B (AKT) and extracellular receptor kinase (ERK) (Balakumar and Sharma 2012) leading to increased myocardial injury following MI.

Fluctuations in glucose level and uptake contribute to cardiovascular disease by directly modifying proteins, DNA and gene expression. In the case of glucose,

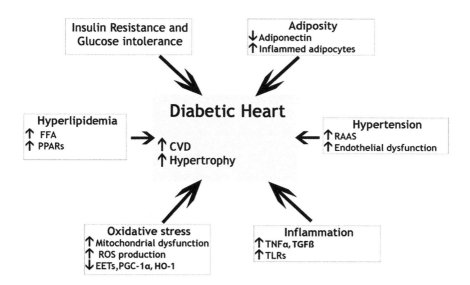

FIGURE 7.1 Pathways and factors leading to the diabetic heart phenotype. Diabetes is characterized by insulin resistance and glucose intolerance that in obese individuals is often accompanied with hyperlipidemia and hypertension due to activation of the RAS. Diabetes may also lead to autonomic nervous dysfunction, increased inflammation, ROS production and increased susceptibility to coronary artery disease. As a result, the diabetic heart suffers from energetic deficiency leading to diastolic dysfunction, cardiac hypertrophy, fibrosis and eventually systolic dysfunction.

clinical studies have shown that increased dietary sugars for healthy individuals or poor glycemic control in diabetic patients further increased CVD risk (Brahma, Pepin et al. 2017). This chapter outlines recent advances in the pathophysiological mechanisms implicated in the development and progression of diabetic cardiomyopathy due to obesity or high glucose levels and current therapeutic strategies and the protective effect of caloric restriction (CR) on the diabetic heart and underlying mechanism.

7.2   METABOLIC AND MOLECULAR CONSEQUENCES OF DIABETES

7.2.1   Oxidative Stress

The heart is capable of metabolizing a wide range of substrates, including fatty acids, glucose, ketone bodies, lactate and amino acids. In the normal heart, insulin stimulates glucose uptake and oxidation. Insulin binding to its surface receptor elicits a signaling cascade that includes activation of PI3K, AKT and PKC. This leads to translocation of glucose transporters to the cardiomyocyte membrane, thus facilitating glucose uptake. Although it increases fatty acid (FA) uptake, it inhibits FA utilization for energy. Insulin resistance in diabetic patients results in impaired cardiac glucose uptake and hyperglycemia, whereas cardiac FA uptake and metabolism increase (Abel, O'Shea et al. 2012). This initial switch towards FA utilization elevates intracellular FA derivatives, such as malondialdehyde (MDA), which activate the peroxisome proliferator-activated receptor α and γ (PPARα/γ) transcription factor. Subsequently, there is an increase in reactive oxygen species (ROS) generation and oxidative damage in the mitochondria from diabetic hearts. Diabetes is characterized by a decreased expression of oxidative phosphorylation genes. Many of these genes are regulated by nuclear respiratory factor 2 (NRF2)-dependent transcription of antioxidant enzymes, such as superoxide dismutase (SOD), catalase, Heme Oxygenase -1 (HO-1), NAD (P) H, quinine oxidoreductase-1 and thioredoxin (Trx-1). The levels of lipid peroxidation and activity of (SOD), glutathione peroxidase (GSH), and catalase, are significantly elevated in tissues and blood in STZ-induced diabetic rats (Kakkar, Kalra et al. 1995). When antioxidant defenses are insufficient to counteract ROS production, the excessive levels of ROS become cytotoxic (Abraham and Kappas 2008; Gao and Mann 2009). HO-1 induction with cobalt protoporphyrin (CoPP), in STZ-diabetic rats enhanced the levels of HO-1 and adiponectin, reduced oxidative stress, and restored eNOS/iNOS expression balance in the heart (Cao, Vecoli et al. 2012; Issan, Kornowski et al. 2014). Resveratrol administration to STZ diabetic myocardium was cardioprotective by the upregulation of Trx-1, NO/HO-1 and VEGF in addition to increased MnSOD activity and reduced blood glucose level (Thirunavukkarasu, Penumathsa et al. 2007).

Heme oxygenase 1 (HO-1) is the inducible enzyme that directly degrades heme in an identical stereospecific manner to biliverdin with the concurrent release of CO and iron (Abraham, Drummond et al. 1996; Abraham, Junge et al. 2016). In mammals, biliverdin is rapidly reduced by biliverdin reductase to bilirubin (Ahmad, Salim et al. 2002; Kapitulnik and Maines 2009). HO-1 and HO-2, the constitutive form of HO, are similar in terms of mechanism, cofactor and substrate requirements.

Both are inhibited by synthetic metalloporphyrins in which the central iron atom is replaced by other elements including tin, zinc, cobalt and chromium (reviewed in Abraham and Kappas 2008). HO-1 expression is regulated by various transcription factors, including NRF2 and a number of intracellular signaling molecules, including the mitogen-activated protein kinase (MAPK), and AKT. HO-1 can be induced by an extraordinarily wide variety of drugs and chemical agents including statins, aspirin, niacin, specific prostaglandins, eicosanoids such as EETs, and free and complexed metals (Abraham, Drummond et al. 1996; Ryter and Choi 2002; Ryter, Otterbein et al. 2002). Iron, bilirubin and CO are the three degradation products of the HO reaction, which have important regulatory functions in cells on their own. Iron is an essential requirement for the synthesis of hemoglobin and ferritin, and HO-1-deficient mice are known to develop anemia (Poss and Tonegawa 1997). HO-1 deficiency leads to reduced stress defense (Poss and Tonegawa 1997) and accelerates the formation of arterial thrombosis (True, Olive et al. 2007). The discovery that HO-1 is a potential target for modulating the inflammatory response (Abraham, Drummond et al. 1996; LaniadoSchwartzman, Abraham et al. 1997; Nicolai, Li et al. 2009) and for diminishing fibrosis (Kie, Kapturczak et al. 2008; Correa-Costa, Semedo et al. 2010) increases interest in HO-1 signaling pathways (Burgess, Li et al. 2010). HO-1 induction in diabetic mice reduces oxidative stress and inflammation following MI and improves heart function. In cardiomyocytes, an increase in HO-1 levels improved mitochondria membrane potential following hypoxia (Issan, Kornowski et al. 2014).

Increased activation of another ROS-generating enzyme, the xanthine oxidase, has also been shown in different organs of diabetic rats or mice (Hotamisligil 2006), including the heart (Rajesh, Mukhopadhyay et al. 2009). Conversely, inhibition of xanthine oxidase (XO) by allopurinol significantly attenuated most pathological features of diabetic cardiomyopathy (e.g. myocardial ROS generation, iNOS expression, cell death and fibrosis) and improved both systolic and diastolic dysfunctions in type 1 diabetic mouse hearts (Rajesh, Mukhopadhyay et al. 2009) or rat hearts (Gao, Xu et al. 2012). In patients with ischemic heart disease, a high dose of allopurinol, an XO inhibitor reduced left ventricular muscle mass and reduced the symptoms of the disease (Rekhraj, Gandy et al. 2013).

A substantial amount of experimental evidence suggests that there is reduced nitric oxide (NO) availability in diabetic tissues. Different mechanisms have been proposed to be responsible for the diabetes-induced dysfunction of NO production, bioavailability and/or signaling. A generally accepted mechanism is the alterations in nitric oxide synthases (NOS), particularly in its endothelial isoform (eNOS), function (Bai and Canto 2012; Johansson, Neovius et al. 2014). The composition of the eNOS complex is critical for the relative formation of NO or superoxide ROS and reactive nitrogen species (RNS). Under pathological conditions oxidative DNA injury may also occur leading to overactivation of the nuclear enzyme poly (ADP-ribose) polymerase 1 (PARP-1), the predominant isoform of the PARP enzyme family, which normally participates in the regulation of DNA repair, cell death, metabolism and inflammatory responses (Bai and Canto 2012). Following binding to damaged DNA, PARP-1 forms homodimers and catalyzes the cleavage of its substrate NAD+ into nicotinamide and ADP-ribose, leading to the formation of long branches of

ADP-ribose polymers on target proteins such as histones and PARP-1. These events result in cellular energetic depletion, mitochondrial dysfunction and ultimately necrosis. Numerous transcription factors involved in controlling inflammation such as NFkB, and various signaling molecules have also been shown to become activated by PARP-1 (Pacher and Szabo 2008). Thus, over-activation of PARP-1 due to ROS/RNS formation not only promotes cell death by ATP and NAD+ depletion, but also stimulates pro-inflammatory mediator production. PARP inhibitors exerted marked tissue protective and anti-inflammatory effects in preclinical models of ischemia-reperfusion injury, endothelial and cardiac dysfunction, circulatory shock, HF and diabetic complications (Pacher and Szabo 2005; Pacher and Szabo 2008). Interestingly, several recent studies have also suggested that PARP-1 and PARP-2 (a minor isoform of the PARP enzyme family) are involved in regulating mitochondrial function/biogenesis and adipogenesis in various organ systems (Bai, Canto et al. 2011a; Bai, Canto et al. 2011; Bai and Canto 2012), including the liver, via the modulation of NAD+ levels and consequent sirtuin 1 activity (Bai and Canto 2012; Mukhopadhyay, Rajesh et al. 2014). Specifically, PARP inhibitors in rodent models of type 1 diabetes were very effective in improving endothelial (Soriano, Pacher et al. 2001) and cardiac function (Pacher, Liaudet et al. 2002). PARP inhibition also prevented the hyperglycemia-induced pathological activation of PKC isoforms, hexosaminase pathway flux and advanced glycosylation end products formation *in vitro*, suggesting its key role in regulating pathological processes promoting the development of all major diabetic complications (Garcia Soriano, Virag et al. 2001).

7.2.2 ENHANCED INFLAMMATORY RESPONSE

Obesity-associated inflammation includes the activation of both the innate and adaptive immune systems (Traba and Sack 2017). Moreover, in contrast to other inflammatory diseases that can be linked to specific organs or locations in the body, obesity results in the infiltration and/or activation of inflammatory cells in various lipid-accumulating organs which in turn result in the production of cytokines and acute-phase reactants that confer systemic effects. The organs most studied include immune activation in adipocytes, the liver, pancreas, muscle and the hypothalamus. Sterile inflammation linked to obesity is mediated in part by the NOD-like receptor family protein 3 (NLRP3) inflammasome pathway (Vandanmagsar, Youm et al. 2011) The activation of this program, as a component of the innate immune system, similarly exacerbates obesity-linked diseases including insulin resistance, diabetes and asthma (Lee, Kim et al. 2013). The biological pathways driving this innate immune program are now well-defined (Sutterwala, Haasken et al. 2014). In the context of obesity, triggers that engage toll-like receptors (TLRs) to initiate transcriptional priming of the NLRP3 inflammasome include: adipose tissue hypertrophy with macrophage infiltration and cytokine secretion; elevated circulating saturated fatty acids; and/or obesity-linked endotoxemia (Osborn and Olefsky 2012). Chronic inflammation is an important pathophysiological factor in the development of diabetes. CVD patients with diabetes and metabolic syndrome demonstrate increased circulating levels of inflammatory cytokines, such as C-reactive protein (CRP), TNFα, and interleukins IL-6 and IL1β (Gao, Belmadani et al. 2007). Increased levels

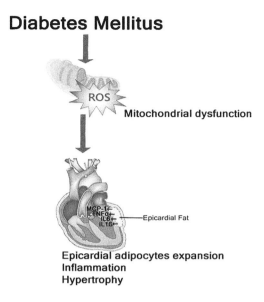

FIGURE 7.2 Illustrates the effect of diabetes on mitochondrial dysfunction through increased mitochondrial ROS production leading to pro-inflammatory cytokines secretion (TNFα, IL1β, MCP-1, IL6) from the epicardial fat as well as the heart, resulting in cardiac hypertrophy, fibrosis and dysfunction.

of TNFα is associated with NF-κB activation signaling, enhancing the expression of pro-oxidant genes and GSH depletion (Rahman, Gilmour et al. 2002; Tripathy, Mohanty et al. 2003; Turkseven, Kruger et al. 2005). Figure 7.2 illustrates the effect of diabetes on mitochondrial dysfunction through increased mitochondrial ROS production leading to pro-inflammatory cytokines secretion (TNFα, IL1β, MCP-1, IL6) from the epicardial fat as well as the heart, resulting in cardiac hypertrophy fibrosis and dysfunction.

The TLR families of receptors are important in the regulation of immune function and inflammation (Trinchieri and Sher 2007). The activation of these receptors in cells of the innate immune system and multiple organs, such as the heart and liver, leads to the production of cytokines and chemokines (Fallach, Shainberg et al. 2010; Hochhauser, Avlas et al. 2013). TLR4 is activated in peripheral monocytes and heart tissue obtained from patients with ischemic heart disease and reduced left ventricular function. Coronary revascularization decreases TLR4 expression (Avlas, Bragg et al. 2015). TLR4 has emerged as a strong candidate for a cellular link between inflammation and insulin resistance. TLR4 is activated by saturated FAs during the hyperlipidemia state associated with obesity and secondary to long-term ingestion of HFD (Shi, Kokoeva et al. 2006; Schaeffler, Gross et al. 2009; Holland, Bikman et al. 2011). In addition, TLR4 expression is primarily upregulated in visceral adipose tissue during insulin resistance (Waller, Huettner et al. 2012). Indeed, its activation has been implicated in the onset of insulin resistance in adipocytes of type 2 diabetic subjects (Song, Kim et al. 2006). Furthermore, humans with TLR4 mutations tend to be protected against developing diabetes (Manolakis, Kapsoritakis et al. 2011).

7.2.3 Enhanced Cardiac Fibrosis

Unlike cardiomyocytes which have, if any, only minimal proliferative capacity beyond the neonatal stage, fibroblasts retain the ability to proliferate and differentiate throughout the human lifespan. Cardiac fibrosis is characterized by excessive production, deposition and contraction of the extracellular matrix (ECM) (Roncarati, Viviani Anselmi et al. 2014). Cardiac fibrosis is the result of chronic arterial hypertension and induces abnormality of cardiac function and arrhythmia (Hasselberg, Edvardsen et al. 2014; Zhang, Zhong et al. 2014). On the other hand, it is accepted that vascular inflammation plays a major role in cardiac fibrosis (Cavalera, Wang et al. 2014). The abnormal proliferation of cardiac fibroblasts and deposition of the ECM proteins and collagens results in the development of cardiac fibrosis, which then adversely affects the performance of the heart (Castaldo, Di Meglio et al. 2013; Hutchinson, Lord et al. 2013). In the heart, pathologic transformation of cardiac fibroblasts (CFs) into activated (profibrogenic) myofibroblasts leads to decreased myocardial compliance, diastolic dysfunction and accompanying HF (Engebretsen, Lunde et al. 2013). Fibroblast collagen synthesis is transcriptionally regulated by fibrogenic growth factors including transforming growth factor β1 (TGF-β1), platelet-derived growth factor (PDGF) and fibroblast growth factor (FGF) (Zou, Jung et al. 2004; Vega-Hernandez, Kovacs et al. 2011; Purnomo, Piccart et al. 2013). The link between an overactive TGF-β cascade and cardiac fibrosis is well-established (Biernacka, Dobaczewski et al. 2011; Dobaczewski, Chen et al. 2011). Fibrogenic actions of TGF-β are primarily mediated through the effects involving Smad-signaling (Dobaczewski, Bujak et al. 2010) although Smad-independent actions have also been implicated. Increased expression of myocardial TGF-β is consistently noted in experimental models of obesity and is associated with cardiac fibrosis (Carroll and Tyagi 2005). Upregulation of TGF-β in obesity-associated cardiomyopathy may be due to angiotensin II signaling (Toblli, Cao et al. 2005), but may also involve angiotensin-independent pathways mediated through the direct stimulatory effects of high glucose and leptin on TGF-β transcription and activation (Ziyadeh, Sharma et al. 1994; Kumpers, Gueler et al. 2007). TGF-β exerts potent pro-hypertrophic actions and promotes a matrix-preserving phenotype in cardiac fibroblasts (Dobaczewski, Chen et al. 2011).

Several pharmacological agents have been specifically designed to inhibit or reverse progression of myocardial fibrosis in HF patients. These therapies primarily act to inhibit the renin–angiotensin–aldosterone system pathway, although agents developed primarily to control blood glucose also have some impact on myocardial fibrosis, as described below.

7.3 MODERN MEDICAL THERAPIES: IMPLICATIONS FOR THE 'DIABETIC' HEART

7.3.1 Pharmacological Approach, Bariatric Surgery, Intragastric Balloon

At present, there is no specific pharmacologic therapy for human diabetic cardiomyopathy (Holscher, Bode et al. 2016). Heart disease is treated according to its clinical phenotype in keeping with evidence-based drug therapies and established cardiology guidelines (Murarka and Movahed 2010; Maisch, Alter et al. 2011). Strict control of cardiovascular risk factors, in particular, hypertension hyperlipidemia and glucose, is a key factor to prevent the progression of coronary disease and to alleviate the hemodynamic load on the diabetic heart.

Lipid-lowering medications: lipid-lowering medications such as statins or PCSK9 inhibitors are used as primary and secondary prevention therapy against atherosclerotic cardiovascular disease. Statins may have a beneficial effect on myocardial fibrosis and inflammation, even in glucose-controlled diabetes (Shin, Min et al. 2017). The newer class of drugs—PCSK9 inhibitors—may be considered in very high risk, statin refractory or intolerant patients. Ironically, both statins and PCSK9 inhibitors slightly increase the prevalence of diabetes, but the benefit from cholesterol reduction greatly exceeds this added risk (Davies, Delfino et al. 2016; Sattar, Preiss et al. 2016).

Several clinical trials of promising pharmaceutical agents for weight loss were published in 2015 (Dixon 2016). In a landmark weight-reduction study with glucagon-like peptide 1 receptor agonist liraglutide, which previously had been primarily used to improve glycemic control, patients lost about 10% of their body weight in a year (Pi-Sunyer, Astrup et al. 2015). Participants receiving liraglutide had greater improvements in glucose metabolism and both weight-specific and generic measures of quality of life, as well as lower cardiovascular risk than patients receiving placebo. The adverse effects of liraglutide, which included gastrointestinal symptoms and an increased risk of symptomatic gallstones, were in line with reported effects of the drug in patients with T2DM (Pi-Sunyer, Astrup et al. 2015). Sodium–glucose co-transporter inhibitors (SGLT2), such as empagliflozin, reduce the risk of cardiovascular events in patients with T2DM, concomitantly leading to sustained weight loss (Zinman, Wanner et al. 2015). Over a median observation time of 3.1 years, empagliflozin reduced cardiovascular mortality, all-cause mortality and hospitalization for HF by 38%, 35% and 32%, respectively. The findings also have implications for managing weight in patients with T2DM, as empagliflozin led to a sustained weight loss of 1.5–2.0 kg over three years. Patients with T2DM who underwent bariatric surgery had sustained weight loss five years after surgery. However, within five years of bariatric surgery, 33–50% of patients again showed signs of diabetes (Mingrone, Panunzi et al. 2015). The first five-year report of health outcomes of bariatric surgery plus conventional medical therapy compared favorably with conventional therapy alone for the management of T2DM (Mingrone, Panunzi et al. 2015). Weight loss is very effective in treating T2DM but unfortunately patients regain weight once their weight loss goal is achieved. While managing obesity, sustaining weight loss in a chronic disease requires the integration of tools that provide excellent short-term weight loss, such as very low energy diets and intragastric balloons, with effective weight maintenance programs. This can include long-term pharmacotherapy and/or intermittent intensification of diets to maintain clinically important weight loss (Johansson, Neovius et al. 2014).

7.4 LIFESTYLE MODIFICATION

There are several modes of therapy for diabetic cardiomyopathy and lifestyle modification comprising caloric restriction and physical activity as the main therapy.

Ultimately, glycemic control over a long period of time ameliorates the hazardous effects of obesity and glucose on atherosclerosis and cardiac muscle (Varga, Giricz et al. 2015; Brahma, Pepin et al. 2017).

First line therapy requires institution of lifestyle modification, specifically the implementation of dietary modification and regular exercise to achieve intentional weight loss. However, HF care guidelines suggest that intentional weight loss is not recommended in HF patients unless the BMI exceed 35–40 kg/m^2 (Riegel, Moser et al. 2009) (i.e. morbid obesity). Physiological studies demonstrated a similar energy intake in HF patients and healthy adults. However, increased energy expenditure in HF patients leads to a negative energy balance associated with increased catecholamines and proinflammatory cytokines levels which may lead to cachexia and increased mortality (Aquilani, Opasich et al. 2003; Curtis, Selter et al. 2005; Lennie 2008; Sandek, Doehner et al. 2009). Otherwise, obesity increases the risk for HF and contributes to its pathophysiology (Kenchaiah, Evans et al. 2002). Thus, there may be a role for encouraging a moderate amount of weight loss in this population. Many different methods have been proposed to incur weight loss. In general, restriction of energy intake and increasing energy expenditure through exercise results in weight loss. A modest 5–10% weight loss improves the cardiovascular risk profile by decreasing hypertension (–9.5/5.3 mmHg) (Miller, Erlinger et al. 2002), dyslipidemias (–9% total cholesterol, –30% triglycerides) (Marckmann, Toubro et al. 1998) and incidence of DM2 (–58%) (Tuomilehto, Lindstrom et al. 2001) and is recommended for obese patients.

7.4.1 CALORIC RESTRICTION (CR)

Caloric restriction (CR) is defined as reducing caloric intake without depriving the individual of essential nutrients, beginning early or in mid-life and sustained over the life span. CR increased longevity and delays or slows the progression of multiple age-related diseases in many, but not all, laboratory animal models (Speakman and Mitchell 2011; Mattison, Roth et al. 2012; Colman, Beasley et al. 2014). Observational studies of individuals who practice long-term CR suggest that it favorably affects chronic disease risk factors and oxidative stress, influencing many transcription factors and genes expression (Weiss and Fontana 2011).

'Calorie restriction' refers to a state in which energy intake in animals or humans is minimized to low-normal levels while adequate intakes of protein and micronutrients are maintained at sufficient levels to avoid malnutrition. CR typically consists of an energy intake that is 30–50% below that which is required to maintain normal body weight and adiposity and thus results in a very lean phenotype. Extensive research over the past seven decades has demonstrated that in animal species ranging from worms to rodents, 30–50% restriction of energy intake without introducing protein or micronutrient malnutrition (i.e. CR) increases life span by 30–50% (Fontana, Partridge et al. 2010). Part of this effect is mediated by preventing or postponing death because of chronic diseases such as cancer (up to 62% reduction in cancer incidence), obesity, type 2 diabetes, as well as autoimmune, cardiovascular, kidney and neurodegenerative diseases (Masoro 2005; Fontana and Klein 2007; Weiss and Fontana 2011). Furthermore, pathological studies have demonstrated that

30% of the rodents undergoing CR die at a very old age without any evidence of lethal pathology (Shimokawa, Higami et al. 1993), suggesting that it is possible to live a long life without overt disease. CR has numerous beneficial effects on the aging cardiovascular system, some of which are likely related to reductions in inflammation and oxidative stress (Weiss and Fontana 2011). In the vasculature, CR appears to protect against endothelial dysfunction and arterial stiffness and attenuates atherogenesis by reducing cardio-metabolic risk factors (Weiss and Fontana 2011) and vascular inflammation. CR attenuates the age-related increase in the expression of vascular adhesion molecules, an effect that coincided with reductions in ROS production (Zou, Jung et al. 2004).

7.4.1.1   CR Intracellular Mode of Action

Animals under CR display a reduction in body weight and fat tissue, reduced body temperature, a state of hunger, and a decrease in leptin levels, reflecting the diminution of white adipose tissue mass. The levels of adiponectin rise under CR concomitant with a large decline in circulating insulin and glucose levels. There are remarkable tissue level changes in the metabolism with a generalized shift from a carbohydrate to a fat metabolism and reduced oxidative stress. Several metabolic regulators/pathways are thought to mediate the CR effect. In particular, the insulin-like growth factor (IGF-1) insulin signaling pathway, AMPK and SIRT1. With CR, adiponectin and phospho-AMPK increase in the heart and they have been involved in cardioprotection (Shinmura, Tamaki et al. 2005; Rohrbach, Aurich et al. 2007; Shinmura, Tamaki et al. 2007; Kondo, Shibata et al. 2009; Edwards, Donato et al. 2010). The beneficial effects in the cardiovascular system of CR appear to be mediated by the prevention of the age-related reductions in NFκB inhibitory complexes, thereby preventing NFκB from entering the nucleus to promote inflammatory gene transcription. CR has also been shown to attenuate age-related increases in prostanoids in both serum (prostaglandin E2) and in the aorta (prostaglandin E2 and thromboxanes A2); this effect appears to be mediated by an attenuation of the age-related increases in cyclooxygenase 2 and cytosolic phospholipase A2, both of which are involved in prostanoid synthesis (Luo, Lin et al. 2008). In the heart, CR attenuates age-related changes in the myocardium (i.e. CR protects against fibrosis, reduces cardiomyocyte apoptosis) and preserves or improves left ventricular diastolic function (Weiss and Fontana 2011). Numerous mechanisms have been suggested to be involved in CR cardio-protection. CR in mice protects the heart from ischemic injury (Levy, Kornowski et al. 2015). A unique metabolic case of spontaneously reduced food intake and cardioprotection is depicted by transgenic αMUPA (Miskin and Masos 1997; Levy, Kornowski et al. 2015). These mice show a set of metabolic changes compared to their control wild type (WT) mice, including reduced food intake when fed at libitum, resistance to obesity, cardioprotection, increased life span and several other benefits seen in calorically restricted mice. The first three changes are caused in αMUPA mice by lifelong increased levels of leptin, an anorectic satiety hormone. In contrast, leptin is strongly reduced after experimentally imposed CR leading to sustained hunger. A lifelong increased ischemic tolerance of αMUPA mice is best demonstrated by the finding that after LAD ligation at 24 months of age, αMUPA mice demonstrated 50% survival after seven ischemic days, while none of the WT mice survived the first ischemic day (Figure 7.3). αMUPA mice also showed increased fractional shortening under ischemic conditions along with reduced infarct size and reduced neutrophil infiltration. Pretreatment with leptin-neutralizing antibodies or with inhibitors for leptin signaling abrogated the αMUPA benefits in the ischemic heart (Figure 7.3). Leptin/JAK2/PI3K/AKT and leptin/JAK2/STAT3 cascades appear to mediate the leptin-induced cardioprotection in αMUPA mice, possibly by activating myocardial survival pathways (Miskin and Masos 1997; Levy, Kornowski et al. 2015). In diabetic mice, CR significantly attenuates the development of cardiomyopathy and improved metabolic markers. We studied the effect of CR on normal and diabetic heart structure and function. Angiotensin was infused to accelerate a cardiomyopathic phenotype in diabetic mice. CR in AT-stressed diabetic mice reduces body weight, improves the metabolic profile and attenuates the cardiomyopathy phenotype (Table 7.1). It attenuates fibrosis and inflammatory cell infiltration (Figure 7.4). CR ameliorates both fibrosis and inflammation, meaning that these markers were eliminated due to the CR regimen. We found that CR is associated with elevated cardiac PPARα which improved the utilization of FAs and a reduction in inflammatory markers, as manifested by reduced levels of TNF-α, TLR2, TLR4 and Fet A, an endogenous ligand of TLR4 and TLR2 (Cohen, Waldman et al. 2017).

7.4.2 SIRTUINS

Sirtuins (SIRTs) are a family of class III histone deacetylases (HDACs), distinguished from other HDAC classes by their requirement for nicotinamide adenine dinucleotide (NAD) in the deacetylation reaction (Mostoslavsky, Chua et al. 2006; Kawahara, Michishita et al. 2009). Seven SIRT homologues exist in the mammalian genome, termed SIRT1 to SIRT7. These proteins regulate a wide range of biological processes including: metabolism (Gerhart-Hines, Rodgers et al. 2007), gene silencing (Vaquero, Scher et al. 2004), aging and lifespan extension (Haigis and Guarente 2006; Bordone, Cohen et al. 2007) and cell survival in response to stress (Outeiro, Kontopoulos et al. 2007; Yuan, Zhang et al. 2007). SIRTs were first discovered in yeast, in which an extra copy of SIR2 was shown to extend the lifespan by 50%, whereas its deletion shortened the lifespan. In addition, SIRT6-deficient mice show the most severe phenotype of all SIRT gene knockouts, with premature aging that includes features of osteoporosis, absence of subcutaneous fat, severe metabolic imbalance, lymphopenia and acute onset hypoglycemia that result in the death of mice within one month (Finkel, Deng et al. 2009). In contrast, transgenic overexpression of SIRT6 protects mice against the adverse consequences of a high-fat diet, which is commonly considered a forerunner of cardiovascular morbidity (Imai and Guarente 2010; Kanfi, Peshti et al. 2010; Zhong, D'Urso et al. 2010; Kanfi, Naiman et al. 2012). Thus, SIRT6 protects against pathological damage caused by hypoxia (Maksin-Matveev, Kanfi et al. 2015) diet-induced obesity and extends lifespan similarly to CR (Bordone, Cohen et al. 2007; Kanfi, Naiman et al. 2012). The protective mechanism stimulated by the overexpression of SIRT6 include the activation of pAMPKα pathway, an increase in protein level of B-cell lymphoma 2 (BCL2), inhibition of nuclear factor kappa-light-chain-enhancer of activated B cells (NFκβ), a decrease in reactive oxygen species (ROS) and reduction in the protein level of

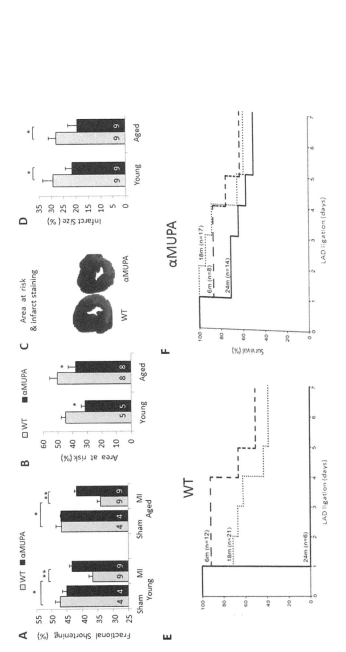

FIGURE 7.3 Functional, histological and survival parameters following MI in αMUPA and WT mice at young and old ages: infarct size examined in the LV area (A–D). Mice were subjected to LAD ligation for 24 h at the ages of six (young) and 18 months and examined for functional and histological parameters as indicated in the Figure. Mice were subjected to LAD ligation for seven days at the ages of six, 18, and 24 months and monitored for survival (E, F). The number of mice in each group is indicated in the Figure. Survival was observed after each day throughout the ischemic period. For each individual mouse, the time of death was plotted against the percentage of mice still alive. Kaplan Meier survival curves are shown. Mouse genotype had a significant effect on survival (p <0.04). The age effect was non-significant (p=0.45) in αMUPA mice, while it was significant (p <0.001) in WT mice. At the youngest age, survival after the entire ischemic period was 50% and 63% in WT and αMUPA mice, respectively (p >0.05). At 18 months, the survival rate was 38% and 59% respectively (p< 0.05). None of the WT mice survived the first ischemic day at 24 months of age while αMUPA mice demonstrated a 50% survival rate after seven ischemic days (p <0.005) (Results taken from Ref. 119). As ischemic tolerance is known to decline with age, these results show that old αMUPA mice exhibit youthful behavior under ischemic conditions. This figure was modified from the previously published manuscript. (From Levy, Kornowski et al. 2015.)

TABLE 7.1
Physiological and Metabolic Markers

	db/db n = 6	*db/db* + AT n = 5	*db/db* + AT+CR n = 5
Body Weight (g)	40.7 ± 9.7	40.3 ± 5.3	33.1 ± 6.7[ab]
Heart Weight (mg)	117 ± 20	163 ± 30[a]	139 ± 20[b]
Glucose (mg/dL)	617 ± 93	658 ± 107	531 ± 127[b]
Cholesterol (mg/dL)	112 ± 21	199 ± 91[a]	118 ± 25[b]
HDL (mg/dL)	112 ± 26	188 ± 74[a]	103 ± 18[b]

[a] $p < 0.05$ vs. db/db, [b]$p < 0.05$ vs.db/db + AT

Results were modified from previously published manuscript (Waldman, Cohen et al. 2018).

protein kinase B (pAkt) during hypoxia. All these pathways preclude necrosis/apoptosis and protect cardiac cells from hypoxic stress and might explain lifespan extension (Maksin-Matveev, Kanfi et al. 2015). SIRT1 is a histone deacetylase which regulates PGC-1α activity. Low insulin levels, such as occur during CR, result in dephosphorylation and deacetylation of FOXO and PPARα, thereby increasing their transcriptional activity (Corton and Brown-Borg 2005; Finck and Kelly 2007).

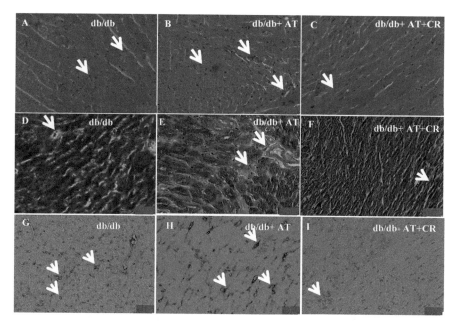

FIGURE 7.4 Diabetic mice (db/db) treated with angiotensin have elevated cardiac fibrosis (D–F) (Masson Trichrome) and cardiac inflammation, as seen by leukocyte infiltration (A–C) using H&E staining. Macrophages infiltration (G–I) using CD68 immunostaining demonstrated that CR reduced the inflammatory response, ($p < 0.05$). Five fields were counted for each slide. CR attenuated fibrosis and cardiac inflammation. This figure was modified from the previously published manuscript. (From Waldman, Cohen et al. 2018.)

7.4.3 PGC-1α

PGC-1α is a potent transcriptional coactivator of nuclear receptors and other transcription factors (King, Aubert et al. 1998; Kleiner, Mepani et al. 2012) that affect metabolic pathways and especially oxidative metabolism, by controlling mitochondrial function and biogenesis. PGC-1 α was first discovered by Bruce Spiegelman's laboratory in a yeast two-hybrid screen designed to discover regulatory proteins that distinguish brown adipose tissue from white adipose tissue (Kondo, Shibata et al. 2009). The constant workload of the heart requires a high capacity mitochondrial system to match ATP production with functional demands. In the adult mammalian heart, ATP synthesis occurs primarily through complete oxidation of fatty acids and glucose in the mitochondrion. In the heart, three major PGC-1α transcription factor partners have been identified. The first cardiac PGC-1α target to be identified was peroxisome proliferatoractivated receptors (PPARs) (Kumpers, Gueler et al. 2007). The major biological role of the PPAR/PGC-1α complex in the myocardium appears to be the transcriptional control of enzymes involved in FA uptake and oxidation (Lagouge, Argmann et al. 2006). PGC1αsignaling plays a role in controlling adipogenesis and mitochondrial function (Puigserver and Spiegelman 2003). Adipose-specific PGC-1α deficiency leads to a decrease in mitochondrial biogenesis and an increase in fatty acid oxidation, glucose, and insulin resistance (Kleiner, Mepani et al. 2012). Induction of PGC-1α is accompanied with a decrease in lipid accumulation and mitochondrial ROS production, increased adiponectin levels and elevation of the expression of genes of the canonical Wnt signaling cascade (Lagouge, Argmann et al. 2006). PGC-1α is activated either by SIRT1 through deacetylation or by AMPK through phosphorylation (Wood, Rogina et al. 2004; Lagouge, Argmann et al. 2006). The main physiological inducers are CR, fasting (Lehman, Barger et al. 2000), exercise (Terjung, Klinkerfuss et al. 1973; Holloszy 1975; Joseph, Pilegaard et al. 2006) and thyroid hormone (Goldenthal, Weiss et al. 2004). Cumulatively, these pathways lead to the metabolic adaptation to glucose and energy deficiency by promoting gluconeogenesis and by increasing FA oxidation (Chalkiadaki and Guarente 2011; Guarente 2013). CR attenuates the development of cardiomyopathy in diabetic mice through SIRT1, HO-1 and PGC-1α. There is a link between SIRT1, PGC-1α and HO-1 signaling in diabetic cardiomyopathy. Targeting PGC-1α and HO-1 may facilitate the development of novel therapies for cardiomyopathy in diabetes (Waldman, Cohen et al. 2018). The postulated mechanisms underlying the effect of CR in diabetes are presented in Figure 7.5. CR results in the NAD+ dependent activation of SIRT1. Either the activation of AMPK or SIRT1 leads to the induction of PGC-1α through phosphorylation and deacetylation, respectively. Increased activity of PGC-1α gene leads to increased expression of HO-1. The activation of the PGC-1α-HO-1 axis improves mitochondrial function, which is impaired in the diabetic heart, and alleviates oxidative stress and inhibits the apoptosis caused by the hyperglycemia.

While the field of pharmaceutical therapies continues to expand, efforts to facilitate weight loss by pharmacological intervention have resulted in rather limited success in protecting the diabetic heart. Rigorous weight and glycemic control along with treatment of concomitant hypertension are very important to try to slow down the progression of the disease. It has become clear that oxidative/nitrosative stress

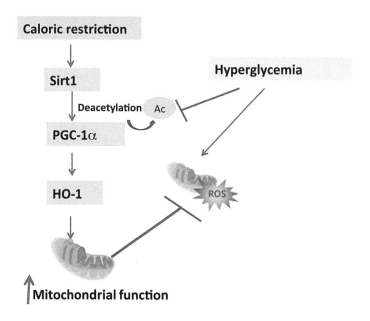

FIGURE 7.5   Postulated mechanisms underlying the effect of CR in diabetes. CR leads to the NAD+ dependent activation of SIRT1. Either the activation of AMPK or SIRT1 lead to the induction of PGC-1α through phosphorylation and deacetylation, respectively. The increased activity of PGC-1α gene leads to augmented expression of HO-1. The activation of the PGC1αHO-1 axis improves mitochondrial function which is impaired in the diabetic heart, alleviates oxidative stress and inhibits apoptosis caused by the hyperglycemia.

and inflammation are central components in triggering and driving the pathological processes associated with diabetic cardiomyopathy. CR has numerous beneficial effects on the aging cardiovascular system, many of which appear to be mediated by reductions in oxidative stress and inflammation in the vasculature and the heart. CR protects against the progressive decline in cardiovascular function and prevents diseases of the cardiovascular system. These latter effects on the cardiovascular system, in combination with other benefits of CR, such as protection against the development of excess adiposity and insulin resistance/type 2 diabetes suggest that CR may be highly beneficial for health, life span and quality of life in humans.

7.5   SUMMARY AND FUTURE PERSPECTIVES

Diabetes mellitus is placing a significant burden on healthcare systems around the globe, largely due to the increased risk of associated cardiovascular disease. These patients suffer from diastolic dysfunction and accelerated atherosclerosis and some develop a distinct form of HF termed diabetic cardiomyopathy that is independent of other comorbidities. Although this area of research is gaining momentum, unraveling the complex molecular and pathophysiological mechanisms that define the disease process is critical in order to develop specific and effective therapies. In particular, the precise mechanisms by which the chronic metabolic disturbances observed in diabetes result in structural and functional changes remain unknown. It is hoped that recent advances in the area of epigenetics, among others, may provide some answers and facilitate drug discovery in the specific context of diabetic cardiomyopathy. Furthermore, the emergence of hypoglycemic agents that confer benefits on cardiovascular outcomes is particularly promising, although mechanistic studies are once again required to elucidate their mechanism(s) of action. Large-scale clinical trials specifically aimed at diabetic patients with cardiovascular disease and HF are required. It is clear that we are entering a new and exciting era in diabetes research that is likely to result in a major clinical impact with regard to improved treatment of cardiovascular complications. Studies with CR identify novel molecular pathways associated with attenuated pathology and improved prognosis in the diabetic heart. Pharmacological agents to activate these pathways and recapitulate the protective effect of CR are being developed to expand the present armamentarium.

ABBREVIATIONS

AKT:	Protein Kinase B
AMPK 5':	AMP-activated Protein Kinase
cGMP:	Cyclic Guanosine Monophosphate
CoPP:	Cobalt Protoporphyrin
CR:	Caloric Restriction
CVD:	Cardiovascular Disease
DM2:	Type 2 Diabetes Mellitus
ECM:	Extracellular Matrix
ERK:	Extracellular Receptor Kinase
FA:	Fatty Acid
FGF:	Fibroblast Growth Factor
GLP-1:	Glucagon Like Peptide
GSH:	Glutathione Peroxidase
HDACs:	Histone Deacetylases
HF:	Heart Failure
HO-1:	Heme Oxygenase
IGF-1:	Insulin-like Growth Factor
IR:	Insulin Resistance
MAPK:	Mitogen-activated Protein Kinase
MDA:	Malondialdehyde
MI:	Myocardial Infarction
NAD (P) H:	Nicotinamide Adenine Dinucleotide Phosphate
NF-κB:	Nuclear Factor Kappa-light-chain-enhancer of Activated B Cells
NLRP3:	Nod-like Receptor Family Protein 3
NO:	Nitric Oxide
eNOS:	Endothelial Nitric Oxide Synthases
NRF2:	Nuclear Respiratory Factor 2
PDGF:	Platelet-Derived Growth Factor
PGC-1α:	Peroxisome Proliferator-Activated Receptor Gamma Coactivator-1α
PKC:	Protein Kinase C

PPARs: Peroxisome Proliferator-Activated Receptors
RAS: Renin-Angiotensin System
RNS: Reactive Nitrogen Species
ROS: Reactive Oxygen Species
SGLT2:   Sodium-Glucose Co-transporter 2
SIRT1:   NAD-dependent Deacetylase Sirtuin-1
SOD: Superoxide Dismutase
TGF-β1:   Transforming Growth Factor β1
TIMP-1:   Metallopeptidase Inhibitor-1
TLR4:   Toll-Like Receptor 4
Trx-1:   Thioredoxin
VEGF: Vascular Endothelial Growth Factor
WAT: White Adipose Tissue
WT: Wild Type
XO: Xanthine Oxidase

REFERENCES

Abel, E. D., K. M. O'Shea, et al. (2012). "Insulin resistance: metabolic mechanisms and consequences in the heart." *Arterioscler Thromb Vasc Biol* 32(9): 2068–2076.

Abraham, N., G. Drummond, et al. (1996). "The biological significance and physiological role of heme oxygenase." *Cell Physiol Biochem* 6(3): 129–168.

Abraham, N. G., J. M. Junge, et al. (2016). "Translational significance of heme oxygenase in obesity and metabolic syndrome." *Trends Pharmacol Sci* 37(1): 17–36.

Abraham, N. G. and A. Kappas (2008). "Pharmacological and clinical aspects of heme oxygenase." *Pharmacol Rev* 60(1): 79–127.

Ahmad, Z., M. Salim, et al. (2002). "Human biliverdin reductase is a leucine zipper-like DNA-binding protein and functions in transcriptional activation of heme oxygenase-1 by oxidative stress." *J Biol Chem* 277(11): 9226–9232.

Amato, L., G. Paolisso, et al. (1997). "Congestive heart failure predicts the development of non-insulin-dependent diabetes mellitus in the elderly. The Osservatorio Geriatrico Regione Campania Group." *Diabetes Metab* 23(3): 213–218.

Ansley, D. M. and B. Wang (2013). "Oxidative stress and myocardial injury in the diabetic heart." *J Pathol* 229(2): 232–241.

Aquilani, R., C. Opasich, et al. (2003). "Is nutritional intake adequate in chronic heart failure patients?" *J Am Coll Cardiol* 42(7): 1218–1223.

Avlas, O., A. Bragg, et al. (2015). "TLR4 expression is associated with left ventricular dysfunction in patients undergoing coronary artery bypass surgery." *PLoS One* 10(6): e0120175.

Bai, P. and C. Canto (2012). "The role of PARP-1 and PARP-2 enzymes in metabolic regulation and disease." *Cell Metab* 16(3): 290–295.

Bai, P., C. Canto, et al. (2011a). "PARP-2 regulates SIRT1 expression and whole-body energy expenditure." *Cell Metab* 13(4): 450–460.

Bai, P., C. Canto, et al. (2011b). "PARP-1 inhibition increases mitochondrial metabolism through SIRT1 activation." *Cell Metab* 13(4): 461–468.

Balakumar, P. and N. K. Sharma (2012). "Healing the diabetic heart: does myocardial preconditioning work?" *Cell Signal* 24(1): 53–59.

Biernacka, A., M. Dobaczewski, et al. (2011). "TGF-beta signaling in fibrosis." *Growth Factors* 29(5): 196–202.

Bordone, L., D. Cohen, et al. (2007). "SIRT1 transgenic mice show phenotypes resembling calorie restriction." *Aging Cell* 6(6): 759–767.

Boudina, S. and E. D. Abel (2007). "Diabetic cardiomyopathy revisited." *Circulation* 115(25): 3213–3223.

Brahma, M. K., M. E. Pepin, et al. (2017). "My sweetheart is broken: role of glucose in diabetic cardiomyopathy." *Diabetes Metab J* 41(1): 1–9.

Burgess, A., M. Li, et al. (2010). "Adipocyte heme oxygenase-1 induction attenuates metabolic syndrome in both male and female obese mice." *Hypertension* 56(6): 1124–1130.

Cao, J., C. Vecoli, et al. (2012). "Cobalt-protoporphyrin improves heart function by blunting oxidative stress and restoring NO synthase equilibrium in an animal model of experimental diabetes." *Front Physiol* 3: 160.

Carroll, J. F. and S. C. Tyagi (2005). "Extracellular matrix remodeling in the heart of the homocysteinemic obese rabbit." *Am J Hypertens* 18(5 Pt 1): 692–698.

Castaldo, C., F. Di Meglio, et al. (2013). "Cardiac fibroblast-derived extracellular matrix (biomatrix) as a model for the studies of cardiac primitive cell biological properties in normal and pathological adult human heart." *Biomed Res Int* 2013: 352370.

Cavalera, M., J. Wang, et al. (2014). "Obesity, metabolic dysfunction, and cardiac fibrosis: pathophysiological pathways, molecular mechanisms, and therapeutic opportunities." *Transl Res* 164(4): 323–335.

Chalkiadaki, A. and L. Guarente (2011). "Metabolic signals regulate SIRT1 expression." *EMBO Rep* 12(10): 985–986.

Cohen, K., M. Waldman, et al. (2017). "Caloric restriction ameliorates cardiomyopathy in animal model of diabetes." *Exp Cell Res* 350(1): 147–153.

Colman, R. J., T. M. Beasley, et al. (2014). "Caloric restriction reduces age-related and all-cause mortality in rhesus monkeys." *Nat Commun* 5: 3557.

Correa-Costa, M., P. Semedo, et al. (2010). "Induction of heme oxygenase-1 can halt and even reverse renal tubule-interstitial fibrosis." *PLoS One* 5(12): e14298.

Corton, J. C. and H. M. Brown-Borg (2005). "Peroxisome proliferator-activated receptor γ coactivator 1 in caloric restriction and other models of longevity." *J Gerontol A Biol Sci Med Sci* 60(12): 1494–1509.

Curtis, J. P., J. G. Selter, et al. (2005). "The obesity paradox: body mass index and outcomes in patients with heart failure." *Arch Intern Med* 165(1): 55–61.

Davies, J. T., S. F. Delfino, et al. (2016). "Current and emerging uses of statins in clinical therapeutics: a review." *Lipid Insights* 9: 13–29.

DeFronzo, R. A. and D. Tripathy (2009). "Skeletal muscle insulin resistance is the primary defect in type 2 diabetes." *Diabetes Care* 32(Suppl 2): S157–S163.

Dei Cas, A., V. Spigoni, et al. (2013). "Diabetes and chronic heart failure: from diabetic cardiomyopathy to therapeutic approach." *Endocr Metab Immune Disord Drug Targets* 13(1): 38–50.

Dixon, J. B. (2016). "Advances in managing obesity." *Nat Rev Endocrinol* 12: 65.

Dobaczewski, M., M. Bujak, et al. (2010). "Smad3 signaling critically regulates fibroblast phenotype and function in healing myocardial infarction." *Circ Res* 107(3): 418–428.

Dobaczewski, M., W. Chen, et al. (2011). "Transforming growth factor (TGF)-β signaling in cardiac remodeling." *J Mol Cell Cardiol* 51(4): 600–606.

Du, X., T. Ninomiya, et al. (2009). "Risks of cardiovascular events and effects of routine blood pressure lowering among patients with type 2 diabetes and atrial fibrillation: results of the ADVANCE study." *Eur Heart J* 30(9): 1128–1135.

Edwards, A. G., A. J. Donato, et al. (2010). "Life-long caloric restriction elicits pronounced protection of the aged myocardium: a role for AMPK." *Mech Ageing Dev* 131(11–12): 739–742.

Eguchi, K., K. Kario, et al. (2005). "Type 2 diabetes is associated with left ventricular concentric remodeling in hypertensive patients." *Am J Hypertens* 18(1): 23–29.

Engebretsen, K. V., I. G. Lunde, et al. (2013). "Lumican is increased in experimental and clinical heart failure, and its production by cardiac fibroblasts is induced by mechanical and proinflammatory stimuli." *FEBS J* 280(10): 2382–2398.

Falcao-Pires, I., N. Hamdani, et al. (2011). "Diabetes mellitus worsens diastolic left ventricular dysfunction in aortic stenosis through altered myocardial structure and cardiomyocyte stiffness." *Circulation* 124(10): 1151–1159.

Fallach, R., A. Shainberg, et al. (2010). "Cardiomyocyte Toll-like receptor 4 is involved in heart dysfunction following septic shock or myocardial ischemia." *J Mol Cell Cardiol* 48(6): 1236–1244.

Fang, C. X., F. Dong, et al. (2008). "Hypertrophic cardiomyopathy in high-fat diet-induced obesity: role of suppression of forkhead transcription factor and atrophy gene transcription." *Am J Physiol Heart Circ Physiol* 295(3): H1206–H1215.

Finck, B. N. and D. P. Kelly (2007). "Peroxisome proliferator-activated receptor γ coactivator-1 (PGC-1) regulatory cascade in cardiac physiology and disease." *Circulation* 115(19): 2540–2548.

Finkel, T., C. X. Deng, et al. (2009). "Recent progress in the biology and physiology of sirtuins." *Nature* 460(7255): 587–591.

Flegal, K. M., M. D. Carroll, et al. (2010). "Prevalence and trends in obesity among US adults, 1999–2008." *JAMA* 303(3): 235–241.

Fontana, L. and S. Klein (2007). "Aging, adiposity, and calorie restriction." *JAMA* 297(9): 986–994.

Fontana, L., L. Partridge, et al. (2010). "Extending healthy life span—from yeast to humans." *Science* 328(5976): 321–326.

Gao, L. and G. E. Mann (2009). "Vascular NAD(P)H oxidase activation in diabetes: a double-edged sword in redox signalling." *Cardiovasc Res* 82(1): 9–20.

Gao, X., S. Belmadani, et al. (2007). "Tumor necrosis factor-α induces endothelial dysfunction in Lepr(db) mice." *Circulation* 115(2): 245–254.

Gao, X., Y. Xu, et al. (2012). "Allopurinol attenuates left ventricular dysfunction in rats with early stages of streptozotocin-induced diabetes." *Diabetes Metab Res Rev* 28(5): 409–417.

Garcia Soriano, F., L. Virag, et al. (2001). "Diabetic endothelial dysfunction: the role of poly(ADP-ribose) polymerase activation." *Nat Med* 7(1): 108–113.

Gerhart-Hines, Z., J. T. Rodgers, et al. (2007). "Metabolic control of muscle mitochondrial function and fatty acid oxidation through SIRT1/PGC-1α." *EMBO J* 26(7): 1913–1923.

Goldenthal, M. J., H. R. Weiss, et al. (2004). "Bioenergetic remodeling of heart mitochondria by thyroid hormone." *Mol Cell Biochem* 265(1–2): 97–106.

Guarente, L. (2013). "Calorie restriction and sirtuins revisited." *Genes Dev* 27(19): 2072–2085.

Haigis, M. C. and L. P. Guarente (2006). "Mammalian sirtuins—emerging roles in physiology, aging, and calorie restriction." *Genes Dev* 20(21): 2913–2921.

Hasselberg, N. E., T. Edvardsen, et al. (2014). "Risk prediction of ventricular arrhythmias and myocardial function in Lamin A/C mutation positive subjects." *Europace* 16(4): 563–571.

Hayat, S. A., B. Patel, et al. (2004). "Diabetic cardiomyopathy: mechanisms, diagnosis and treatment." *Clin Sci (Lond)* 107(6): 539–557.

Hochhauser, E., O. Avlas, et al. (2013). "Bone marrow and nonbone marrow Toll like receptor 4 regulate acute hepatic injury induced by endotoxemia." *PLoS One* 8(8): e73041.

Holland, W. L., B. T. Bikman, et al. (2011). "Lipid-induced insulin resistance mediated by the proinflammatory receptor TLR4 requires saturated fatty acid-induced ceramide biosynthesis in mice." *J Clin Invest* 121(5): 1858–1870.

Holloszy, J. O. (1975). "Adaptation of skeletal muscle to endurance exercise." *Med Sci Sports* 7(3): 155–164.

Holscher, M. E., C. Bode, et al. (2016). "Diabetic cardiomyopathy: does the type of diabetes matter?" *Int J Mol Sci* 17(12).

Hotamisligil, G. S. (2006). "Inflammation and metabolic disorders." *Nature* 444(7121): 860–867.

Hutchinson, K. R., C. K. Lord, et al. (2013). "Cardiac fibroblast-dependent extracellular matrix accumulation is associated with diastolic stiffness in type 2 diabetes." *PLoS One* 8(8): e72080.

Iltis, I., F. Kober, et al. (2005). "Noninvasive characterization of myocardial blood flow in diabetic, hypertensive, and diabetic-hypertensive rats using spin-labeling MRI." *Microcirculation* 12(8): 607–614.

Imai, S. and L. Guarente (2010). "Ten years of NAD-dependent SIR2 family deacetylases: implications for metabolic diseases." *Trends Pharmacol Sci* 31(5): 212–220.

Issan, Y., R. Kornowski, et al. (2014). "Heme oxygenase-1 induction improves cardiac function following myocardial ischemia by reducing oxidative stress." *PLoS One* 9(3): e92246.

Johansson, K., M. Neovius, et al. (2014). "Effects of anti-obesity drugs, diet, and exercise on weight-loss maintenance after a very-low-calorie diet or low-calorie diet: a systematic review and meta-analysis of randomized controlled trials." *Am J Clin Nutr* 99(1): 14–23.

Joseph, A. M., H. Pilegaard, et al. (2006). "Control of gene expression and mitochondrial biogenesis in the muscular adaptation to endurance exercise." *Essays Biochem* 42: 13–29.

Kakkar, R., J. Kalra, et al. (1995). "Lipid peroxidation and activity of antioxidant enzymes in diabetic rats." *Mol Cell Biochem* 151(2): 113–119.

Kanfi, Y., S. Naiman, et al. (2012). "The sirtuin SIRT6 regulates lifespan in male mice." *Nature* 483: 218.

Kanfi, Y., V. Peshti, et al. (2010). "SIRT6 protects against pathological damage caused by diet-induced obesity." *Aging Cell* 9(2): 162–173.

Kapitulnik, J. and M. D. Maines (2009). "Pleiotropic functions of biliverdin reductase: cellular signaling and generation of cytoprotective and cytotoxic bilirubin." *Trends Pharmacol Sci* 30(3): 129–137.

Kawahara, T. L., E. Michishita, et al. (2009). "SIRT6 links histone H3 lysine 9 deacetylation to NF-κB-dependent gene expression and organismal life span." *Cell* 136(1): 62–74.

Kenchaiah, S., J. C. Evans, et al. (2002). "Obesity and the risk of heart failure." *N Engl J Med* 347(5): 305–313.

Kie, J. H., M. H. Kapturczak, et al. (2008). "Heme oxygenase-1 deficiency promotes epithelial-mesenchymal transition and renal fibrosis." *J Am Soc Nephrol* 19(9): 1681–1691.

King, H., R. E. Aubert, et al. (1998). "Global burden of diabetes, 1995–2025: prevalence, numerical estimates, and projections." *Diabetes Care* 21(9): 1414–1431.

Kleiner, S., R. J. Mepani, et al. (2012). "Development of insulin resistance in mice lacking PGC-1α in adipose tissues." *Proc Natl Acad Sci USA* 109(24): 9635–9640.

Kondo, M., R. Shibata, et al. (2009). "Caloric restriction stimulates revascularization in response to ischemia via adiponectin-mediated activation of endothelial nitric-oxide synthase." *J Biol Chem* 284(3): 1718–1724.

Kumpers, P., F. Gueler, et al. (2007). "Leptin is a coactivator of TGF-beta in unilateral ureteral obstructive kidney disease." *Am J Physiol Renal Physiol* 293(4): F1355–F1362.

Lagouge, M., C. Argmann, et al. (2006). "Resveratrol improves mitochondrial function and protects against metabolic disease by activating SIRT1 and PGC-1α." *Cell* 127(6): 1109–1122.

Laniado-Schwartzman, M., N. G. Abraham, et al. (1997). "Heme oxygenase induction with attenuation of experimentally induced corneal inflammation." *Biochem Pharmacol* 53(8): 1069–1075.

Lean, M. E., J. K. Powrie, et al. (1990). "Obesity, weight loss and prognosis in type 2 diabetes." *Diabet Med* 7(3): 228–233.

Lee, H. M., J. J. Kim, et al. (2013). "Upregulated NLRP3 inflammasome activation in patients with type 2 diabetes." *Diabetes* 62(1): 194–204.

Lehman, J. J., P. M. Barger, et al. (2000). "Peroxisome proliferator-activated receptor γ coactivator-1 promotes cardiac mitochondrial biogenesis." *J Clin Invest* 106(7): 847–856.

Lennie, T. A. (2008). "Nutrition self-care in heart failure: state of the science." *J Cardiovasc Nurs* 23(3): 197–204.

Levy, E., R. Kornowski, et al. (2015). "Long-lived alphaMUPA mice show attenuation of cardiac aging and leptin-dependent cardioprotection." *PLoS One* 10(12): e0144593.

Lindman, B. R., S. V. Arnold, et al. (2011). "The adverse impact of diabetes mellitus on left ventricular remodeling and function in patients with severe aortic stenosis." *Circ Heart Fail* 4(3): 286–292.

Luo, S. F., C. C. Lin, et al. (2008). "Involvement of MAPKs, NF-κB and p300 co-activator in IL-1β-induced cytosolic phospholipase A2 expression in canine tracheal smooth muscle cells." *Toxicol Appl Pharmacol* 232(3): 396–407.

MacDonald, M. R., M. C. Petrie, et al. (2008). "Impact of diabetes on outcomes in patients with low and preserved ejection fraction heart failure: an analysis of the Candesartan in Heart failure: assessment of Reduction in Mortality and morbidity (CHARM) programme." *Eur Heart J* 29(11): 1377–1385.

Maisch, B., P. Alter, et al. (2011). "Diabetic cardiomyopathy—fact or fiction?" *Herz* 36(2): 102–115.

Maksin-Matveev, A., Y. Kanfi, et al. (2015). "Sirtuin 6 protects the heart from hypoxic damage." *Exp Cell Res* 330(1): 81–90.

Manolakis, A. C., A. N. Kapsoritakis, et al. (2011). "TLR4 gene polymorphisms: evidence for protection against type 2 diabetes but not for diabetes-associated ischaemic heart disease." *Eur J Endocrinol* 165(2): 261–267.

Marckmann, P., S. Toubro, et al. (1998). "Sustained improvement in blood lipids, coagulation, and fibrinolysis after major weight loss in obese subjects." *Eur J Clin Nutr* 52(5): 329–333.

Masoro, E. J. (2005). "Overview of caloric restriction and ageing." *Mech Ageing Dev* 126(9): 913–922.

Mattison, J. A., G. S. Roth, et al. (2012). "Impact of caloric restriction on health and survival in rhesus monkeys from the NIA study." *Nature* 489(7415): 318–321.

Miller, E. R., 3rd, T. P. Erlinger, et al. (2002). "Results of the Diet, Exercise, and Weight Loss Intervention Trial (DEW-IT)." *Hypertension* 40(5): 612–618.

Mingrone, G., S. Panunzi, et al. (2015). "Bariatric-metabolic surgery versus conventional medical treatment in obese patients with type 2 diabetes: 5 year follow-up of an open-label, single-centre, randomised controlled trial." *Lancet* 386(9997): 964–973.

Miskin, R. and T. Masos (1997). "Transgenic mice overexpressing urokinase-type plasminogen activator in the brain exhibit reduced food consumption, body weight and size, and increased longevity." *J Gerontol A Biol Sci Med Sci* 52(2): B118–B124.

Mostoslavsky, R., K. F. Chua, et al. (2006). "Genomic instability and aging-like phenotype in the absence of mammalian SIRT6." *Cell* 124(2): 315–329.

Mukhopadhyay, P., M. Rajesh, et al. (2014). "Poly (ADP-ribose) polymerase-1 is a key mediator of liver inflammation and fibrosis." *Hepatology* 59(5): 1998–2009.

Murarka, S. and M. R. Movahed (2010). "Diabetic cardiomyopathy." *J Card Fail* 16(12): 971–979.

Nicolai, A., M. Li, et al. (2009). "Heme oxygenase-1 induction remodels adipose tissue and improves insulin sensitivity in obesity-induced diabetic rats." *Hypertension* 53(3): 508–515.

Ogden, C. L., M. D. Carroll, et al. (2010). "Prevalence of high body mass index in US children and adolescents, 2007–2008." *JAMA* 303(3): 242–249.

Osborn, O. and J. M. Olefsky (2012). "The cellular and signaling networks linking the immune system and metabolism in disease." *Nat Med* 18(3): 363–374.

Ouchi, N., J. L. Parker, et al. (2011). "Adipokines in inflammation and metabolic disease." *Nat Rev Immunol* 11(2): 85–97.

Outeiro, T. F., E. Kontopoulos, et al. (2007). "Sirtuin 2 inhibitors rescue alpha-synuclein-mediated toxicity in models of Parkinson's disease." *Science* 317(5837): 516–519.

Pacher, P., L. Liaudet, et al. (2002). "The role of poly(ADP-ribose) polymerase activation in the development of myocardial and endothelial dysfunction in diabetes." *Diabetes* 51(2): 514–521.

Pacher, P. and C. Szabo (2005). "Role of poly(ADP-ribose) polymerase-1 activation in the pathogenesis of diabetic complications: endothelial dysfunction, as a common underlying theme." *Antioxid Redox Signal* 7(11–12): 1568–1580.

Pacher, P. and C. Szabo (2008). "Role of the peroxynitrite-poly(ADP-ribose) polymerase pathway in human disease." *Am J Pathol* 173(1): 2–13.

Patel, B. M. and A. A. Mehta (2012). "Aldosterone and angiotensin: role in diabetes and cardiovascular diseases." *Eur J Pharmacol* 697(1–3): 1–12.

Paul, S. (2003). "Ventricular remodeling." *Crit Care Nurs Clin N Am* 15(4): 407–411.

Pi-Sunyer, X., A. Astrup, et al. (2015). "A randomized, controlled trial of 3.0 mg of liraglutide in weight management." *N Engl J Med* 373(1): 11–22.

Poss, K. D. and S. Tonegawa (1997a). "Heme oxygenase 1 is required for mammalian iron reutilization." *Proc Natl Acad Sci USA* 94(20): 10919–10924.

Poss, K. D. and S. Tonegawa (1997b). "Reduced stress defense in heme oxygenase 1-deficient cells." *Proc Natl Acad Sci USA* 94(20): 10925–10930.

Puigserver, P. and B. M. Spiegelman (2003). "Peroxisome proliferator-activated receptor-gamma coactivator 1 alpha (PGC-1 alpha): transcriptional coactivator and metabolic regulator." *Endocr Rev* 24(1): 78–90.

Purnomo, Y., Y. Piccart, et al. (2013). "Oxidative stress and transforming growth factor-β1-induced cardiac fibrosis." *Cardiovasc Hematol Disord Drug Targets* 13(2): 165–172.

Rahman, I., P. S. Gilmour, et al. (2002). "Oxidative stress and TNF-alpha induce histone acetylation and NF-kappaB/AP-1 activation in alveolar epithelial cells: potential mechanism in gene transcription in lung inflammation." *Mol Cell Biochem* 234–235(1–2): 239–248.

Rajesh, M., P. Mukhopadhyay, et al. (2009). "Xanthine oxidase inhibitor allopurinol attenuates the development of diabetic cardiomyopathy." *J Cell Mol Med* 13(8B): 2330–2341.

Rekhraj, S., S. J. Gandy, et al. (2013). "High-dose allopurinol reduces left ventricular mass in patients with ischemic heart disease." *J Am Coll Cardiol* 61(9): 926–932.

Riegel, B., D. K. Moser, et al. (2009). "State of the science: promoting self-care in persons with heart failure: a scientific statement from the American Heart Association." *Circulation* 120(12): 1141–1163.

Rohrbach, S., A. C. Aurich, et al. (2007). "Age-associated loss in adiponectin-activation by caloric restriction: lack of compensation by enhanced inducibility of adiponectin paralogs CTRP2 and CTRP7." *Mol Cell Endocrinol* 277(1–2): 26–34.

Roncarati, R., C. Viviani Anselmi, et al. (2014). "Circulating miR-29a, among other up-regulated microRNAs, is the only biomarker for both hypertrophy and fibrosis in patients with hypertrophic cardiomyopathy." *J Am Coll Cardiol* 63(9): 920–927.

Ryter, S. W. and A. M. Choi (2002). "Heme oxygenase-1: molecular mechanisms of gene expression in oxygen-related stress." *Antioxid Redox Signal* 4(4): 625–632.

Ryter, S. W., L. E. Otterbein, et al. (2002). "Heme oxygenase/carbon monoxide signaling pathways: regulation and functional significance." *Mol Cell Biochem* 234(1): 249–263.

Sam, S. and T. Mazzone (2014). "Adipose tissue changes in obesity and the impact on metabolic function." *Transl Res* 164(4): 284–292.

Sandek, A., W. Doehner, et al. (2009). "Nutrition in heart failure: an update." *Curr Opin Clin Nutr Metab Care* 12(4): 384–391.

Sattar, N., D. Preiss, et al. (2016). "Lipid-lowering efficacy of the PCSK9 inhibitor evolocumab (AMG 145) in patients with type 2 diabetes: a meta-analysis of individual patient data." *Lancet Diabetes Endocrinol* 4(5): 403–410.

Schaeffler, A., P. Gross, et al. (2009). "Fatty acid-induced induction of Toll-like receptor-4/nuclear factor-kappaB pathway in adipocytes links nutritional signalling with innate immunity." *Immunology* 126(2): 233–245.

Schannwell, C. M., M. Schneppenheim, et al. (2002). "Left ventricular diastolic dysfunction as an early manifestation of diabetic cardiomyopathy." *Cardiology* 98(1–2): 33–39.

Schilling, J. D. and D. L. Mann (2012). "Diabetic cardiomyopathy: bench to bedside." *Heart Fail Clin* 8(4): 619–631.

Shi, H., M. V. Kokoeva, et al. (2006). "TLR4 links innate immunity and fatty acid-induced insulin resistance." *J Clin Invest* 116(11): 3015–3025.

Shimokawa, I., Y. Higami, et al. (1993). "Diet and the suitability of the male Fischer 344 rat as a model for aging research." *J Gerontol* 48(1): B27–B32.

Shin, Y. H., J. J. Min, et al. (2017). "The effect of fluvastatin on cardiac fibrosis and angiotensin-converting enzyme-2 expression in glucose-controlled diabetic rat hearts." *Heart Vessels* 32(5): 618–627.

Shinmura, K., K. Tamaki, et al. (2005). "Short-term caloric restriction improves ischemic tolerance independent of opening of ATP-sensitive K+ channels in both young and aged hearts." *J Mol Cell Cardiol* 39(2): 285–296.

Shinmura, K., K. Tamaki, et al. (2007). "Cardioprotective effects of short-term caloric restriction are mediated by adiponectin via activation of AMP-activated protein kinase." *Circulation* 116(24): 2809–2817.

Song, M. J., K. H. Kim, et al. (2006). "Activation of Toll-like receptor 4 is associated with insulin resistance in adipocytes." *Biochem Biophys Res Commun* 346(3): 739–745.

Soriano, F. G., P. Pacher, et al. (2001). "Rapid reversal of the diabetic endothelial dysfunction by pharmacological inhibition of poly(ADP-ribose) polymerase." *Circ Res* 89(8): 684–691.

Speakman, J. R. and S. E. Mitchell (2011). "Caloric restriction." *Mol Aspects Med* 32(3): 159–221.

Sutterwala, F. S., S. Haasken, et al. (2014). "Mechanism of NLRP3 inflammasome activation." *Ann NY Acad Sci* 1319: 82–95.

Swan, J. W., S. D. Anker, et al. (1997). "Insulin resistance in chronic heart failure: relation to severity and etiology of heart failure." *J Am Coll Cardiol* 30(2): 527–532.

Terjung, R. L., G. H. Klinkerfuss, et al. (1973). "Effect of exhausting exercise on rat heart mitochondria." *Am J Physiol* 225(2): 300–305.

Thirunavukkarasu, M., S. V. Penumathsa, et al. (2007). "Resveratrol alleviates cardiac dysfunction in streptozotocin-induced diabetes: role of nitric oxide, thioredoxin, and heme oxygenase." *Free Radic Biol Med* 43(5): 720–729.

Thom, G. and M. Lean (2017). "Is there an optimal diet for weight management and metabolic health?" *Gastroenterology* 152(7): 1739–1751.

Toblli, J. E., G. Cao, et al. (2005). "Reduced cardiac expression of plasminogen activator inhibitor 1 and transforming growth factor beta1 in obese Zucker rats by perindopril." *Heart* 91(1): 80–86.

Tocci, G., S. Sciarretta, et al. (2008). "Development of heart failure in recent hypertension trials." *J Hypertens* 26(7): 1477–1486.

Traba, J. and M. N. Sack (2017). "The role of caloric load and mitochondrial homeostasis in the regulation of the NLRP3 inflammasome." *Cell Mol Life Sci* 74(10): 1777–1791.

Trinchieri, G. and A. Sher (2007). "Cooperation of Toll-like receptor signals in innate immune defence." *Nat Rev Immunol* 7(3): 179–190.

Tripathy, D., P. Mohanty, et al. (2003). "Elevation of free fatty acids induces inflammation and impairs vascular reactivity in healthy subjects." *Diabetes* 52(12): 2882–2887.

True, A. L., M. Olive, et al. (2007). "Heme oxygenase-1 deficiency accelerates formation of arterial thrombosis through oxidative damage to the endothelium, which is rescued by inhaled carbon monoxide." *Circ Res* 101(9): 893–901.

Tuomilehto, J., J. Lindstrom, et al. (2001). "Prevention of type 2 diabetes mellitus by changes in lifestyle among subjects with impaired glucose tolerance." *N Engl J Med* 344(18): 1343–1350.

Turkseven, S., A. Kruger, et al. (2005). "Antioxidant mechanism of heme oxygenase-1 involves an increase in superoxide dismutase and catalase in experimental diabetes." *Am J Physiol Heart Circ Physiol* 289(2): H701–H707.

Vandanmagsar, B., Y. H. Youm, et al. (2011). "The NLRP3 inflammasome instigates obesity-induced inflammation and insulin resistance." *Nat Med* 17(2): 179–188.

Vaquero, A., M. Scher, et al. (2004). "Human SirT1 interacts with histone H1 and promotes formation of facultative heterochromatin." *Mol Cell* 16(1): 93–105.

Varga, Z. V., Z. Giricz, et al. (2015). "Interplay of oxidative, nitrosative/nitrative stress, inflammation, cell death and autophagy in diabetic cardiomyopathy." *Biochim Biophys Acta* 1852(2): 232–242.

Vega-Hernandez, M., A. Kovacs, et al. (2011). "FGF10/FGFR2b signaling is essential for cardiac fibroblast development and growth of the myocardium." *Development* 138(15): 3331–3340.

Waldman, M., K. Cohen, et al. (2018). "Regulation of diabetic cardiomyopathy by caloric restriction is mediated by intracellular signaling pathways involving 'SIRT1 and PGC-1α'." *Cardiovasc Diabetol* 17(1): 111.

Waller, A. P., L. Huettner, et al. (2012). "Novel link between inflammation and impaired glucose transport during equine insulin resistance." *Vet Immunol Immunopathol* 149(3–4): 208–215.

Weiss, E. P. and L. Fontana (2011). "Caloric restriction: powerful protection for the aging heart and vasculature." *Am J Physiol Heart Circ Physiol* 301(4): H1205–H1219.

Wood, J. G., B. Rogina, et al. (2004). "Sirtuin activators mimic caloric restriction and delay ageing in metazoans." *Nature* 430(7000): 686–689.

Yuan, Z., X. Zhang, et al. (2007). "SIRT1 regulates the function of the Nijmegen breakage syndrome protein." *Mol Cell* 27(1): 149–162.

Zhang, H., H. Zhong, et al. (2014). "Blockade of A2B adenosine receptor reduces left ventricular dysfunction and ventricular arrhythmias 1 week after myocardial infarction in the rat model." *Heart Rhythm* 11(1): 101–109.

Zhong, L., A. D'Urso, et al. (2010). "The histone deacetylase Sirt6 regulates glucose homeostasis via Hif1α." *Cell* 140(2): 280–293.

Zinman, B., C. Wanner, et al. (2015). "Empagliflozin, cardiovascular outcomes, and mortality in type 2 diabetes." *N Engl J Med* 373(22): 2117–2128.

Ziyadeh, F. N., K. Sharma, et al. (1994). "Stimulation of collagen gene expression and protein synthesis in murine mesangial cells by high glucose is mediated by autocrine activation of transforming growth factor-beta." *J Clin Invest* 93(2): 536–542.

Zou, Y., K. J. Jung, et al. (2004). "Alteration of soluble adhesion molecules during aging and their modulation by calorie restriction." *FASEB J* 18(2): 320–322.

8 Personalized Nutrition in Children with Crohn Disease

Andrew S. Day

CONTENTS

8.1 INTRODUCTION

Crohn disease (CD) is one of the main subtypes of conditions known as the inflammatory bowel diseases (IBD) (Day and Brown 2017; Sairenji et al. 2017). Nutrition and nutritional factors are critical elements in the pathogenesis of CD, are often adversely affected in individuals with CD and are increasingly seen as essential components of comprehensive disease management. Although much is known about the epidemiology and pathogenesis of CD, there remain many aspects yet to be clarified and understood. An overview of these aspects is important to provide a framework around which nutritional influences can then be considered.

Nutritional factors contribute to the pathogenesis of CD in various fashions. Although data is continuing to expand, again there remain gaps in current understanding. Furthermore, nutritional factors almost certainly have contributed to changing patterns of the epidemiology of CD both in parts of the world where CD has traditionally been diagnosed more commonly and also in other regions where CD has been increasing in the last decade or so (Bernstein 2017).

Gastrointestinal dysfunction, as evident in active CD, commonly leads to nutritional changes—in children and adolescents this manifests as poor weight gains (or weight loss), impaired linear growth and disruption to typical pubertal patterns. Micronutrient deficiencies, such as with vitamin D, are also commonly seen (Day and Brown 2017; Massironi et al. 2013; Weisshof and Chermesh 2015).

In addition, nutrition is increasingly recognized as an essential specific component of the management of individuals with CD; generic aspects of nutritional monitoring are also very important. Again, these are paramount issues in the management of children or adolescents with CD.

Building from the foundation of these various elements of the interactions between nutrition and CD are novel options to personalize and individualize nutritional aspects, with the potential to greatly enhance outcomes in the short- and longer-term. Although many of the underlying concepts outlined are also relevant to adults diagnosed with CD, this discussion will focus upon children and adolescents, as nutrition and nutritional management are especially important and applicable to this population.

8.2 CROHN DISEASE: AN INFLAMMATORY BOWEL DISEASE

8.2.1 CROHN DISEASE AND ULCERATIVE COLITIS

The conditions known as the inflammatory bowel diseases (IBD) mainly comprise CD and ulcerative colitis (UC) (Day and Brown 2017). These two types of chronic gut inflammation share some features but are distinguished by other specific findings. A hallmark of IBD is active and chronic inflammatory changes present in the gut. Extra-intestinal manifestations, such as skin or joint features, may also occur in both CD and UC.

CD features discontinuous inflammation that may involve any segment of the gastrointestinal tract, with inflammation typically extending through the wall of the bowel (with involvement from epithelium to serosa). The finding of mucosal granulomata is specific to CD. Involvement of the mouth or lips and fistulizing involvement of the peri-anal region are also characteristic to CD.

In contrast, UC typically features superficial inflammatory changes that begin in the rectum and extend in a continuous fashion around a variable length of the colon, without involvement of more proximal segments of the gut. Discrete patterns include very distal inflammation (proctitis) and changes extending from rectum to cecum (pan-colitis).

Whilst it may be convenient to think of these two classifications, some individuals will have clear features of IBD, but not clearly be UC or CD; this is termed IBD

Unclassified (IBDU). Most of those labelled as IBDU at diagnosis will, in time, develop further changes that enable re-classification as clearly CD or UC.

8.2.2 PATHOGENESIS OF IBD

Although the exact cause of CD and UC is not fully established, the current best understanding is that an environmental change triggers the onset of dysregulated immune responses in an individual possessing one or more at risk genes (Ye and McGovern 2016). Environmental factors include a discrete infectious process (such as IBD beginning after an isolated enteric infection like *Campylobacter*) and dietary factors (Ananthakrishnan et al. 2018; Bernstein 2017; Rogler and Vavricka 2015). Complicit also in the development of IBD are changes in the intestinal microbiota: so-called dysbiosis (Imhann et al. 2018; Sartor and Wu 2017). This may include a reduction in the diversity of organisms present or (at least in adults) loss of protective factors, such as *Fusobacterium prausnitzii*. Factors that influence the development of the intestinal microbiota early in life, such as method of feeding or place of residence, are relevant also. In summary, although the precise pathogenesis of IBD is not completely understood, it is clear that the microbiota, the environment, innate immune responses and genetic risk profiles are all involved in the development of IBD.

A number of environmental factors have been investigated to elucidate their potential roles (Ananthakrishnan et al. 2018; Klement et al. 2004). Breastfeeding appears to provide protection against the development of IBD, whilst early and repeated antibiotic exposure increases risk. Rural residence is generally protective compared to urban location. Pet animal ownership appears to be protective as does having a family vegetable garden in the early years of life.

Although more than 200 genes are associated with the development of IBD, most provide only small increases in risk. Mutations in the CARD 15/NOD 2 gene that encodes an intracellular bacterial receptor (Nod 2) are present in up to 30% of Western populations with CD, but are not associated with any risk in Japanese individuals (Ahuja and Tandon 2010; Ye and McGovern 2016). Polymorphisms in the CARD 15/NOD 2 gene are also linked with disease phenotype; ileal involvement is seen more commonly in such individuals.

8.2.3 PRESENTATION AND EPIDEMIOLOGY OF IBD

IBD may have onset at any age, from infancy to the elderly. The highest rates of IBD are seen in the early adult years, between 15 and 35 years of age (Day and Brown 2017; Sairenji et al. 2017). From the time of onset of symptoms, diagnosis may take months or even years, during which time symptoms, nutritional consequences and related morbidities may be ongoing.

Typical symptoms include abdominal pain, diarrhea, hematochezia, anorexia and weight loss. More colonic involvement may lead to more prominent bloody diarrhea. In children and adolescents, weight loss or poor weight gains, impaired linear growth and delayed pubertal development are commonly seen.

During the 20th century, IBD was considered to be predominantly a condition seen in Western countries, such as in Europe, North America and Australasia (Benchimol et al. 2011). Since the turn of the century, however, IBD has been increasingly recognized in the East, with increasing rates of diagnosis in China, Taiwan and Japan for instance (Ahuja and Tandon 2010; Ng et al. 2018). Furthermore, rates of IBD have steadily increased over the last several decades in almost all countries with serial records of incidence. For instance, high rates of IBD (especially CD) were noted in the Canterbury region of New Zealand in 2004 (Gearry et al. 2006). A decade later, almost twice as many people were being diagnosed in the same region, with an increase in the incidence to 26 per 100,000 in 2014 (Su et al. 2016). Furthermore, over a 25-year period, rates of IBD have increased many-fold in children aged less than 16 years in this same region (Lopez et al. 2018). This trend of increasing incidence and particularly of increased rates in younger populations has been observed in many locations.

8.2.4 DIAGNOSIS OF CD

The diagnosis of CD is based primarily upon endoscopic, histologic and radiological findings (Lemberg and Day 2015). Standard serum-based tests indicating an inflammatory response (e.g. elevated C-reactive protein or platelet count) along with increased fecal calprotectin levels (a more specific indicator of gut inflammation) in an individual with relevant symptoms should prompt further investigations to establish the diagnosis and to define the phenotype of disease fully. The Porto criteria guiding the diagnosis of IBD in children specify a requirement to perform upper gastrointestinal endoscopy and ileocolonoscopy (with mucosal biopsies) and (in most cases) an assessment of the small bowel (IBD Working Group 2005). Magnetic resonance enterography would be preferred in most centers as it provides high-quality cross-sectional imaging of the small bowel without radiation exposure.

8.3 RELEVANCE OF NUTRITION AND DIETARY FACTORS IN THE DEVELOPMENT OF IBD

As mentioned above, a number of environmental factors may contribute to the pathogenesis of IBD; these include various dietary factors. Breastfeeding is considered protective against the development of IBD. This is most likely secondary to the key interactions between breast milk exposure and early development of the intestinal microbiota.

Westernization of the diet, with increasing reliance upon processing and high food miles, has been attributed to be one of the factors implicated in the recent increasing incidence of IBD. There are likely multiple aspects to these observations, with complex interactions between diet, nutrition, the intestinal microbiota and the gut. Farming practices, the use of preservatives and food additives, changing biodiversity of crops and processing methodologies may all contribute.

Western diet patterns typically have high fat and high carbohydrate content. In animal models, such high fat/high sugar diets are associated with changes in the intestinal microbiota and contribute to the development of gut inflammation. In a murine model, this diet change leads to dysbiosis, defined as a reduction in beneficial organisms and an increase in proinflammatory species (Agus et al. 2016).

Consequent to these changes, a marked reduction in fecal short-chain fatty acid (SCFA) levels was demonstrated. These investigators also showed reduced expression of a specific receptor for SCFA.

The same group also showed, in a separate set of experiments, that a high fat/high sugar diet also led to a change in barrier function in the treated animals (Martinez-Medina et al. 2014). The mucus layer was thinner, intestinal permeability was increased, and proinflammatory responses were demonstrated. Furthermore, these changes resulted in higher rates of colonization of specific gram-negative bacteria.

A separate group has also shown that a high fat diet resulted in increased ileal inflammation and a change in barrier function in a murine model of ileitis (Gruber et al. 2013). In the diet-treated mice, ileal inflammation was increased, whilst expression of occludin (a tight junction protein) was reduced, along with additional epithelial changes evident.

A Danish case-control study evaluated dietary and environmental factors related to the risk of developing IBD (Hansen et al. 2011). Two hundred and sixty patients with IBD were matched to controls without IBD. In this evaluation breastfeeding was protective, and high sugar intake and low fiber intakes were associated with increased risk of developing IBD. Furthermore, daily consumption of fruit, vegetables and wholemeal breads were associated with a reduced risk of developing IBD. A report of dietary and environmental factors in a large case-control study involving Asian and Australian patients concluded that daily tea consumption and prior breastfeeding for more than 12 months were protective for both CD and UC (Ng et al. 2015).

Numerous studies have evaluated the impact of breastfeeding. Meta-analysis of the evidence arising from these reports concludes that breastfeeding is protective (Klement et al. 2004).

As noted above, a number of the important dietary factors relate to the early years of life; these and other dietary exposures likely contribute to altered risk of IBD through modulation of the intestinal microbiota and altered interactions between the host and the microbiota. Further, these dietary factors comprise just some of the identified environmental factors, and complex interactions are likely between these dietary and non-dietary aspects.

Another diet element that has been considered is microparticles (Powell et al. 2007; Zhao et al. 2014). These sub-micron particles are present in many processed foods and comprise excipients or food additives. One of these is titanium dioxide. Butler et al. (2007) evaluated the impact of microparticles upon macrophage function and host/bacterial interactions. Although macrophages did not appear to be affected directly, the microparticles did act as adjuvants that aggravated bacterial responses. A recent report indicated that the administration of titanium dioxide particles to mice lead to worsening of colitis (Ruiz et al. 2017). These authors also showed detectable levels of titanium in the blood of patients with UC.

Two other dietary additives are carrageenan and carboxymethyl cellulose (Martino et al. 2017). In various animal studies, these additives lead to inflammatory changes, induce dysbiosis and trigger changes in intestinal barrier function. *In vitro* and human studies support these animal data. Together, this evidence suggests that these additives are further potentially implicated in the development of IBD.

8.4   NUTRITIONAL IMPACTS OF CD IN CHILDREN

Typical presentation of CD in children and adolescents includes weight loss, or a reduction in weight gain (Moeeni and Day 2011). This is predominantly triggered by early satiety, food-induced pain or other gut symptoms and mediated (in part) by the anorexic effects of several of the key pro-inflammatory cytokines produced in the inflamed gut.

In addition to effects upon weight gain, many children may also have impaired linear growth (Gasparetto and Guariso 2014; Walters and Griffiths 2009). This again is predominantly mediated by the effects of pro-inflammatory cytokines (especially interleukin 6) upon growth plates and upon the production of insulin-like growth factor 1. Caloric restriction (malnutrition) may also contribute to this adverse effect.

The effects upon linear growth may be particularly prominent in adolescents entering or going through their pubertal years. Interruption to this normal developmental phase may lead to ultimate impaired adult height acquisition, as well as the adverse social and psychological impacts secondary to delayed puberty.

In addition to the altered anthropometry commonly evident in active CD, many children with this condition will also have micronutrient deficiencies (Day and Brown 2017; Lemberg and Day 2015; Martino et al. 2017). Iron and vitamin D deficiency are most common, but deficiencies of other vitamins, folate or zinc can also be seen.

8.5   NUTRITIONAL THERAPY FOR CHILDREN WITH ACTIVE CD

The most common nutritional intervention in children with active CD is exclusive enteral nutrition (EEN) (Critch et al. 2012; Otley et al. 2010). This therapeutic approach involves the administration of a liquid formula with the exclusion of normal solid foods for a period of time—typically eight weeks. This therapy leads to high rates of remission with minimal side-effects and a multitude of other benefits as well. These include nutritional corrections, improved bone health and correction of micronutrient deficiencies. In addition to high rates of induction of clinical and biochemical remission, EEN also leads to mucosal healing (MH). In an Italian study, EEN resulted in MH in 75% of a group of children, contrasting with a MH rate of 33% in a comparative group treated with corticosteroids (Berni Canani et al. 2006).

Consequent to the numerous benefits, EEN is now accepted as the primary therapy to induce remission in children diagnosed with CD, and endorsed in several clinical guidelines in this fashion (Critch et al. 2012; Ruemmele et al. 2014). EEN may also have a role in the reinduction of remission in individuals with established CD who are suffering an exacerbation.

Although partial enteral nutrition has a lower response rate compared to EEN in the induction of remission of active disease, ongoing maintenance enteral nutrition in addition to normal diet has been shown to assist in the prevention of relapse, as well as various nutritional benefits (Johnson et al. 2006; Nakahigashi et al. 2016).

EEN may exert its benefits in various ways (Nahidi et al. 2014). Although the standard commercial products do not include the various additives mentioned earlier, such as carrageenan, it is not clear how important avoidance of these is within a short time period. EEN has been demonstrated to lead to significant changes in the intestinal microbiota, with clear patterns that are maintained for a limited time after the end of EEN (Gatti et al. 2017). EEN also has been shown to lead to direct anti-inflammatory effects with suppression of intracellular signal transduction pathways, for example (Alhagamhmad et al. 2017). Furthermore, the enteral formulae used for EEN enhance barrier function, through redistribution and upregulation of tight junction proteins (Nahidi et al. 2013).

8.5.1 INDUCTION OF REMISSION WITH EXCLUSIVE ENTERAL NUTRITION

EEN is considered the most appropriate therapy for the induction of remission in all children diagnosed with CD, regardless of age and disease phenotype (Critch et al. 2012; Ruemmele et al. 2014). Although traditionally considered for luminal CD, EEN may be beneficial in children with peri-oral CD, peri-anal CD and even as a component in the management of fistulizing/perforating CD (Day and Brown 2017).

Typically, EEN would involve the administration of a polymeric formula for a period of six to eight weeks as the sole nutritional source (Critch et al. 2012). Most children in our experience would be able to drink sufficient volume orally; a small number of children will require placement of a naso-gastric tube to facilitate the required volumes.

Prior to commencing EEN, preparation should include time to outline the rationale and requirements of this therapy along with adequate time to clarify or answer questions. During EEN, close support and regular follow-up should be instituted. Local practice includes clinical review every two weeks, with close phone contact from members of the multidisciplinary IBD team between visits. At review visits, adherence to EEN should be assessed, issues or concerns resolved and response to therapy assessed.

Monitoring during EEN should include repeat inflammatory markers (assuming that were elevated at diagnosis) to confirm improvements, serial measurement of patient's weight and ascertainment of clinical improvements (resolution in symptoms). Follow-up measurement of fecal calprotectin may also be helpful.

At the completion of the treatment course of EEN (typically eight weeks), normal diet should then be introduced following a standardized protocol, with a view to progressive readjustment to normal diet. General dietary advice and recommendations should be given at that time, with subsequent further clarification and reiteration of these.

In addition, after the completion of a course of EEN, ongoing daily formula intake in addition to normal diet (maintenance enteral nutrition: MEN) should be considered. This strategy may assist in the maintenance of remission and also provide specific nutritional benefits (Nakahigashi et al. 2016). Japanese data also suggests that ongoing MEN also protects from post-operative recurrence after surgically induced remission (resectional surgery) (Yamamoto et al. 2013). Other reports indicate that ongoing nutritional support may also prevent secondary loss of efficacy of biologic therapies (Hirai et al. 2013).

Other nutritional interventions that may have roles in the management of CD are the specific carbohydrate diet (SCD) and the CD exclusion diet (CDED). Although the SCD was developed more than five decades ago as a way to manage coeliac disease, in the last decade or so there has been increasing interest in its potential role in the management of IBD, especially in children. A retrospective assessment of the outcomes from SCD in seven children with CD showed that all the children had resolution of symptoms within three months of starting the diet (Suskind et al. 2014). Along with the clinical response, standard markers of inflammation were also noted to normalize or improve over this time period. Similar improvements were demonstrated in a second cohort of patients—these data were, however, collected prospectively (Cohen et al. 2014). Further, another retrospective assessment including 11 children with SCD who started the diet at diagnosis or at the time of a disease exacerbation (Burgis et al. 2016). During the period of having a strict SCD, most of this group of children had enhanced growth parameters and improved inflammatory markers.

Suskind and colleagues (2016) conducted an online survey of patients and parent experiences with SCD. The investigators received 417 responses to their survey, with almost half of the individuals having CD. These respondents reported their perception of clinical improvement with SCD, with more than half reporting a response within the first three months of therapy. This data was likely biased towards those respondents who were more positive towards the therapy, and was uncontrolled.

More recently, a group of investigators from two North American centers has reported their prospective assessment of the SCD in 12 patients with CD or UC (Suskind et al. 2018). Overall, disease activity scores fell, as did serum markers of inflammation (CRP). Analysis of the patterns of the fecal microbiota in serial stool samples demonstrated marked changes consequent to the SCD intervention.

The SCD involves limitations in intake, with various foods allowed and disallowed. The inclusion of disallowed foods along with the SCD, resulting in the so-called modified SCD, was assessed in a small group of 11 subjects (Wahbeh et al. 2017). This report evaluated mucosal healing rates with follow-up endoscopic assessments. Although patients had normalization of serum-based markers of inflammation and clinical improvements, none of the children taking a modified SCD had complete mucosal healing.

Overall the data available from recent focused studies of the role of the SCD in children provide some support for this intervention. Most of the data has derived from two discrete centers, so additional reports from other centers would provide further support. In addition, the reports to date have included relatively small cohorts of patients. Furthermore, it appears that the SCD does need to be maintained strictly to ensure full benefits. Data is not yet available to be able to provide patient-specific prognostication or to establish the longer term impacts of this intervention.

The other major dietary intervention that has been considered for the management of CD is the CDED. This diet was developed by Israeli investigators and incorporates the key concepts of excluding various foods or food groups that may be associated with inflammation, affect the microbiota or alter intestinal permeability. Although a full list of foods has not been published or made available, the diet involves the exclusion of diary and gluten and the avoidance of processed foods. Two reports of the responses to the CDED have been published to date, but further evaluations are ongoing.

In the initial report of the use of the CDED, 47 individuals were managed with the CDED in combination with an enteral formula for a period of six weeks (Sigall-Boneh et al. 2014). Almost 80% of the subjects had a response, and remission (judged by disease activity scores) was seen in 70%. The mean pediatric CD Activity Index fell overall (p<0.001) and CRP levels fell (in 70% of those in remission). Further, a subset of 7 patients utilized the CDED without concurrent enteral formula. This report was uncontrolled and conducted in multiple centers in Israel. The long-term outcomes of this intervention have not been evaluated.

More recently, this team of investigators has evaluated the role of CDED in the management of individuals who have had secondary loss of response to a biologic therapy (Sigall Boneh et al. 2017). Twenty-one subjects (ten children) were assessed prospectively to receive CDED with supplementary enteral nutrition for 12 weeks. Some subjects received an initial two weeks of EEN prior to starting CDED. Clinical remission (assessed by the physician global assessment and the Harvey Bradshaw index) was seen in 13 of the 21 subjects. This study did not contain a control comparative group and was conducted for just 12 weeks of observation. Nonetheless, these data suggest that the CDED may be of relevance in this setting where even biologic therapies have failed.

At present the data for the SCD and for the CDED provide support for these interventions. Further data, including prospective controlled studies and assessments over longer periods of time are required to be able to establish the role(s) of these interventions.

8.6 GENERAL ASPECTS OF NUTRITIONAL MANAGEMENT IN CHILDREN WITH IBD

8.6.1 OVERALL NUTRITIONAL CARE

General nutritional management principles should include full nutritional assessment and anthropometry at the time of diagnosis in all children with CD. Anthropometry of current status should be augmented with all available information about historical growth patterns and trajectory. Parental height measurement and calculation of mid-parental height may guide assessment of adequacy of the child's current height. Estimation of bone age and assessment of Tanner staging are also relevant.

Diagnostic baseline nutritional testing to screen for micronutrient deficiency should include iron studies and ferritin, vitamin D, vitamin B12 and folate. Assessment of zinc, vitamins A, vitamin E, vitamin C, selenium and magnesium also may be indicated. Micronutrient deficiencies should be corrected with follow-up measurement to confirm response. Measurement of insulin-like growth factor 1 (IGF-1) may be helpful if there is significant linear growth impairment at diagnosis.

Subsequent to diagnosis, serial measurement of anthropometry is essential, along with regular dietetic review. Measurement of micronutrient levels may be considered as part of annual blood test monitoring.

General dietary recommendations have typically focused upon the goal of a well balanced healthy diet, with inclusion of all food groups. Unnecessary dietary restrictions expose the child to increased risk of overall caloric compromise (leading to malnutrition), or a lack of specific required nutrients. Furthermore, general dietary recommendations would include guidance for food choices during a flare of the disease; this might include recommendations to consume high-calorie foods or simple foods that are easily tolerated during a short-term period of disease exacerbation.

Some children, particularly entering puberty or during pubertal growth spurts, will require calories in addition to their usual intake to support this period of increased growth. This could take the form of ongoing maintenance enteral nutrition (as detailed above), or as intermittent supplementary enteral nutrition products. Some children will require the commencement of supplementary enteral nutrition delivered overnight by nasogastric tube, and a few children will require the creation of a gastrostomy to facilitate additional feed delivery (overnight or as boluses through each day).

8.6.2 Dietary Management of Functional Symptoms in Children with CD

Many individuals with a background of CD develop functional symptoms over time. It can often be difficult to elucidate whether symptoms are related to active disease or are of a functional nature. Determination of inflammatory markers and potential repeat endoscopic assessments may permit differentiation between these causes and guide subsequent management steps. In the context of ongoing symptoms without any features of active inflammation, further nutritional changes could then be considered. One such intervention is the removal of small fermentable carbohydrates (FODMAPS) from the diet (Halmos et al. 2014). The introduction of a FODMAP-free diet should be undertaken only with the guidance and supervision of a dietitian experienced in the use of this diet. This diet change may greatly improve symptoms and assist in ongoing management (Gearry et al. 2009; Gibson 2017). Although there are not yet specific data of the use of this intervention in children or adolescents with CD, the adult data does support this in selected patients.

8.7 APPLICATION OF CURRENT UNDERSTANDING TO THE DEVELOPMENT OF PERSONALIZED NUTRITION FOR CHILDREN WITH CD

8.7.1 Concepts to Guide Personalized Nutrition

Although there are no current well-established evidence-based guidelines to guide a personalized approach to nutrition in CD, there are clearly some key concepts and premises that are based upon current understanding of the importance of diet and nutrition in the context of CD. General aspects include the avoidance of preservatives and other food additives, avoiding highly processed foods and a preference towards whole foods. However, although these food elements may contribute to the development of IBD, it is not yet fully clear whether avoidance after diagnosis is beneficial or not in terms of the course of the disease. It may be too late in the course of the disease to effect any real benefit at that time.

8.7.2 PERSONALIZATION OF NUTRITIONAL THERAPY TO INDUCE REMISSION

At present most evaluations of EEN in children with CD have included relatively small cohorts with variations in disease severity and location, with the same protocol utilized regardless of these other variables. Consequently, there is not yet data that supports individualizing an EEN protocol to suit particular circumstances. For instance, it may be appropriate to adjust the duration of EEN according to disease activity, with a shorter duration in a child with mild disease activity and a longer duration for children with moderate–severe disease activity.

Some work also suggests that early changes in inflammatory markers could assist in guiding the duration of therapy (Gerasimidis et al. 2011). For instance, if the fecal calprotectin was to normalize completely within two weeks of commencing EEN, EEN may only be need for a short time afterwards, rather that routine completion of the full eight-week period. Further focused clinical studies are required to answer these questions more fully: data arising from such work would be greatly received as it would enable further personalization of EEN in children with CD.

As above, the SCD and the CDED are two further dietary interventions aimed at inducing and/or maintaining remission of CD. Rather than these being considered as competitors to EEN as options for therapy, it would be ideal to be able to direct a patient to one of these options over another based upon key disease-characterizing factors or disease location. Current evidence does not assist in this process, and further data is required to enable a personalized approach to selecting a nutritional intervention.

8.7.3 PATIENT AND DISEASE-SPECIFIC FACTORS TO GUIDE DIETARY CHOICES

Although patients diagnosed with CD are all classified with the one label (namely CD), in practice, all individuals with CD do have different characteristics. These include factors such as the extent of disease, specific disease location, variations in pro-inflammatory cytokine profiles, variations in patterns of intestinal microbiota and differences in genetic risk factors. The impact of these variations upon dietary recommendations in the individual patient have not been established. Data that might guide the development of algorithms to guide more individualized management is required.

8.7.4 NUTRIGENOMICS, NUTRIGENETICS AND NUTRIEPIGENETICS

While nutrigenomics refers to the impact of tolerance of diet due to underlying genetic determinants, nutrigenetics refers to the impacts of dietary choices upon genetic modulation and nutriepigenetics upon epigenetic modulation. Given the importance of genetic polymorphisms on the pathogenesis of IBD, it has been hypothesized that these genetic variations could be relevant to tolerance of various foods and/or assist in guiding dietary advice (Ferguson 2015). A recent consensus statement has established a framework for the evaluation of specific findings and to ensure that consequent personal dietary advice is well-founded and reliable (Grimaldi et al. 2017).

Although there are not yet clear findings arising from these works that have day-to-day applicability to the patient with IBD, this avenue of work may well be useful in the future. These applications may, for instance, help to guide the use of one specific nutritional intervention over another.

8.8 CONCLUSIONS

It is very clear that nutrition plays important roles in both the development of IBD and that IBD often impacts adversely upon nutrition. Hence, nutrition assessment and nutritional interventions are clearly very critical aspects of the management of IBD, especially in children and adolescents. At present there are various aspects of nutrition that are important to all patients. There is a burgeoning role for individualization of nutritional interventions and of dietary guidance that will lead on to a personalized approach to nutrition for individuals with IBD. At present, based upon available evidence, there are limited aspects that can be applied. Further work is necessary to better understand the interactions between genes, the environment, the intestinal microbiota and nutrition or diet. These remain keys to advancing the management of IBD in children.

ABBREVIATIONS

CD:	Crohn disease
CDED:	CD exclusion diet
CARD 15:	Caspase recruitment domain-containing protein 15
CRP:	C-reactive protein
EEN:	Enteral nutrition
FODMAPS:	Small fermentable carbohydrates
IBD:	Inflammatory bowel diseases
IBDU:	IBD Unclassified
IGF-1:	Insulin-like growth factor 1
MH:	Mucosal healing
Nod 2:	Nucleotide-binding oligomerization domain-containing protein 2
SCFA:	Short-chain fatty acid
UC:	Ulcerative colitis

REFERENCES

Agus, A., et al., Western diet induces a shift in microbiota composition enhancing susceptibility to adherent-invasive *E. coli* infection and intestinal inflammation. *Sci Rep*, 2016. **6**: p. 19032.

Ahuja, V. and R.K. Tandon, Inflammatory bowel disease in the Asia-Pacific area: a comparison with developed countries and regional differences. *J Dig Dis*, 2010. **11**(3): p. 134–147.

Alhagamhmad, M.H., et al., Exploring and enhancing the anti-inflammatory properties of polymeric formula. *JPEN J Parenter Enteral Nutr*, 2017. **41**(3): p. 436–445.

Ananthakrishnan, A.N., et al., Environmental triggers in IBD: a review of progress and evidence. *Nat Rev Gastroenterol Hepatol*, 2018. **15**(1): p. 39–49.

Benchimol, E.I., et al., Epidemiology of pediatric inflammatory bowel disease: a systematic review of international trends. *Inflamm Bowel Dis*, 2011. **17**(1): p. 423–439.

Berni Canani, R., et al., Short- and long-term therapeutic efficacy of nutritional therapy and corticosteroids in paediatric Crohn's disease. *Dig Liver Dis*, 2006. **38**(6): p. 381–387.

Bernstein, C.N., Review article: changes in the epidemiology of inflammatory bowel disease-clues for aetiology. *Aliment Pharmacol Ther*, 2017. **46**(10): p. 911–919.

Burgis, J.C., et al., Response to strict and liberalized specific carbohydrate diet in pediatric Crohn's disease. *World J Gastroenterol*, 2016. **22**(6): p. 2111–2117.

Butler, M., et al., Dietary microparticles implicated in Crohn's disease can impair macrophage phagocytic activity and act as adjuvants in the presence of bacterial stimuli. *Inflamm Res*, 2007. **56**(9): p. 353–361.

Cohen, S.A., et al., Clinical and mucosal improvement with specific carbohydrate diet in pediatric Crohn disease. *J Pediatr Gastroenterol Nutr*, 2014. **59**(4): p. 516–521.

Critch, J., et al., Use of enteral nutrition for the control of intestinal inflammation in pediatric Crohn disease. *J Pediatr Gastroenterol Nutr*, 2012. **54**(2): p. 298–305.

Day, A.S. and S.C. Brown, The adjunctive role of nutritional therapy in the management of phlegmon in two children with Crohn's disease. *Front Pediatr*, 2017. **5**: p. 199.

Ferguson, L.R., Nutritional modulation of gene expression: might this be of benefit to individuals with Crohn's disease? *Front Immunol*, 2015. **6**: p. 467.

Gasparetto, M. and G. Guariso, Crohn's disease and growth deficiency in children and adolescents. *World J Gastroenterol*, 2014. **20**(37): p. 13219–13233.

Gatti, S., et al., Effects of the exclusive enteral nutrition on the microbiota profile of patients with Crohn's disease: a systematic review. *Nutrients*, 2017. **9**(8).

Gearry, R.B., et al., High incidence of Crohn's disease in Canterbury, New Zealand: results of an epidemiologic study. *Inflamm Bowel Dis*, 2006. **12**(10): p. 936–943.

Gearry, R.B., et al., Reduction of dietary poorly absorbed short-chain carbohydrates (FODMAPs) improves abdominal symptoms in patients with inflammatory bowel disease-a pilot study. *J Crohns Colitis*, 2009. **3**(1): p. 8–14.

Gerasimidis, K., et al., Serial fecal calprotectin changes in children with Crohn's disease on treatment with exclusive enteral nutrition: associations with disease activity, treatment response, and prediction of a clinical relapse. *J Clin Gastroenterol*, 2011. **45**(3): p. 234–239.

Gibson, P.R., Use of the low-FODMAP diet in inflammatory bowel disease. *J Gastroenterol Hepatol*, 2017. **32**(Suppl 1): p. 40–42.

Grimaldi, K.A., et al., Proposed guidelines to evaluate scientific validity and evidence for genotype-based dietary advice. *Genes Nutr*, 2017. **12**: p. 35.

Gruber, L., et al., High fat diet accelerates pathogenesis of murine Crohn's disease-like ileitis independently of obesity. *PLoS One*, 2013. **8**(8): p. e71661.

Halmos, E.P., et al., A diet low in FODMAPs reduces symptoms of irritable bowel syndrome. *Gastroenterology*, 2014. **146**(1): p. 67–75 e5.

Hansen, T.S., et al., Environmental factors in inflammatory bowel disease: a case-control study based on a Danish inception cohort. *J Crohns Colitis*, 2011. **5**(6): p. 577–584.

Hirai, F., et al., Effectiveness of concomitant enteral nutrition therapy and infliximab for maintenance treatment of Crohn's disease in adults. *Dig Dis Sci*, 2013. **58**(5): p. 1329–1334.

Imhann, F., et al., Interplay of host genetics and gut microbiota underlying the onset and clinical presentation of inflammatory bowel disease. *Gut*, 2018. **67**(1): p. 108–119.

Johnson, T., et al., Treatment of active Crohn's disease in children using partial enteral nutrition with liquid formula: a randomised controlled trial. *Gut*, 2006. **55**(3): p. 356–361.

Klement, E., et al., Breastfeeding and risk of inflammatory bowel disease: a systematic review with meta-analysis. *Am J Clin Nutr*, 2004. **80**(5): p. 1342–1352.

Lemberg, D.A. and A.S. Day, Crohn disease and ulcerative colitis in children: an update for 2014. *J Paediatr Child Health*, 2015. **51**(3): p. 266–270.

Lopez, R.N., et al., Rising incidence of paediatric inflammatory bowel disease in Canterbury, New Zealand, 1996–2015. *J Pediatr Gastroenterol Nutr*, 2018. **66**(2): p. e45–e50.

Martinez-Medina, M., et al., Western diet induces dysbiosis with increased *E coli* in CEABAC10 mice, alters host barrier function favouring AIEC colonisation. *Gut*, 2014. **63**(1): p. 116–124.

Martino, J.V., J. Van Limbergen, and L.E. Cahill, The role of carrageenan and carboxymethylcellulose in the development of intestinal inflammation. *Front Pediatr*, 2017. **5**: p. 96.

Massironi, S., et al., Nutritional deficiencies in inflammatory bowel disease: therapeutic approaches. *Clin Nutr*, 2013. **32**(6): p. 904–910.

Moeeni, V. and A.S. Day, Impact of inflammatory bowel disease upon growth in children and adolescents. *ISRN Pediatr*, 2011. **2011**: p. 365712.

Nahidi, L., et al., Inflammatory bowel disease therapies and gut function in a colitis mouse model. *Biomed Res Int*, 2013. **2013**: p. 909613.

Nahidi, L., et al., Paediatric inflammatory bowel disease: a mechanistic approach to investigate exclusive enteral nutrition treatment. *Scientifica (Cairo)*, 2014. **2014**: p. 423817.

Nakahigashi, M., et al., Enteral nutrition for maintaining remission in patients with quiescent Crohn's disease: current status and future perspectives. *Int J Colorectal Dis*, 2016. **31**(1): p. 1–7.

Ng, S.C., et al., Environmental risk factors in inflammatory bowel disease: a population-based case-control study in Asia-Pacific. *Gut*, 2015. **64**(7): p. 1063–1071.

Ng, S.C., et al., Worldwide incidence and prevalence of inflammatory bowel disease in the 21st century: a systematic review of population-based studies. *Lancet*, 2018. **390**(10114): p. 2769–2778.

Otley, A.R., R.K. Russell, and A.S. Day, Nutritional therapy for the treatment of pediatric Crohn's disease. *Expert Rev Clin Immunol*, 2010. **6**(4): p. 667–676.

Powell, J.J., V. Thoree, and L.C. Pele, Dietary microparticles and their impact on tolerance and immune responsiveness of the gastrointestinal tract. *Br J Nutr*, 2007. **98**(Suppl 1): p. S59–S63.

Rogler, G. and S. Vavricka, Exposome in IBD: recent insights in environmental factors that influence the onset and course of IBD. *Inflamm Bowel Dis*, 2015. **21**(2): p. 400–408.

Ruemmele, F.M., et al., Consensus guidelines of ECCO/ESPGHAN on the medical management of pediatric Crohn's disease. *J Crohns Colitis*, 2014. **8**(10): p. 1179–1207.

Ruiz, P.A., et al., Titanium dioxide nanoparticles exacerbate DSS-induced colitis: role of the NLRP3 inflammasome. *Gut*, 2017. **66**(7): p. 1216–1224.

Sairenji, T., K.L. Collins, and D.V. Evans, An update on inflammatory bowel disease. *Prim Care*, 2017. **44**(4): p. 673–692.

Sartor, R.B. and G.D. Wu, Roles for intestinal bacteria, viruses, and fungi in pathogenesis of inflammatory bowel diseases and therapeutic approaches. *Gastroenterology*, 2017. **152**(2): p. 327–339 e4.

Sigall-Boneh, R., et al., Partial enteral nutrition with a Crohn's disease exclusion diet is effective for induction of remission in children and young adults with Crohn's disease. *Inflamm Bowel Dis*, 2014. **20**(8): p. 1353–1360.

Sigall Boneh, R., et al., Dietary therapy with the Crohn's disease exclusion diet is a successful strategy for induction of remission in children and adults failing biological therapy. *J Crohns Colitis*, 2017. **11**(10): p. 1205–1212.

Su, H.Y., et al., Rising incidence of inflammatory bowel disease in Canterbury, New Zealand. *Inflamm Bowel Dis*, 2016. **22**(9): p. 2238–2244.

Suskind, D.L., et al., Nutritional therapy in pediatric Crohn disease: the specific carbohydrate diet. *J Pediatr Gastroenterol Nutr*, 2014. **58**(1): p. 87–91.

Suskind, D.L., et al., Patients perceive clinical benefit with the specific carbohydrate diet for inflammatory bowel disease. *Dig Dis Sci*, 2016. **61**(11): p. 3255–3260.

Suskind, D.L., et al., Clinical and fecal microbial changes with diet therapy in active inflammatory bowel disease. *J Clin Gastroenterol*, 2018. **52**(2): p. 155–163.

Wahbeh, G.T., et al., Lack of mucosal healing from modified specific carbohydrate diet in pediatric patients with Crohn disease. *J Pediatr Gastroenterol Nutr*, 2017. **65**(3): p. 289–292.

Walters, T.D. and A.M. Griffiths, Mechanisms of growth impairment in pediatric Crohn's disease. *Nat Rev Gastroenterol Hepatol*, 2009. **6**(9): p. 513–523.

Weisshof, R. and I. Chermesh, Micronutrient deficiencies in inflammatory bowel disease. *Curr Opin Clin Nutr Metab Care*, 2015. **18**(6): p. 576–581.

Yamamoto, T., et al., Enteral nutrition to suppress postoperative Crohn's disease recurrence: a five-year prospective cohort study. *Int J Colorectal Dis*, 2013. **28**(3): p. 335–340.

Ye, B.D. and D.P. McGovern, Genetic variation in IBD: progress, clues to pathogenesis and possible clinical utility. *Expert Rev Clin Immunol*, 2016. **12**(10): p. 1091–1107.

Zhao, R., et al., Intestinal microparticles and inflammatory bowel diseases: incidental or pathogenic? *Inflamm Bowel Dis*, 2014. **20**(4): p. 771–775.

9 Personalized Nutrition in Chronic Kidney Disease

A New Challenge

Giorgina Barbara Piccoli

CONTENTS

9.1　CHRONIC KIDNEY DISEASE: AN EPIDEMIC THAT REFLECTS SOCIO-ECONOMIC DIFFERENCES

Chronic kidney disease (CKD) is presently defined as every persistent (lasting for at least three months) alteration in renal function or structure, or in urine composition, regardless of kidney function (Figure 9.1) (2002; Inker, Astor et al. 2014). This broad definition, which has replaced previously used ones (i.e. 'kidney insufficiency' and 'renal insufficiency'), shows how nephrology, formerly seen as a branch of medicine dealing with rare and complex diseases, has come to be viewed in the new millennium. The changes reflected in this new outlook, and the dogmas they have replaced, will be considered and analyzed in the discussion that follows.

9.1.1　KIDNEY DISEASES ARE NOT RARE: A PROBLEM OF MEASURE

Previous dogma: kidney diseases are rare.
Present perspective: kidney diseases are frequent.

GFR Categories (ml/min/1.73 m²) Stage. Description. Range			Persistent Albuminuria Categories. Description. Range		
			Normal to mildly increased	Moderately increased	Severely increased
			< 30 mg/g (< 3 mg/mmol)	30-300 mg/g (3-30 mg/mmol)	> 300 mg/g (> 30 mg/mmol)
1	Normal or high	≥ 90	1 if CKD	1	2
2	Mildly decreased	60-89	1 if CKD	1	2
3A	Mildly to moderately decreased	45-59	1	2	3
3B	Moderately to severely decreased	30-44	2	3	3
4	Severely decreased	15-29	3	3	4+
5	Kidney failure	< 15	4+	4+	4+

FIGURE 9.1　CKD stages: Classification and stratification.

If kidney diseases are considered as synonymous with a reduction in the kidney function, their prevalence may be as low as 5–10 : 1,000 individuals, while the lives of about 1 : 1,000 individuals are sustained by dialysis and transplantation (Inker, Astor et al. 2014; Drawz and Rahman 2015; Zoccali, Vanholder et al. 2017).

This definition fails, however, to identify not only the initial phases of several important, treatable but quite rare diseases, such as the glomerulonephritides, but also does not take into consideration a reduction of less than 50% of the kidney tissue, which roughly represents what is called the renal functional reserve, i.e. the amount of the kidney tissue that can be damaged without causing a reduction in the kidney function, at least as currently measured (Bauer, Brooks et al. 1982; Rodriguez-Iturbe 1986).

Renal functional reserve is also what makes possible living-donor kidney transplantation, aimed at guaranteeing a healthy life to the donor and 'functionally healing' the recipient. While renal functional reserve is 'good news' for kidney transplantation, its presence makes it difficult to capture the initial phases of renal damage, when the baseline disease is often amenable to complete healing (Tapson, Mansy et al. 1986).

Studies on kidney donors, the prototype of a 'healthy' reduction in the kidney parenchyma, are accumulating. They show the importance of a reduction in the kidney tissue in particular situations, such as pregnancy, in which the renal function is physiologically stressed (Bauer 1982). In fact, the incidence of pre-eclampsia and the hypertensive disorders of pregnancy is significantly higher after kidney donation, underlining the importance of not neglecting even the first signs of kidney disease (Ibrahim, Akkina et al. 2009; Reisaeter, Roislien et al. 2009; Garg, McArthur et al. 2015).

When the new definitions are used (Figure 9.1), the prevalence of kidney diseases rises to at least 10% in virtually all populations, with different distributions in type of disease in wealthy and developing countries, involving complex interactions between several factors, including life expectancy and competitive mortality (Garcia-Garcia and Jha 2015; Norton, Moxey-Mims et al. 2016; Webster, Nagler et al. 2017).

As discussed below, the problem of defining the presence of CKD merges with the critical review of the role of serum creatinine as a marker of kidney disease.

9.1.2 THE MAIN MARKERS OF KIDNEY FUNCTION ARE ALSO BASIC NUTRITIONAL MARKERS

Previous dogma: urea is not a relevant marker of kidney function.
Present perspective: urea is a basic marker in the dietary management of CKD patients.

The kidney function is measured by the level of two substances derived from protein catabolism: urea and creatinine (Kassirer 1978; Bostom, Kronenberg et al. 2002; Kumar and Mohan 2017).

Urea levels reflect exogenous protein intake, while creatinine is an indicator of muscle mass. More specifically, urea levels are the result of the integration between kidney function, protein intake and, to a lesser degree, fluid intake (Harvey,

Blumenkrantz et al. 1980). In a healthy individual, the intake of proteins and the type of proteins are the main modulators of kidney function. In fact, in humans as well as in other omnivorous and carnivorous animals, protein intake elicits a sharp rise in kidney function (renal plasma flow, glomerular filtration rate, intraglomerular pressure), and the rise is steeper if the proteins are of animal origin (Hirschberg, Rottka et al. 1985; Wiseman, Hunt et al. 1987; Kontessis, Bossinakou et al. 1995; Moe, Zidehsarai et al. 2011; So, Song et al. 2016).

There are several corollaries of this physiological characteristic, in health and disease:

- In the case of protein-rich diets (such as those followed by some athletes), normal kidneys increase the glomerular filtration rate (GFR), thus keeping urea and creatinine at normal levels, increasing urinary urea and creatinine; these changes are fully captured only by analysis of 24-hour urine collection (Schwingshackl and Hoffmann 2014; Lovshin, Skrtic et al. 2018).
- Two diseases may lead to hyperfiltration in normal kidneys: diabetes and nephrotic syndrome. Both are 'perceived' by kidney sensors as meals 'rich' in calories and proteins, and in both cases dietary approaches may be useful way to reduce kidney damage and control symptoms, as will be discussed later (Hilton, Roth et al. 1969; Cameron, Maisey et al. 1970; Levey, Greene et al. 1999).
- In CKD, the hyperfiltration response is blunted, urea levels increase and become roughly proportional to protein intake. Once more, the effect is higher for animalderived proteins; the hyperfiltration response is, however, partially preserved in the diseased kidney and a CKD patient on a protein restricted-diet has lower clearances and lower urea levels, compared to a patient on an unrestricted diet (Bosch, Saccaggi et al. 1983).
- While individuals on vegetarian diets, and especially those on vegan diets, which ban all animal-derived food, may have fully normal kidney tissue and function, they have lower urea levels and renal clearances (Kolpek, Ott et al. 1989; Brandle, Sieberth et al. 1996; So, Song et al. 2016).
- Hypercatabolic patients have disproportionally high urea levels, since protein catabolism mimics high exogenous protein intake; urea levels rise more in younger well-nourished patients (Stellato, Rhodes et al. 1980; Dickerson, Tidwell et al. 2005).
- The presence of blood in the gastrointestinal tract has a similar effect, and a sharp rise in the urea level in a CKD patient should prompt a search for gastrointestinal bleeding (Masud, Manatunga et al. 2002).

Urea is the most important determinant of the countercurrent osmotic gradient; hence, urea clearance is also dependent upon fluid intake, highly variable not only in different individuals but also in the same individual in different conditions. Consequently, urea levels increase in patients needing high-dose diuretics, out of proportion to kidney function reduction.

Due to its intrinsic variability, blood urea is not employed in screening for kidney diseases; however, urea is pivotal in the monitoring of CKD patients, where

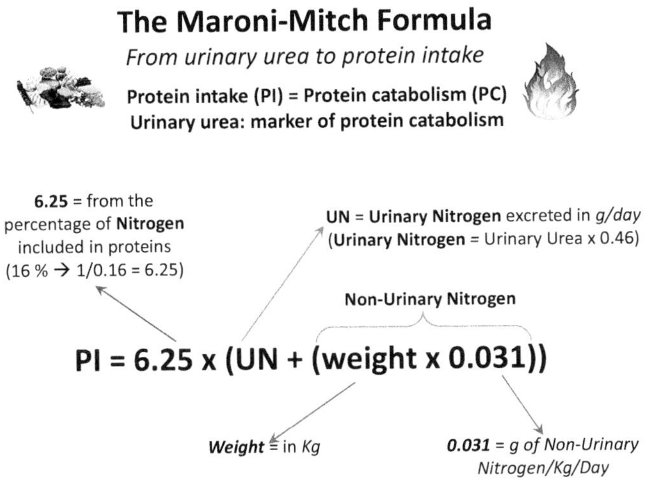

The Maroni-Mitch Formula
From urinary urea to protein intake

Protein intake (PI) = Protein catabolism (PC)
Urinary urea: marker of protein catabolism

6.25 = from the
percentage of **Nitrogen**
included in proteins
(16 % → 1/0.16 = 6.25)

UN = Urinary Nitrogen excreted in *g/day*
(**Urinary Nitrogen** = Urinary Urea x 0.46)

Non-Urinary Nitrogen

$$PI = 6.25 \times (UN + (weight \times 0.031))$$

Weight = in *Kg*

0.031 = g of Non-Urinary
Nitrogen/Kg/Day

FIGURE 9.2 The Maroni-Mitch formula for protein intake assessment from 24-hour urine collection.

it continues to be a marker of protein intake, and is one of the elements on which the decision to start dialysis is based, particularly in the context of protein restriction. Furthermore, in stable patients, the only validated method for assessing protein intake is the analysis of 24-hour urinary urea (the Maroni-Mitch formula, Figure 9.2) (Maroni, Steinman et al. 1985; Oscanoa, Amado et al. 2018).

> *Previous dogma: kidney insufficiency is diagnosed on the basis of an increase in serum creatinine.*
> *Present perspective: serum creatinine has to be contextualized at least to age, sex and ethnicity to reflect the kidney function.*

As previously mentioned, in healthy individual's serum creatinine levels are linked to muscular mass, and practically unaffected by protein intake. Protein intake regulates (increases) glomerular filtration; high muscle mass is usually associated with high protein (and calorie) intake, and hyperfiltration is the main reason why athletes' urea and creatinine levels do not increase or increase only marginally (Hirschberg, Rottka et al. 1985; Kontessis, Bossinakou et al. 1995; Bernstein, Treyzon et al. 2007; Schwingshackl and Hoffmann 2014; Lovshin, Skrtic et al. 2018).

Individuals of African origin, who have a higher muscle mass compared to Caucasians or Orientals, have 10–20% higher creatinine levels; this difference is accounted for in the main formulae for indirect assessment of the glomerular filtration rate (GFR) (Figure 9.3) (Madero and Sarnak 2011; Delanaye and Mariat 2013; Levey, Inker et al. 2014; Filler and Lee 2018).

Relying on 'absolute' values of serum creatinine for the definition of chronic kidney disease was customary before the new millennium, and the fact that over time different cut-points were employed to define kidney insufficiency has been an important source of heterogeneity in CKD research.

CKD-EPI Equation
For Use with Standardized Serum Creatinine

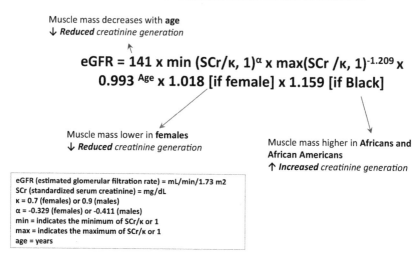

Muscle mass decreases with **age**
↓ *Reduced* creatinine generation

$$eGFR = 141 \times min\ (SCr/κ, 1)^α \times max(SCr\ /κ, 1)^{-1.209} \times 0.993\ ^{Age} \times 1.018\ [\text{if female}] \times 1.159\ [\text{if Black}]$$

Muscle mass lower in **females**
↓ *Reduced* creatinine generation

Muscle mass higher in **Africans and African Americans**
↑ *Increased* creatinine generation

eGFR (estimated glomerular filtration rate) = mL/min/1.73 m2
SCr (standardized serum creatinine) = mg/dL
κ = 0.7 (females) or 0.9 (males)
α = -0.329 (females) or -0.411 (males)
min = indicates the minimum of SCr/κ or 1
max = indicates the maximum of SCr/κ or 1
age = years

FIGURE 9.3 'Anatomy' of one of the main formulae for the assessment of GFR. (Adapted from Levey AS, Schoolwerth AC, Burrows NR, et al. (2009). "Comprehensive public health strategies for preventing the development, progression, and complications of CKD: Report of an expert panel convened by the Centers for Disease Control and Prevention." *Am J Kidney Dis.* 53: 522–535.)

The 'normal' values of serum creatinine were assessed in a particular population: young adult males without apparent diseases (usually at medical checkups for men being drafted into the army). Since muscle mass is higher in young individuals and in males, serum creatinine is a very poor marker of CKD in females and the elderly (and an especially poor marker in elderly females), in which a 'normal' serum creatinine level may correspond to a GFR as low as 20–30 mL/min (Levey, Inker et al. 2014; Filler and Lee 2018).

The KDOQI (Kidney Disease Outcomes and Quality Initiative) guidelines mark a systematic shift towards measuring kidney function via the evaluation of GFR, which can be directly assessed in a number of ways: measuring radio isotopic clearances or inulin clearance (the gold standard, although time-consuming and expensive); using 24-hour urine collection to calculate creatinine clearance (not very practical and subject to relevant pre-analytical error); or using indirect formulae that take age, sex and ethnicity into account (Inker, Astor et al. 2014; Drawz and Rahman 2015).

Regrouping CKD patients in broad categories (CKD stages) is of pragmatic interest, but leaves many questions unanswered:

- Different formulae are available, and none of them is perfect.
- Standardization for body surface is not always done, and does not capture obesity or may underestimate kidney function reduction in small patients or in patients with lowbody weight.

- The creatinine-based formulae which consider body weight (i.e. the Cockcroft and Gault formula, still widely used in Europe) are based on the assumption that individuals (and patients) have a coherent distribution of lean and fat body mass; they therefore overestimate GFR in obese patients (and to a greater degree if they are sarcopenic), and may underestimate it in thin, athletic patients (Delanaye and Mariat 2013; Levey, Inker et al. 2014).
- The GFR may be overestimated in patients with low muscle mass with respect to body weight, for example, in the case of neoplastic diseases or diabetes, categories in which iatrogenic damage is frequent and drug administration is often modulated by GFR (Alper, Yi et al. 2011; Madero and Sarnak 2011; Delanaye and Mariat 2013; Levey, Inker et al. 2014).
- There is no validated formula for the assessment of GFR in pregnancy, a physiological cause of hyperfiltration and a challenge for the diseased kidney, as will shortly be discussed (Smith, Moran et al. 2008; Koetje, Spaan et al. 2011; BustosGuadano, Martin-Calderon et al. 2017).
- There is no validated formula for elderly patients, and the definition of old age differs markedly, from over 65 to over 80 (Rosansky 2011; Kilbride, Stevens et al. 2013; Delanaye, Cavalier et al. 2014).

Isotope dilution mass spectrometry (IDMS), the new gold standard method for assessing serum creatinine, has not completely replaced the older Jaffe test, which has higher laboratory variability, and by using colorimetric interference can also detect other substances. Systematic differences are observed between Jaffe and enzymatic methods, and some manufacturers subtract 0.30 mg/dl from Jaffe results to match enzymatic results ('the compensated Jaffe method'). In fact, the formula presently most widely used (CKD-EPI, Table 9.1) requires creatinine dosing by IDMS (Sood, Manns et al. 2014).

TABLE 9.1
The Main GFR Formulae Used in Research and Clinical Practice

Formulae	Features	Cons
Cockcroft-Gault Formula	• The only formula that includes body weight	• Estimates Creatinine Clearance (ClCr) rather than GFR • Based on older assessment • Method for serum creatinine • Overestimates ClCr in obese or edematous patients, and underestimates ClCr in thin athletic ones
MDRD *Modification of Diet in Renal Disease Study* **Equation**	• Originally expressed with six variables, later expressed with four (creatinine, age, sex, race)	• Not validated in elderly patients (> 85 years), pregnant women and children
CKD-EPI Equation	• Same variables as MDRD • More accurate than MDRD in higher GFR levels • Requires creatinine dosing by IDMS	• Not validated in children and pregnant women

While replacing GFR with the value of serum creatinine has greatly increased the sensitivity of CKD diagnosis, a rigid interpretation of the formulae, and consequently of the stratification of actions for CKD stages, may even be misleading. This has been fully acknowledged in the most critical CKD phase (renal replacement therapy (RRT)), in which clinical reasoning is presently considered the main, albeit not the sole criterion for RRT initiation (McIntyre and Rosansky 2012; Erkan 2013; Sood, Manns et al. 2014).

9.1.3   CKD Assessment: What Is Missed?

Simplistic dogma: CKD can be diagnosed on the basis of GFR, integrated by proteinuria.

Complex reality: hematuria, a possible sign of CKD, imaging for urologic diseases and integrated blood and urine evaluation for interstitial disorders have to be considered.

Proteinuria is a sign of kidney disease. While low-grade proteinuria may be a sign of interstitial damage, proteinuria over 1 g per day is highly likely to be of glomerular origin. Proteinuria is both a sign of kidney disease and a progression factor, and its presence and degree is a therapeutic target in CKD (Kincaid-Smith and Fairley 2005; Cirillo 2010; Gorriz and Martinez-Castelao 2012; Cyriac, Holden et al. 2017).

There are three categories of kidney diseases that elude detection based solely upon GFR levels and proteinuria (Figure 9.1).

Hematuria of glomerular origin is a sign that the glomerular architecture of at least some glomeruli is damaged; while macroscopic hematuria (red or 'black' urines) is usually immediately reported, microscopic hematuria is still often overlooked. It may, however, be an indication of progressive diseases, such as IgA nephropathy, one of the most common glomerulonephritides worldwide. Furthermore, at least in IgA nephropathy, the entity of hematuria does not correlate with the risk of end-stage kidney disease (Yadin 1994; Hodgkins and Schnaper 2012).

Urologic anomalies, including reflux nephropathy, renal malformations and chronic pyelonephritis, may go undetected until more than 50% of the renal parenchyma is damaged, when serum creatinine increases, or proteinuria develops. Easily detected using ultrasounds, these diseases respond well to nutritional management, in the absence of urinary tract infections.

The kidney is the most sophisticated metabolic machine in the body. It is the main regulator of acid base, calcium phosphate, sodium, water and potassium homeostasis (Raghavan and Eknoyan 2014; Fogo, Lusco et al. 2017). Ultimately, urine composition affects body composition, and urine composition is determined at the tubule-interstitial level, where reabsorption of about 180 liters of pre-urine occurs every day. Interstitial diseases affect the metabolic core of the kidney, and, while in advanced stages they may be revealed by proteinuria (usually low-grade) and kidney function impairment, in the early stages signs may be elusive, including impaired growth in children, kidney stone disease, nephrocalcinosis or osteoporosis in adults (Piccoli, De Pascale et al. 2016; Piccoli, Alrukhaimi et al. 2018). While included in the broad CKD definition, these diseases may be diagnosed only if specifically searched for.

9.1.4 THE EPIDEMIOLOGY OF KIDNEY DISEASES FOLLOWS WEALTH DISTRIBUTION

Paradoxically, kidney diseases are overall more frequent in rich countries, where people live longer.

However, kidney diseases in younger individual have a higher incidence in poor countries, where competitive causes of death reduce the overall burden of CKD.

CKD have been called a hidden epidemic; it is estimated that about 10% of the world's population suffer from some form of kidney disease; these diseases are among the 20 most common causes of death worldwide and represent a negative prognostic factor in patients affected by other diseases (cardiovascular ailments, diabetes, etc.) (Martin-Cleary and Ortiz 2014; Garcia-Garcia and Jha 2015; De Broe, Gharbi et al. 2017; Carrero, Hecking et al. 2018; Piccoli, Alrukhaimi et al. 2018).

The kidneys are the most vascularized organs in the human body and are therefore involved in vascular diseases, which are an important part of ageing; hence, the graying of the population is increasing the burden of CKD (Kurella, Covinsky et al. 2007; Stevens, Li et al. 2010; Carrero, Hecking et al. 2018).

Improvement in the care of the elderly, diabetic patients and patients with diffuse atherosclerosis has reduced mortality and increased life expectancy. As a consequence of the decrease in competitive mortality, chronic diseases such as CKD are increasing. Hence, the prevalence of CKD is higher in populations with a longer life expectancy, i.e. people in highincome countries (Kurella, Covinsky et al. 2007).

Conversely, almost all kidney diseases of the young are more common in medium and low-income countries; interstitial and infectious diseases are more prevalent where sanitary conditions are suboptimal; diseases with toxic and environmental causes (e.g. Mesoamerican nephropathy) are more frequent in the developing world; immunologic diseases occur more frequently in some ethnic groups (such as Latin-Americans), and may be triggered by infectious or inflammatory diseases (Hoy, White et al. 2015; Benghanem Gharbi, Elseviers et al. 2016; Lozano-Kasten, Sierra-Diaz et al. 2017). The obesity and diabetes epidemics have not spared low-income countries, where preventive measures and early diagnosis of CKD are less available. As a consequence, age-adjusted incidence of CKD is higher in low-income countries, even when the overall burden of disease may be lower, due to the presence of competitive mortality (Figure 9.4) (Kovesdy, Furth et al. 2017a).

9.2 NUTRITIONAL APPROACH TO CKD: TEN REASONS WHY ONE SIZE DOES NOT FIT ALL

1. Kidney diseases are different.
2. CKD phases are different.
3. Comorbidity is different.
4. Cultures are different.
5. Dietary habits are different.
6. Availability of nutritional support is different.
7. Nephrologists (and dietitians) are different.

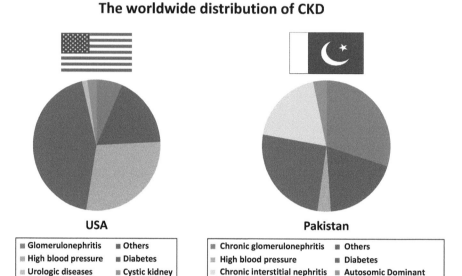

FIGURE 9.4 Two examples of differences in the worldwide distribution of CKD. (From: For Pakistan: Jha, Vivekanand (2013). "Current status of end-stage renal disease care in India and Pakistan." *Kidney Int. Suppl.* 3: 157–160. 10.1038/kisup.2013.3; For USA: Health, United States, 2011: Table 51. End-stage renal disease patients, by selected characteristics: United States, selected years 1980–2010. Centers for Disease Control and Prevention website. www.cdc.gov/nchs/data/hus/2011/051.pdff (PDF, 25 KB). Updated 2011. Accessed December 20, 2013.)

8. Life phases are different.
9. Patients are different.
10. Most of the previous points may change.

9.2.1 KIDNEY DISEASES ARE DIFFERENT

All the major structures of the kidney can be affected by CKD: glomerular involvement is usually signaled by hematuria and proteinuria; tubular disorders lead to disturbances in handling sodium, water and electrolytes; vascular diseases are usually only detected when the kidney function decreases; and the same is true in the case of diseases that disrupt a part of the kidney parenchyma (such as neoplasia, infection or malformations). It is therefore clear that nutrition needs to focus on specific issues (protein metabolism, water and electrolytes, kidney function).

9.2.2 CKD PHASES ARE DIFFERENT

Kidney diseases are often asymptomatic, at least in the early phases, meriting, as previously mentioned, the definition of silent killers. Severe kidney disease can impair virtually all body functions; however, end-stage disease of this vital organ is the only one that can be treated using substitutive treatments, i.e. dialysis and transplantation (ESKD). Nutritional needs change in each of the three main phases of the disease (CKD, dialysis and transplantation) and are modulated by the specific causes of CKD, type of dialysis and by antirejection treatment after kidney transplantation (Cupisti, D'Alessandro et al. 2016; Jonker, de Heer et al. 2018).

9.2.3 Comorbidity Is Different

When CKD is the only disease a patient has, nutritional issues are focused on type and stage of CKD. This may not be the case in diabetic or obese patients, in which the choice between privileging the management of CKD or that of the associated condition modulates the nutritional prescriptions, or in elderly patients in which protein energy wasting goes along with diffuse vascular disease, and a mediation between risk of malnutrition and 'wise restriction' has to be pursued (de Roij van Zuijdewijn, Grooteman et al. 2016; Kang, Chang et al. 2017; Munoz-Perez, Espinosa-Cuevas et al. 2017).

9.2.4 Cultures Are Different

The importance of eating together differs in different part of the world.

In large families, changing dietary habits may be difficult and will require a whole-family commitment; in cultures in which the same meal is shared from a common pot, it may be difficult to calculate individual portions; in settings in which the habit of cooking has been lost and most meals are bought pre-prepared, nutritional counseling may be even more challenging.

Conversely, modifying choices in the context of a healthy nutritional culture, for example, in Mediterranean countries, represents an enormous advantage and facilitates compliance.

The recent BMC Nephrology series, dedicated to different approaches to low-protein diets worldwide demonstrates the important role culture plays in determining compliance with dietary prescriptions (Piccoli and Cupisti 2017).

9.2.5 Dietary Habits Are Different

Within each culture, every individual has different habits, dictated both by lifestyle (working or retired, sedentary or travelling, self-employed or employee, etc.) and personal preferences (vegan vs. omnivorous, fast food vs. carefully prepared meals, sweets vs. cheese, etc.), which have to be taken into consideration when prescribing a diet to attain good compliance (Bellizzi, Cupisti et al. 2016; D'Alessandro, Piccoli et al. 2016; Fouque, Chen et al. 2016; Garneata, Stancu et al. 2016; Piccoli, Nazha et al. 2016; Piccoli and Cupisti 2017).

9.2.6 Availability of Nutritional Support Is Different

The availability of protein-free food, supplements rich in proteins and calories, or other nutritional supplements differs widely even in neighboring countries, and the prices charged for the same preparations can differ considerably from one country to another. Thus, the decision to prescribe certain diets, such as complex very low-protein diets, will depend to some extent on the availability of protein-free food and nutritional supplements, as well as on whether the patient will be reimbursed for purchasing the items on the diet (D'Alessandro, Piccoli et al. 2016; Fouque, Chen et al. 2016; Garneata, Stancu et al. 2016; Piccoli and Cupisti 2017).

9.2.7 NEPHROLOGISTS (AND DIETITIANS) ARE DIFFERENT

The fact that each physician (or dietitian) reflects his/her beliefs in the major decisions they take is well known; nutrition is in the core habits of each individual, so it is little wonder that diets prescribed differ from one person to another.

9.2.8 LIFE PHASES ARE DIFFERENT

Children and young adults eat more than their grandparents; a trend towards reducing protein intake is often observed across age; pregnancy and lactation are life phases in which diet needs adaptation, but that are also sensitive to cultural influences. The effect of protein restriction in children is less constant than in adults (Ornish 2012).

While in our sedentary society hard manual work is increasingly unusual, energy expenditure is profoundly different in different individuals.

9.2.9 PATIENTS ARE DIFFERENT

It may seem too obvious to even mention, but each individual is a puzzle of genetic background, cultural and personal preferences, lifestyle, work and leisure, habits and changes. In such a context, personalized nutrition is probably the only approach able to ensure compliance from patients needing life-long changes in their dietary habits so they can cope with kidney disease.

9.2.10 MOST OF THE PREVIOUS POINTS MAY CHANGE

In the Western world, the media repeatedly call for changes in dietary habits. One example is the campaign against red meat presently being carried out in many European countries. Such positions change both the perception of what healthy habits are and modify habits at the community level, since the advice given is followed by food providers in public services, such as schools, hospitals and company canteens (Gorelik, Kanner et al. 2012; Ornish 2012; Hermans, de Bruin et al. 2017; Lv, Yu et al. 2017; Pettigrew, Talati et al. 2018). The impact is enormous; for instance, the changes in the daily allowances for proteins from 1.0 to 0.8 g/Kg/day deeply affected the prescription of 'normal' diets, in health and disease (World Health Organization 2007; Millward 2012; Wolfe, Cifelli et al. 2017).

Furthermore, independently of cultural pressure, dietary habits may change over the course of an individual's life, and according to the availability of *ad libitum* food and restrictions.

9.3   NUTRITIONAL ISSUES IN DIFFERENT PHASES OF CKD

9.3.1   EARLY CKD PHASES: NEPHRON SPARING

The idea that CKD can be modulated by diet, in particular by a reduction in protein intake, is as old as nephrology itself. In fact, the first convincing data on a reduction in protein intake as a tool to slow the progression of kidney disease came from one of the pioneers of our discipline, Thomas Addis. He was the first to describe nephron hypertrophy as a response to irreversible loss of kidney tissue, and as a further element for progression, almost 30 years before the 'Brenner Theory' became the basis for the analysis of CKD progression. In extreme synthesis, Addis's working hypothesis was that diseased kidneys should be given 'functional rest' and, since protein intake is one of the main modulators of the renal function, proteins in the diet should be reduced (Brenner, Lawler et al. 1996; Piccoli 2010; Boulton 2011).

Since the publication of Addis's book *Glomerular Nephritis, Diagnosis and Treatment* in 1948, low-protein diets in the treatment of CKD have encountered different fortunes. While they were initially widely used to prolong survival, in the 1970s, when dialysis became widely available in most Western countries, very few doctors continued to prescribe diets. This trend continued in the 1980s, due to the increased availability of dialysis and transplantation. Then, in the decades that followed, with the progressive aging of the dialysis population, nutritional issues were once more recognized as crucial for patient survival (Giovannetti and Maggiore 1964; Berlyne, Janabi et al. 1966; Kerr, Robson et al. 1967; Bergstrom 1984; Walser, Mitch et al. 1999; Mitch and Remuzzi 2004; Piccoli, Ferraresi et al. 2013; Mitch and Remuzzi 2016; Piccoli, Ventrella et al. 2016).

The difficulty of demonstrating that low-protein diets are able to reduce CKD progression, due to the heterogeneity of kidney diseases, too short follow-up, poor patient compliance and inadequately powered trials played against the use of these diets. Notwithstanding these issues, a widely cited systematic Cochrane review suggests that a reduction in protein intake is associated with lower risk to patients starting dialysis (retarded progression) (Fouque and Laville 2009).

These findings are confirmed by the two best-designed recent trials, on two different populations: one, from Brunori, in Italy, focused on diet and dialysis in the elderly, while the other, from Garneata, in Romania, examined the effects of diet on younger nondiabetic patients with good compliance with moderate protein restriction. Both studies elegantly demonstrated that a reduction in protein intake, to a target as low as 0.3 g/Kg/day, was associated with longer 'renal survival' without an increase in mortality (Brunori, Viola et al. 2007; Garneata, Stancu et al. 2016).

There are many reasons why low-protein diets are now encountering a resurgence of interest: the dialysis population is getting older and dialysis-related iatrogenicity is felt to be too high in patients with high comorbidity, leading some authors to 'contraindicate' dialysis in elderly, high-comorbidity individuals; the economic crisis is deeply affecting health services in the Western world, so reducing the costs of dialysis (which can be as high as €50,000–75,000 per year) is becoming a healthcare priority (Meier-Kriesche and Schold 2005; Vanholder, Lameire et al. 2016). In this context, a low-protein diet can delay dialysis start, or be included in 'supportive,' or 'conservative' care for patients for whom dialysis is not planned. No 'best diet' has been identified, however, and different levels of protein restriction, with different distributions of animal and plant-derived proteins are employed worldwide. A recent attempt to describe some of the different approaches to these diets is available in a special issue of *BMC Nephrology* (Piccoli and Cupisti 2017). The main characteristics of the 'low-protein diet menu' are reported in Table 9.2.

A detailed description of the different diets proposed is beyond the scope of a general review, as it would risk oversimplifying such a complex issue; an example of an approach employing all the major diet options, modulating choice according to the patient's preferences and characteristics is reported in Figure 9.5. It should be noted that the diet is modulated to the patient's account of their eating habits and on the availability of protein-free food.

TABLE 9.2
The Main Low-Protein Diet Menu

type of diet	protein restriction (g/Kg/bw)	main features	main advantages	main disadvantages
traditional	0.6-0.8 g/Kg/day; mixed proteins	where traditional cuisine is more plant based, and returning to the roots may be useful	a natural approach, doesn't require special food	demanding: requires attention to quantity and quality of food
vegan	0.6-0.8 g/Kg/day; vegetable proteins	unrestricted vegan diets usually 0.7-0.9 g/Kg/day proteins, due to different bio-availability, a 0.7 diet corresponds to 0.6 mixed protein diet	"trendy" approach, a natural diet that may also have other favourable effects on health	demanding: requires attention to quality and integration of legumes and cereals. Risk of B12, vit D and iron deficits
vegan supplemented	0.6 g/Kg/day; vegetable proteins, supplemented with amino- and keto-acids (1: 8-10 Kg)	based upon forbidden (animal) and allowed (plant based) food. Animal-derived food allowed in "unrestricted meals"	supplements avoid the need to integrate legumes and cereals, reducing the risk of nutritional deficits	adding pills to usual demanding drug list. Expensive if supplements are not reimbursed
protein-free food	0.6 g/Kg/day; mixed proteins	Protein-free pasta, bread and other carbohydrates	allows a reduction of proteins without changing eating habits	protein-free food may not be "tasty", expensive if not reimbursed
very low-protein supplemented (with or without protein-free food)	0.3 g/Kg/day; vegetable proteins, supplemented with amino- and keto-acids (1 tablet : 5 Kg)	Animal-derived food is allowed only in "free meals" (usually no more than 1 per week)	the most effective approach for delaying dialysis start	adding pills to the drug list. Difficult if protein- free food is not available. expensive if supplements and protein-free foods not reimbursed

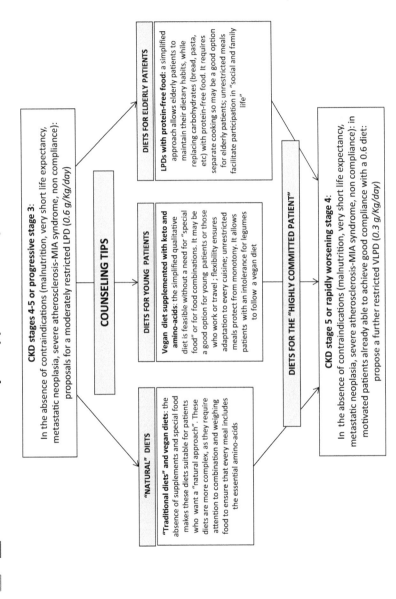

A tailored "Italian-style" approach to low protein diet in CKD patients

CKD stages 4-5 or progressive stage 3:
In the absence of contraindications (malnutrition, very short life expectancy, metastatic neoplasia, severe atherosclerosis-MIA syndrome, non compliance): proposals for a moderately restricted LPD (*0.6 g/Kg/day*)

COUNSELING TIPS

DIETS FOR ELDERLY PATIENTS

LPDs with protein-free food: a simplified approach allows elderly patients to maintain their dietary habits, while replacing carbohydrates (bread, pasta, etc) with protein-free food. It requires separate cooking so may be a good option for elderly patients; unrestricted meals facilitate participation in "social and family life"

DIETS FOR YOUNG PATIENTS

Vegan diet supplemented with keto and amino-acids: the simplified qualitative diet is feasible without a need for "special food" or for food combinations. It may be a good option for young patients or those who work or travel ; flexibility ensures adaptation to every cuisine; unrestricted meals protect from monotony. It allows patients with an intolerance for legumes to follow a vegan diet

"NATURAL" DIETS

"Traditional diets" and vegan diets: the absence of supplements and special food makes these diets suitable for patients who want a "natural approach". These diets are more complex, as they require attention to combination and weighing food to ensure that every meal includes the essential amino-acids

DIETS FOR THE "HIGHLY COMMITTED PATIENT"

CKD stage 5 or rapidly worsening stage 4:
In the absence of contraindications (malnutrition, very short life expectancy, metastatic neoplasia, severe atherosclerosis-MIA syndrome, non compliance): in motivated patients already able to achieve good compliance with a 0.6 diet: propose a further restricted VLPD (*0.3 g/Kg/day*)

FIGURE 9.5   Examples of multiple-choice diet options: Italian and French styles.

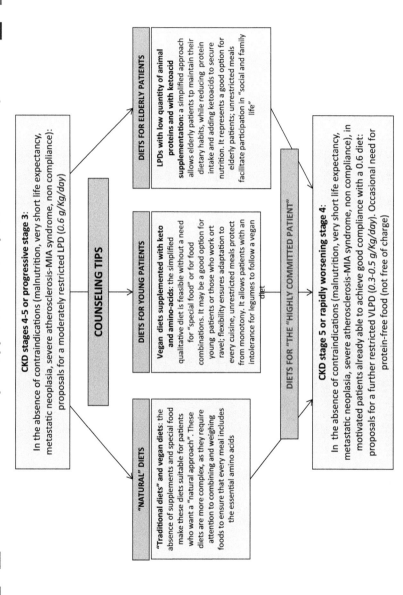

A tailored "French style" approach to low-protein diet in CKD patients

CKD stages 4-5 or progressive stage 3:
In the absence of contraindications (malnutrition, very short life expectancy, metastatic neoplasia, severe atherosclerosis-MIA syndrome, non compliance): proposals for a moderately restricted LPD (*0.6 g/Kg/day*)

COUNSELING TIPS

DIETS FOR ELDERLY PATIENTS

LPDs with low quantity of animal proteins and with ketoacid supplementation: a simplified approach allows elderly patients tp maintain their dietary habits, while reducing protein intake and adding ketoacids to secure nutrition. It represents a good option for elderly patients; unrestricted meals facilitate participation in "social and family life"

DIETS FOR YOUNG PATIENTS

Vegan diets supplemented with keto and amino-acids: the simplified qualitative diet is feasible without a need for "special food" or for food combinations. It may be a good option for young patients or those who work ort ravel; flexibility ensures adaptation to every cuisine, unrestricted meals protect from monotony. It allows patients with an intolerance for legumes to follow a vegan diet

"NATURAL" DIETS

"Traditional diets" and vegan diets: the absence of supplements and special food make these diets suitable for patients who want a "natural approach". These diets are more complex, as they require attention to combining and weighing foods to ensure that every meal includes the essential amino acids

DIETS FOR "THE "HIGHLY COMMITTED PATIENT"

CKD stage 5 or rapidly worsening stage 4:
In the absence of contraindications (malnutrition, very short life expectancy, metastatic neoplasia, severe atherosclerosis-MIA syndrome, non compliance), in motivated patients already able to achieve good compliance with a 0.6 diet: proposals for a further restricted VLPD (*0.3-0.5 g/Kg/day*). Occasional need for protein-free food (not free of charge)

FIGURE 9.5 (CONTINUED)   Examples of multiple-choice diet options: Italian and French styles.

9.3.2 LATE CKD PHASES: RETARDING DIALYSIS START

Over ten years ago, a brilliant editorial by Remuzzi and Mitch, two giants of nephrology, first underlined that the issue of retarding the progression of the kidney disease should be separated from the goal of attaining a stable metabolic balance that, in the last CKD stages, could make it possible to delay dialysis start (Mitch and Remuzzi 2016).

This goal is presently considered to be of prime importance in an aging pre-dialysis population. Low-protein diets were in fact first used on the basis of the hypothesis that since uremia could be described as a state of protein intoxication, reducing protein intake could reduce symptoms and allow better end-of-life quality. The finding that reducing protein intake was associated with prolonged survival was almost forgotten until we began seeing a sharp increase in octogenarians and nonagenarians with end-stage kidney disease (Giovannetti and Maggiore 1964; Berlyne, Janabi et al. 1966; Kerr, Robson et al. 1967; Bergstrom 1984; Walser, Mitch et al. 1999; Mitch and Remuzzi 2004; Piccoli, Ferraresi et al. 2013).

Conversely, in younger patients, pre-emptive kidney transplantation is increasingly considered to be highly advantageous in terms of survival, and retarding the need for dialysis start while waiting for kidney transplant is a short-term goal, at least for some patients (Meier-Kriesche and Schold 2005; Grams, Chen et al. 2013; Amaral, Sayed et al. 2016; Goto, Okada et al. 2016; Jay, Washburn et al. 2017).

Furthermore, the finding that early 'healthy' dialysis start (at 10–15 mL/min of glomerular filtration rate (GFR)) does not bring any survival advantage compared to lower GFR (6–8 mL/min), and may even be correlated with higher mortality, notwithstanding higher costs, focused attention on 'dialysis shock,' and the need for smoother dialysis start (Harris, Cooper et al. 2011; Tattersall, Dekker et al. 2011; Mehrotra, Rivara et al. 2013; Rosansky, Cancarini et al. 2013; Watnick 2013).

In this regard, the option of incremental dialysis, in which dialysis sessions are progressively increased from one to three per week, while relying on moderately restricted diets to control symptoms and metabolic derangements, is presently gaining attention (Basile, Casino et al. 2017; Chin, Appasamy et al. 2017; Golper 2017; Locatelli, Del Vecchio et al. 2017; Obi and Kalantar-Zadeh 2017).

Once more, there are, and there probably will always be, several different ways to face this problem, according to dietary patterns, patient's age, reasons for delaying dialysis and individual preferences (Hanafusa, Lodebo et al. 2017; Locatelli, Del Vecchio et al. 2017) (Table 9.2).

9.3.3 DIALYSIS: PRESERVING NUTRITIONAL STATUS

The close relationship between nutrition and dialysis is both obvious and complex. It is obvious since it is known that renal replacement therapy (RRT) substitutes the kidneys to eliminate metabolic waste products from food introduced into the body, and some of the well-known dietary restrictions (potassium, phosphate) are established to counter the limits of the usual RRT policies (Silvererg and Hunt 1966; Kopple, Swendseid et al. 1975; Feinstein and Kopple 1985; Shinaberger, Greenland et al. 2008; D'Alessandro, Piccoli et al. 2015; Sherman 2016). It is complex because dialysis efficiency is a key modulator of a patient's nutritional status, which is, in turn, probably the most important element associated with survival (Lindsay and Henderson 1988; Raja, Ijelu et al. 1992;

Lindsay and Bergstrom 1994; Chauveau, Naret et al. 1996; Stenvinkel, Heimburger et al. 1999; Kalantar-Zadeh, Stenvinkel et al. 2003; St-Jules, Goldfarb et al. 2016; Biruete, Jeong et al. 2017; Ishii, Takahashi et al. 2017; Kang, Chang et al. 2017; Takahashi, Inoue et al. 2017). Therefore, preserving good nutritional status is one of the main goals in dialysis patients. Hence, the classical indication that protein and calorie intakes should be high (respectively 1.2–1.4 g/Kg/day of proteins and30-35 Kcal/Kg/day) clashes with the need to restrict phosphates (poorly dialyzed, especially in the past) and potassium (for the cardiovascular risks of hyperkalemia), and monitor the patient's lipid and glucose metabolism. In fact, the relationship between dialysis efficiency, nutritional restrictions, nutritional status and outcomes produces a series of paradoxes that ultimately impact on dialysis choices (Cuppari and Kamimura 2011; KalantarZadeh, Tortorici et al. 2015; St-Jules, Goldfarb et al. 2016; Biruete, Jeong et al. 2017; Piccoli, Moio et al. 2017) (Table 9.3). While it is generally acknowledged that malnutrition plays a prominent role in determining survival on dialysis, the best way to assess nutritional status has not yet been established. Several nutritional markers and assessments have been proposed. The most comprehensive ones integrate inflammation, atherosclerosis (MIA, the malnutrition inflammation atherosclerosis syndrome) and/or various comorbidities (Pollock, Ibels et al. 1995; Fein, Gundumalla et al. 2002; Pecoits-Filho, Lindholm et al. 2002; Kalantar-Zadeh, Ikizler et al. 2003; Wang, Sea et al. 2005; Panichi, Cupisti et al. 2014; Carrero and Avesani 2015; Komatsu, Okazaki et al. 2015).

Furthermore, many of the studies carried out to identify the most reliable biomarkers of malnutrition in dialysis were performed in populations on thrice-weekly 'conventional' hemodialysis and, while different strategies, including incremental dialysis, daily dialysis and nocturnal dialysis have been reported to improve nutritional status, the assessment of nutritional status is still based on conventional thrice-weekly hemodialysis (Spanner, Suri et al. 2003; Galland and Traeger 2004; Ipema, Struijk et al. 2016). Peritoneal dialysis and hemodiafiltration pose a further challenge to nutritional evaluation because of the loss of albumin in the dialysate (Guest 2013; Ikizler 2014; Johansson 2015; Vega, Quiroga et al. 2015; Rippe and Oberg 2016; Morena, Jaussent et al. 2017) .

The customary nutritional prescriptions for patients on dialysis generally employed assessments derived from studies from the 1970s and 1980s, targeted to a

TABLE 9.3

Open Issues in Dietary Management in Kidney Transplantation

Nutrient	Indication	main advantages	main disadvantages
Proteins	1.2 g/Kg/day	Protects form protein waste	In contrast with what is suggested to reduce progression of kidney disease
Energy	Not clear	Normal intake is needed for equilibrate nutrition	Reduction of caloric intake is needed to control overweight;
Lipids	Not clear	"safe" oils, such as olive oil, may help conserving nutritional status without increasing proteins	High in calories, may be reduced for weight control
Carbohydrates	Limitation in sugar	Reducing diabetes risk, controlling weight	None, unless needed to increase low energy intake
Salt	Normal or reduced	Better blood pressure control	An overzealous reduction may induce AKI if combined with ACE inhibitors, dehydration

completely different population (younger, lower comorbidity), treated with much less effective methods (without ultrapure water, employing acetate dialysis, associated with lower tolerance, hypercatabolism and lower efficiency, etc.).

Little wonder, therefore, that most of the nutritional indications for dialysis patients are presently being challenged and are undergoing critical revision (Table 9.3) (Cuppari and Kamimura 2011; Kalantar-Zadeh, Tortorici et al. 2015; St-Jules, Goldfarb et al. 2016; Biruete, Jeong et al. 2017; Piccoli, Moio et al. 2017). Once more, the different prescriptions should be adapted to individual patients. Preserving energy intake is a must at all ages; attention to the calcium-phosphate balance may be a priority in younger patients for longterm preservation of their vascular status, while adequate protein intake may need to be age-adjusted and subordinated to attainment of adequate calorie input, to limit acidosis and hyperphosphatemia.

9.3.4 Kidney Transplantation: Preserving Nutritional Status and Slowing CKD Progression

Kidney transplantation merges several different and sometimes incompatible issues: in case of steroid treatment, the main focus is on avoiding obesity, diabetes and metabolic syndrome. Avoidance of protein wasting, which may be enhanced by steroids, requires a higher protein intake, which on the contrary, is not advised for preservation of the kidney function, even if the experience with low-protein diets post-transplant is limited (el-Agroudy, Wafa et al. 2004; Kent 2007; Teplan 2009; Teplan, Valkovsky et al. 2009; Ward 2009; Chakkera, Weil et al. 2013; Wissing and Pipeleers 2014; Deetman, Said et al. 2015; Centenaro, Pedrollo et al. 2018).

Other issues are the reduction of salt intake (especially in countries like Japan, in which the diet is rich in sodium), and of animal-derived fat, which can negatively impact on the cardiovascular profile (Ichimaru, Nakazawa et al. 2016; Nagaoka, Onda et al. 2016; Pedrollo, Nicoletto et al. 2017).

The presence of an inverse correlation between urea excretion, mortality and graft failure underlines the importance of preserving the nutritional status of kidney transplant recipients, as well as in dialysis patients (Deetman, Said et al. 2015; Said, Deetman et al. 2015; Eisenga, Kieneker et al. 2016).

As previously mentioned in the case of dialysis, most of the indications on protein intake for transplanted patients date back to decades ago, when steroid treatment was the basis of therapy, with the well-known effects on protein wasting. In the absence of specific directions, in particular, we suggest that the issues reported in Table 9.4 should be considered in the nutritional assessment and counseling of patients with failing kidneys or a significant reduction in kidney function.

9.4 SOME SUGGESTIONS ON SPECIFIC NUTRITIONAL ISSUES IN DIFFERENT DISEASES

9.4.1 Nephrotic Syndrome and the Case of Focal Segmental Glomerulosclerosis

Protein loss is the hallmark of the nephrotic syndrome, leading to hypoalbuminemia and hypercholesterolemia, hypercoagulation and protein wasting. Loss of

TABLE 9.4
Nutrition on Dialysis: Some Open Issues

THE QUESTIONS	Pros	Cons	Comments
Are the nutritional requirements usually cited (calories: 30–35 kcal/kg; proteins > 1.2 g/kg) still relevant?	International standard followed for more than 40 years	The requirements were assessed in a different dialysis population, and may not be relevant to the present one; they do not consider the changes in the indications given to the overall population	There is a need for a re-assessment of the requirements of elderly dialysis patients
Are the present standards of 'adequate nutrition' applicable to intensive dialysis schedules, and to hemodiafiltration?	Simple markers such as albumin level make it possible to compare results, and are robust enough to maintain a constant predictive value	Sensibility may be lower in nonconventional dialysis techniques, and can be affected by albumin losses in hemodiafiltration	None of the proposed evaluations of malnutrition is clearly superior or selfstanding; results of studies depend in part on the definition-diagnoses chosen
Processed and preserved food may be significantly different from untreated food. What are we eating?	Nutritional approaches have to be simple and basing them on quantity and quality may not be feasible	Processed foods may be rich in rapidly absorbable phosphate and potassium	Not acknowledging the importance of additives in processed and preserved foods can lead to unnecessary restrictions
Is malnutrition a single disease or the result of several diseases?	The clinical signs of malnutrition are universal and do not depend on pathogenesis	If malnutrition is not linked to poor intake but to poor clinical conditions, it will not respond to therapy	Differentiation may allow setting attainable goals according to the individual patient's comorbidity
THE PARADOXES	The 'logic' (overall population or general data in the dialysis population)	The findings (in the dialysis population or in specific dialysis populations)	Comments
Obesity and survival	Obesity is associated with lower survival in the overall population	Obesity is associated with higher survival in dialysis patients; losing weight is associated with higher mortality on dialysis	Obesity is often a contraindication for kidney transplantation

(Continued)

TABLE 9.4 (CONTINUED)
Nutrition on Dialysis: Some Open Issues

THE QUESTIONS	Pros	Cons	Comments
High protein intake and phosphate control	A high protein diet is indicated after dialysis start	Reduction of phosphate intake is not compatible with a high-protein diet	Plant-derived phosphate may be less well-absorbed; acidosis induced by catabolism is often a missing element in hyperphosphatemia
Albumin level, Kt/V and survival	Low serum albumin and low dialysis efficiency are associated with reduced survival	In hemodiafiltration, high efficiency is coupled with significant albumin losses	Albumin losses are incompletely quantified; nutrition is probably more important than high efficiency in elderly or fragile sarcopenic patients
Potassium and vascular health	Since dialysis patients are at risk for hyperkalemia, potassium is often restricted	Banning plant derived food to avoid hyperkalemia limits consumption of 'vascular healthy' food in a high-risk population	Hyperkalemia is still a rare, but possible cause of death

immunoglobulins increases the risk of infections. In the long run, this unfavorable metabolic pattern leads to an increase in atherosclerotic vascular disease. Nutritional management in the context of nephrotic syndrome is complex: increasing protein intake has the effect of increasing GFR and proteinuria, while decreasing protein intake is feared to exacerbate malnutrition (Mansy, Goodship et al. 1989). Very low-protein diets have occasionally been employed with success in the management of nephrotic syndrome resistant to other treatments; some animal models suggest that the addiction of ketoacid and amino acids can play a protective role, possibly through an antioxidant effect (Zhang, Xie et al. 2016; Zhang, Yin et al. 2016). The effect of low-protein diets on nephrotic syndrome is doubtless complex and not merely hemodynamic. Interestingly, there have been found to be several points in common between experiences with very low-protein diets, which are necessarily plant-based, and our experience with vegan diets in pregnancy in focal segmental glomerulosclerosis, suggesting that future studies should also focus on the role of nutritional antigens in the pathogenesis of these diseases (Attini, Leone et al. 2017).

9.4.2 Obesity-Related Kidney Disease and Diabetic Nephropathy

Obesity poses multiple challenges to the kidney function. While in most cases the damage is multifactorial, and is associated with other components of the metabolic syndrome, themselves associated with CKD (such as hypertension, hyperuricemia, diabetes and dyslipidemia), obesity-related kidney disease is now widely considered as a self-standing disease in which the increased renal workload due to the high body mass, and to over-rich nutrition induces hyperfiltration, ultimately resulting in glomerular sclerosis (Lastra, Manrique et al. 2006; Eknoyan 2011; Ritz, Koleganova et al. 2011; Amann and Benz 2013; Kovesdy, S et al. 2017b; Jonker, de Heer et al. 2018; Nehus 2018).

A specific form of focal segmental glomerulosclerosis is associated with morbid obesity, with nephrotic proteinuria and slow progression towards end-stage kidney disease (Ritz, Koleganova et al. 2011; Amann and Benz 2013). This condition often merges with proteinuric diabetic nephropathy. The latter, at least in Western countries, is also increasingly being found in the absence of proteinuria and its hallmark has shifted to diffuse vascular disease (Packham, Ivory et al. 2011; Amann and Benz 2013; de Vries, Ruggenenti et al. 2014; Robles, Villa et al. 2015; Bolignano and Zoccali 2017; Friedman, Wahed et al. 2018; Jonker, de Heer et al. 2018).

While the results obtained with moderately restricted low-protein diets in diabetic patients (mainly affected by type 2 diabetes) are similar to those of nondiabetic patients of a similar age, in both diseases an attempt to correct obesity should probably come first (Bolignano and Zoccali 2013; Chang, Chen et al. 2016; Piccoli, Capizzi et al. 2016).

While the advantages of weight loss are intuitive, the modality is not, and although different dietary approaches have been reported to decrease body weight, with a significant effect on kidney disease progression, the long-term efficacy of weight loss in this population meets with the usual problems that characterize the management of obese patients (Bolignano and Zoccali 2013; Capizzi, Teta et al. 2017; Kalantar-Zadeh, Rhee et al. 2017).

Furthermore, while obesity has a negative effect on kidney disease and survival in CKD and after kidney transplantation, the high mortality linked to malnutrition on dialysis gives a survival advantage to overweight and obese dialysis patients, a phenomenon called 'the obesity paradox' (Salahudeen 2003; Kalantar-Zadeh, Abbott et al. 2005).

9.4.3 Autosomal Dominant Polycystic Kidney Disease (ADPKD)

ADPKD was formerly generally considered refractory to nutritional interventions with lowprotein diets, since the natural history of the disease is genetically determined and linked to the progressive growth of renal cysts, and to complications such as infection, hemorrhage or obstruction of the urinary tract. The few studies done that specifically address the role of low-protein diets in ADPKD progression show a small but positive effect in the retardation of CKD progression (Di Iorio, Cupisti et al. 2018; Jankowska, Qureshi et al. 2018).

Interest in nutrition in ADPKD has recently increased, as a result of evidence showing the positive effect of maintaining very high diuresis (over 4 liters per day), high enough to inhibit vasopressin, or of treatment with specific antagonists of vasopressin receptors (Tolvaptan and related drugs) (Taylor, Hamilton-Reeves et al. 2017; Tran, Huynh et al. 2017; Visconti, Cernaro et al. 2018).

The focus in dietary management is therefore moving from reducing protein intake to low-sodium, high-fluid intake to retard CKD progression in ADPKD patients (Taylor, Hamilton-Reeves et al. 2017).

9.4.4 INTERSTITIAL DISORDERS AND KIDNEY STONES

Interstitial diseases and recurrent kidney stones are often associated with subtle derangements in the complex metabolic machine of tubular reabsorption. While kidney function is often normal, these diseases are seen as forms of CKD (usually stages 1–2) because of alterations they produce in the electrolyte composition of blood or urine. Kidney stones are the most common symptom leading to diagnosis. Correcting a metabolic derangement is not simple or always feasible, and diet alone may not be sufficient; an extensive discussion of these complex alterations is beyond the scope of this review, but some suggestions regarding a pragmatic nutritional approach to kidney stones is reported in Table 9.5 (Gambaro, Croppi et al. 2016; Scales, Tasian et al. 2016).

TABLE 9.5
Open Issues in Dietary Management in Kidney Transplantation

Kidney stones

One size fits all		
↓ Salt	↑ Water	Normalize protein intake to **0.8** g/Kg/day

One size does NOT fit all			
To add ↑		*To reduce ↓*	
Mg	They may be added to protect from lithiasis, especially in cases of low urinary excretion	**Ca**	Control calcium intake, reduce salt, evaluate thiazide diuretics
K		**Phosphate**	Control protein intake and search for tubular deficits
Citric acid		**Uric acid**	Increase urinary pH and evaluate allopurinol

9.4.5 KIDNEY DONATION

Although kidney donation should not be considered a disease, the absence of a kidney falls into the CKD category of 'permanent alteration at imaging' even when kidney function is strictly normal. Protection of the remnant kidney is mandatory, and most transplant centers offer systematic nephrology follow-up to donors. In this context, a balanced diet, 'normal' protein intake (0.8 g/Kg/day), avoidance of obesity, early correction of hypertension and of any metabolic derangements that exist are the basis for preserving kidney health in the long term (Nogueira, Weir et al. 2010; van Ginhoven, de Bruin et al. 2011; Bergen, Reese et al. 2014; Firat Kaya, Sayin et al. 2017).

9.5 NUTRITIONAL ISSUES IN PARTICULAR CKD POPULATIONS

9.5.1 PREGNANT WOMEN WITH CKD

A healthy diet during pregnancy is vital for the well-being of both mother and fetus; in the Western world, the concept of healthy diet has shifted from protection from malnutrition to protection from overnutrition (Kaiser and Allen 2008; Ramakrishnan, Grant et al. 2012; Koletzko, Bauer et al. 2013; Kaiser and Campbell 2014; Procter and Campbell 2014). The rediscovery of the Mediterranean diet and of plant-based diets points to their being able to play a role in preventing many of the 'overeating' diseases, including metabolic syndrome and obesity (Widmer, Flammer et al. 2015). While the meaning of plant-based diet differs in poorly- and highly resourced settings, in the latter a well-balanced plant-based diet makes it possible to avoid excessive weight gain, and related risks, in pregnancy (Jebeile, Mijatovic et al. 2016; Ruchat, Allard et al. 2016).

The literature on vegan, vegetarian or plant-based diets in pregnancy is limited, and heterogeneous. A systematic review based on seven papers supports the position of the American Dietetic Association: 'well-planned vegetarian diets are appropriate for individuals during all stages of the lifecycle, including pregnancy, lactation, infancy, childhood' (Craig and Mangels 2009). A second review, targeting plant-based diets outside the context of limited resources, analyzed 29 papers. None of the studies suggested a higher risk for severe adverse pregnancy-related events in vegan/vegetarian mothers (Piccoli, Clari et al. 2015). A slightly shorter duration of gestation, and lower infant weight, both in the normal and 'atterm' range, were reported in some studies. However, it is not clear if this reflects protection against large for gestational age babies, or a higher prevalence of small for gestational age babies (Piccoli, Clari et al. 2015). The only relevant note of caution was the finding in one large study of a higher incidence of hypospadias in the children of vegan mothers, which has remained unexplained and unconfirmed (North and Golding 2000). It is crucial to prevent nutritional deficits, in particular of vitamin B12, vitamin D, iron and zinc, and their levels need to be regularly monitored in our on-diet patients (Foster, Herulah et al. 2015; Piccoli, Clari et al. 2015).

Available data supported the decision of a few groups, including ours, to offer a plantbased option to CKD patients with good compliance and no history of nutritional disorders (anorexia in particular); this choice was associated with better intrauterine growth and protection from pre-term delivery. Interestingly, while proteinuria increases in gestation in patients with glomerular diseases, this increase was not observed in on-diet patients. This may have indirectly contributed to maintaining stable albumin levels, and promoting fetal growth. None of the patients suffered a rapid reduction in renal function after delivery, differently from many cases in the literature (Piccoli, Leone et al. 2014; Attini, Leone et al. 2016).

9.5.2 CHILDREN WITH CKD

Nutrition is an obvious and essential element for normal growth, and children with CKD face multiple challenges, linked not only to dietary restrictions, but also to the profound endocrinological derangements that accompany uremia, especially a deficit in growth hormone (GH). Hormonal supplementation with GH, correction of anemia and of the calciumphosphate balance are a priority and dietary restrictions are usually limited to the correction of severe derangements in potassium or the calcium-phosphate balance (Graf, Candelaria et al. 2007; Harambat and Cochat 2009; Rees and Mak 2011; Ingulli and Mak 2014; Kim, Lim et al. 2014; Gupta, Mantan et al. 2016; Pugh, Hemingway et al. 2018).

Moderately restricted diets, however, are employed in the clinical practice in some settings, more as a result of following consolidated practice than because this practice is supported by specific studies. Extension of the interesting results obtained when plant-based diets are followed by pregnant women with CKD to children with kidney disease represents an interesting subject for future research.

9.6 WHAT THIS REVIEW DID NOT DISCUSS: SUGGESTIONS FOR FURTHER RESEARCH

The high heterogeneity and metabolic complexity of the kidney function, the challenges involved in its measurement and the peculiarities of the different phases of CKD make it impossible to discuss in a single chapter anything more than some general insights.

We hope that this review will be useful first of all for acknowledging the complexity of CKD, the first step for guiding a dietary management approach that is tailored to the disease, its phases and, most importantly, to individual patients.

Several issues that have emerged merit discussion, or should be better studied in this context. We have summarized them in the last table of this review, also as a call to promote discussion and exchanges on these questions (Table 9.6).

TABLE 9.6

Crucial Issues and Suggestions for Research

Open issue	Crucial points	Suggestions for research
Low -protein diets and CKD progression	Compliance may lead to an important attrition bias; difficult long-term randomization of the diet; ethical problems in intrusive randomization1	Patient-preference randomized trials may account for preferences and at least partially account for biases
	Diets are prescribed in the context of global patient care; distinguishing between different aspects of care may be difficult. Randomization usually requires standardization of care, which is against the current trend towards personalization	Analysis based on 'system of care' and not simple protocol could make it possible to study the effect of personalized approaches to nutrition in CKD
Definition of the outcomes	The hard 'renal' outcome (dialysis start) depends on indications and modalities of dialysis or access to preemptive transplantation; it may not uniformly reflect the effects of diet.	Integration of diet effect, modality of dialysis start, residual clearance at six months and one year could add to the present knowledge
	Doubling of serum creatinine and halving of the clearance are poor markers, due to the difference in 'absolute loss of GFR' for different levels of creatinine or GFR itself	Analysis of trajectories and different trends, classical analysis (on creatinine and GFR) combined with loss of GFR
Assessment of nutritional status	Complex, difficult different in the various CKD phases, according to age and disease; overhydration is a major limit to usual evaluations	Analysis should probably be focused on integration of different evaluations more than on the selection of the 'best way'
Assessment of proteinuria	Difficulties similar to those encountered when assessing GFR; no fully satisfactory assessment on random samples	Integration of 24-hour assessment in patients with severe CKD, also allowing detailed metabolic study (urea, calcium, phosphate, Na, etc.)
Other treatments— indications	There a number of topics we did not discuss. These include Na restriction in CKD, phosphate intake, type of proteins, the role of additives and food preservatives, the role of probiotics. All these issues merit attention in the future.	

9.7 CONCLUDING REMARKS

There are several reasons why a dietary approach to CKD should be personalized: the heterogeneity of kidney diseases, their different phases, the wide variety of therapies and patients' personal preferences have to modulate nutritional approaches in patients with CKD, a disease that will accompany them throughout their lives.

Defining dietary indications requires not only establishing a program and modifying it over time, but also acknowledging the many gaps in present knowledge, and being ready to reconsider dietary indications periodically on the basis of new advances in nutrition in CKD.

Diet in CKD is an evolving issue; in this regard acknowledging the limits, and the need for personalization may help establish a cooperative relationship between caregivers and patients, which is the basis for a durable therapeutic alliance.

ABBREVIATIONS

ADPKD: Autosomal dominant polycystic kidney disease
CKD: Chronic kidney disease
ESKD: End Stage Kidney Disease
GFR: Glomerular filtration rate
GH: Growth hormone
IDMS: Isotope dilution mass spectrometry
Kcal: Kilocalorie
KDOQI: Kidney Disease Outcomes and Quality Initiative
Kg: Kilogram
RRT: Renal replacement therapy
MIA: Malnutrition inflammation atherosclerosis syndrome

REFERENCES

(1990). *Clinical Methods: The History, Physical, and Laboratory Examinations.* rd, H. K. Walker, W. D. Hall and J. W. Hurst, Boston, Butterworth Publishers, A Division of Reed Publishing.

(2002). "K/DOQI clinical practice guidelines for chronic kidney disease: evaluation, classification, and stratification." *Am J Kidney Dis* 39(2 Suppl 1): S1–S266.

Alper, A. B., Y. Yi, et al. (2011). "Performance of estimated glomerular filtration rate prediction equations in preeclamptic patients." *Am J Perinatol* 28(6): 425–430.

Amann, K. and K. Benz (2013). "Structural renal changes in obesity and diabetes." *Semin Nephrol* 33(1): 23–33.

Amaral, S., B. A. Sayed, et al. (2016). "Preemptive kidney transplantation is associated with survival benefits among pediatric patients with end-stage renal disease." *Kidney Int* 90(5): 1100–1108.

Attini, R., F. Leone, et al. (2016). "Vegan-vegetarian low-protein supplemented diets in pregnant CKD patients: fifteen years of experience." *BMC Nephrol* 17(1): 132.

Attini, R., F. Leone, et al. (2017). "Pregnancy, proteinuria, plant-based supplemented diets and focal segmental glomerulosclerosis: a report on three cases and critical appraisal of the literature." *Nutrients* 9(7).

Basile, C., F. G. Casino, et al. (2017). "Is incremental hemodialysis ready to return on the scene? From empiricism to kinetic modelling." *J Nephrol* 30(4): 521–529.

Bauer, J. H., C. S. Brooks, et al. (1982). "Clinical appraisal of creatinine clearance as a measurement of glomerular filtration rate." *Am J Kidney Dis* 2(3): 337–346.

Bellizzi, V., A. Cupisti, et al. (2016). "Low-protein diets for chronic kidney disease patients: the Italian experience." *BMC Nephrol* 17(1): 77.

Benghanem Gharbi, M., M. Elseviers, et al. (2016). "Chronic kidney disease, hypertension, diabetes, and obesity in the adult population of Morocco: how to avoid 'over'- and 'under'-diagnosis of CKD." *Kidney Int* 89(6): 1363–1371.

Bergen, C. R., P. P. Reese, et al. (2014). "Nutrition assessment and counseling of the medically complex live kidney donor." *Nutr Clin Pract* 29(2): 207–214.

Bergstrom, J. (1984). "Discovery and rediscovery of low protein diet." *Clin Nephrol* 21(1): 29–35.

Berlyne, G. M., K. M. Janabi, et al. (1966). "Dietary treatment of chronic renal failure." *Proc R Soc Med* 59(7): 665–667.

Bernstein, A. M., L. Treyzon, et al. (2007). "Are high-protein, vegetable-based diets safe for kidney function? A review of the literature." *J Am Diet Assoc* 107(4): 644–650.

Biruete, A., J. H. Jeong, et al. (2017). "Modified nutritional recommendations to improve dietary patterns and outcomes in hemodialysis patients." *J Ren Nutr* 27(1): 62–70.

Bolignano, D. and C. Zoccali (2013). "Effects of weight loss on renal function in obese CKD patients: a systematic review." *Nephrol Dial Transplant* 28(Suppl 4): iv82–iv98.

Bolignano, D. and C. Zoccali (2017). "Non-proteinuric rather than proteinuric renal diseases are the leading cause of end-stage kidney disease." *Nephrol Dial Transplant* 32(Suppl 2): ii194–ii199.

Bosch, J. P., A. Saccaggi, et al. (1983). "Renal functional reserve in humans. Effect of protein intake on glomerular filtration rate." *Am J Med* 75(6): 943–950.

Bostom, A. G., F. Kronenberg, et al. (2002). "Predictive performance of renal function equations for patients with chronic kidney disease and normal serum creatinine levels." *J Am Soc Nephrol* 13(8): 2140–2144.

Boulton, F. E. (2011). "Thomas Addis, MD (1881–1949): Scottish-American clinical laboratory researcher, social activist and pioneer of renal medicine." *J Nephrol* 24(Suppl 17): S62–S65.

Brandle, E., H. G. Sieberth, et al. (1996). "Effect of chronic dietary protein intake on the renal function in healthy subjects." *Eur J Clin Nutr* 50(11): 734–740.

Brenner, B. M., E. V. Lawler, et al. (1996). "The hyperfiltration theory: a paradigm shift in nephrology." *Kidney Int* 49(6): 1774–1777.

Brunori, G., B. F. Viola, et al. (2007). "Efficacy and safety of a very-low-protein diet when postponing dialysis in the elderly: a prospective randomized multicenter controlled study." *Am J Kidney Dis* 49(5): 569–580.

Bustos-Guadano, F., J. L. Martin-Calderon, et al. (2017). "Glomerular filtration rate estimation in people older than 85: comparison between CKD-EPI, MDRD-IDMS and BIS1 equations." *Nefrologia* 37(2): 172–180.

Cameron, J. S., M. N. Maisey, et al. (1970). "Creatinine clearance in patients with proteinuria." *Lancet* 1(7643): 424–425.

Capizzi, I., L. Teta, et al. (2017). "Weight loss in advanced chronic kidney disease: should we consider individualised, qualitative, ad libitum diets? a narrative review and case study." *Nutrients* 9(10).

Carrero, J. J. and C. M. Avesani (2015). "Pros and cons of body mass index as a nutritional and risk assessment tool in dialysis patients." *Semin Dial* 28(1): 48–58.

Carrero, J. J., M. Hecking, et al. (2018). "Sex and gender disparities in the epidemiology and outcomes of chronic kidney disease." *Nat Rev Nephrol* 14(3): 151–164.

Centenaro, A., E. F. Pedrollo, et al. (2018). "Different dietary patterns and new-onset diabetes mellitus after kidney transplantation: a cross-sectional study." *J Ren Nutr* 28(2): 110–117.

Chakkera, H. A., E. J. Weil, et al. (2013). "Can new-onset diabetes after kidney transplant be prevented?" *Diabetes Care* 36(5): 1406–1412.

Chang, A. R., Y. Chen, et al. (2016). "Bariatric surgery is associated with improvement in kidney outcomes." *Kidney Int* 90(1): 164–171.

Chauveau, P., C. Naret, et al. (1996). "Adequacy of haemodialysis and nutrition in maintenance haemodialysis patients: clinical evaluation of a new on-line urea monitor." *Nephrol Dial Transplant* 11(8): 1568–1573.

Chin, A. I., S. Appasamy, et al. (2017). "Feasibility of incremental 2-times weekly hemodialysis in incident patients with residual kidney function." *Kidney Int Rep* 2(5): 933–942.

Cirillo, M. (2010). "Evaluation of glomerular filtration rate and of albuminuria/proteinuria." *J Nephrol* 23(2): 125–132.

Craig, W. J. and A. R. Mangels (2009). "Position of the American Dietetic Association: vegetarian diets." *J Am Diet Assoc* 109(7): 1266–1282.

Cupisti, A., C. D'Alessandro, et al. (2016). "Nutritional support in the tertiary care of patients affected by chronic renal insufficiency: report of a step-wise, personalized, pragmatic approach." *BMC Nephrol* 17(1): 124.

Cuppari, L. and M. A. Kamimura (2011). "Dialysis: dietary phosphorus restriction: changing the paradigm?" *Nat Rev Nephrol* 7(5): 252–253.

Cyriac, J., K. Holden, et al. (2017). "How to use... urine dipsticks." *Arch Dis Child Educ Pract Ed* 102(3): 148–154.

D'Alessandro, C., G. B. Piccoli, et al. (2015). "The 'phosphorus pyramid': a visual tool for dietary phosphate management in dialysis and CKD patients." *BMC Nephrol* 16: 9.

D'Alessandro, C., G. B. Piccoli, et al. (2016). "'Dietaly': practical issues for the nutritional management of CKD patients in Italy." *BMC Nephrol* 17(1): 102.

De Broe, M. E., M. B. Gharbi, et al. (2017). "Why overestimate or underestimate chronic kidney disease when correct estimation is possible?" *Nephrol Dial Transplant* 32(Suppl 2): ii136–ii141.

de Roij van Zuijdewijn, C. L., M. P. Grooteman, et al. (2016). "Comparing tests assessing protein-energy wasting: relation with quality of life." *J Ren Nutr* 26(2): 111–117.

de Vries, A. P., P. Ruggenenti, et al. (2014). "Fatty kidney: emerging role of ectopic lipid in obesity-related renal disease." *Lancet Diabetes Endocrinol* 2(5): 417–426.

Deetman, P. E., M. Y. Said, et al. (2015). "Urinary urea excretion and long-term outcome after renal transplantation." *Transplantation* 99(5): 1009–1015.

Delanaye, P., E. Cavalier, et al. (2014). "Calibration and precision of serum creatinine and plasma cystatin C measurement: impact on the estimation of glomerular filtration rate." *J Nephrol* 27(5): 467–475.

Delanaye, P. and C. Mariat (2013). "The applicability of eGFR equations to different populations." *Nat Rev Nephrol* 9(9): 513–522.

Di Iorio, B. R., A. Cupisti, et al. (2018). "Nutritional therapy in autosomal dominant polycystic kidney disease." *J Nephrol* 31(5): 635–643.

Dickerson, R. N., A. C. Tidwell, et al. (2005). "Predicting total urinary nitrogen excretion from urinary urea nitrogen excretion in multiple-trauma patients receiving specialized nutritional support." *Nutrition* 21(3): 332–338.

Drawz, P. and M. Rahman (2015). "Chronic kidney disease." *Ann Intern Med* 162(11): ITC1-16.

Eisenga, M. F., L. M. Kieneker, et al. (2016). "Urinary potassium excretion, renal ammoniagenesis, and risk of graft failure and mortality in renal transplant recipients." *Am J Clin Nutr* 104(6): 1703–1711.

Eknoyan, G. (2011). "Obesity and chronic kidney disease." *Nefrologia* 31(4): 397–403.

el-Agroudy, A. E., E. W. Wafa, et al. (2004). "Weight gain after renal transplantation is a risk factor for patient and graft outcome." *Transplantation* 77(9): 1381–1385.

Erkan, E. (2013). "Proteinuria and progression of glomerular diseases." *Pediatr Nephrol* 28(7): 1049–1058.

Fein, P. A., G. Gundumalla, et al. (2002). "Usefulness of bioelectrical impedance analysis in monitoring nutrition status and survival of peritoneal dialysis patients." *Adv Perit Dial* 18: 195–199.

Feinstein, E. I. and J. D. Kopple (1985). "Severe wasting and malnutrition in a patient undergoing maintenance dialysis." *Am J Nephrol* 5(5): 398–405.

Filler, G. and M. Lee (2018). "Educational review: measurement of GFR in special populations." *Pediatr Nephrol* 33(11): 2037–2046.

Firat Kaya, D., B. Sayin, et al. (2017). "Obesity and loss of kidney function: two complications to face for older living kidney donors." *Exp Clin Transplant* 15(Suppl 1): 136–138.

Fogo, A. B., M. A. Lusco, et al. (2017). "AJKD atlas of renal pathology: nephrocalcinosis and acute phosphate nephropathy." *Am J Kidney Dis* 69(3): e17–e18.

Foster, M., U. N. Herulah, et al. (2015). "Zinc status of vegetarians during pregnancy: a systematic review of observational studies and meta-analysis of zinc intake." *Nutrients* 7(6): 4512–4525.

Fouque, D., J. Chen, et al. (2016). "Adherence to ketoacids/essential amino acids-supplemented low protein diets and new indications for patients with chronic kidney disease." *BMC Nephrol* 17(1): 63.

Fouque, D. and M. Laville (2009). "Low protein diets for chronic kidney disease in non diabetic adults." *Cochrane Database Syst Rev* (3): CD001892.

Friedman, A. N., A. S. Wahed, et al. (2018). "Effect of bariatric surgery on CKD risk." *J Am Soc Nephrol* 29(4): 1289–1300.

Galland, R. and J. Traeger (2004). "Short daily hemodialysis and nutritional status in patients with chronic renal failure." *Semin Dial* 17(2): 104–108.

Gambaro, G., E. Croppi, et al. (2016). "Metabolic diagnosis and medical prevention of calcium nephrolithiasis and its systemic manifestations: a consensus statement." *J Nephrol* 29(6): 715–734.

Garcia-Garcia, G. and V. Jha (2015). "World Kidney Day 2015: CKD in disadvantaged populations." *Am J Kidney Dis* 65(3): 349–353.

Garg, A. X., E. McArthur, et al. (2015). "Gestational hypertension and preeclampsia in living kidney donors." *N Engl J Med* 372(15): 1469–1470.

Garneata, L., A. Stancu, et al. (2016). "Ketoanalogue-supplemented vegetarian very low-protein diet and CKD progression." *J Am Soc Nephrol* 27(7): 2164–2176.

Giovannetti, S. and Q. Maggiore (1964). "A low-nitrogen diet with proteins of high biological value for severe chronic uraemia." *Lancet* 1(7341): 1000–1003.

Golper, T. A. (2017). "Incremental dialysis: review of recent literature." *Curr Opin Nephrol Hypertens* 26(6): 543–547.

Gorelik, S., J. Kanner, et al. (2012). "Additional ways to diminish the deleterious effects of red meat." *Arch Intern Med* 172(18): 1424–1425; author reply 1425.

Gorriz, J. L. and A. Martinez-Castelao (2012). "Proteinuria: detection and role in native renal disease progression." *Transplant Rev (Orlando)* 26(1): 3–13.

Goto, N., M. Okada, et al. (2016). "Association of dialysis duration with outcomes after transplantation in a Japanese cohort." *Clin J Am Soc Nephrol* 11(3): 497–504.

Graf, L., S. Candelaria, et al. (2007). "Nutrition assessment and hormonal influences on body composition in children with chronic kidney disease." *Adv Chronic Kidney Dis* 14(2): 215–223.

Grams, M. E., B. P. Chen, et al. (2013). "Preemptive deceased donor kidney transplantation: considerations of equity and utility." *Clin J Am Soc Nephrol* 8(4): 575–582.

Guest, S. (2013). "Hypoalbuminemia in peritoneal dialysis patients." *Adv Perit Dial* 29: 55–60.

Gupta, A., M. Mantan, et al. (2016). "Nutritional assessment in children with chronic kidney disease." *Saudi J Kidney Dis Transpl* 27(4): 733–739.

Hanafusa, N., B. T. Lodebo, et al. (2017). "Current uses of dietary therapy for patients with far-advanced CKD." *Clin J Am Soc Nephrol* 12(7): 1190–1195.

Harambat, J. and P. Cochat (2009). "Growth after renal transplantation." *Pediatr Nephrol* 24(7): 1297–1306.

Harris, A., B. A. Cooper, et al. (2011). "Cost-effectiveness of initiating dialysis early: a randomized controlled trial." *Am J Kidney Dis* 57(5): 707–715.

Harvey, K. B., M. J. Blumenkrantz, et al. (1980). "Nutritional assessment and treatment of chronic renal failure." *Am J Clin Nutr* 33(7): 1586–1597.

Hermans, R. C. J., H. de Bruin, et al. (2017). "Adolescents' responses to a school-based prevention program promoting healthy eating at school." *Front Public Health* 5: 309.

Hilton, P. J., Z. Roth, et al. (1969). "Creatinine clearance in patients with proteinuria." *Lancet* 2(7632): 1215–1216.

Hirschberg, R., H. Rottka, et al. (1985). "Effect of an acute protein load on the creatinine clearance in healthy vegetarians." *Klin Wochenschr* 63(5): 217–220.

Hodgkins, K. S. and H. W. Schnaper (2012). "Tubulointerstitial injury and the progression of chronic kidney disease." *Pediatr Nephrol* 27(6): 901–909.

Hoy, W. E., A. V. White, et al. (2015). "The multideterminant model of renal disease in a remote Australian Aboriginal population in the context of early life risk factors: lower birth weight, childhood post-streptococcal glomerulonephritis, and current body mass index influence levels of albuminuria in young Aboriginal adults." *Clin Nephrol* 83(7 Suppl 1): 75–81.

Ibrahim, H. N., S. K. Akkina, et al. (2009). "Pregnancy outcomes after kidney donation." *Am J Transplant* 9(4): 825–834.

Ichimaru, N., S. Nakazawa, et al. (2016). "Adherence to dietary recommendations in maintenance phase kidney transplant patients." *Transplant Proc* 48(3): 890–892.

Ikizler, T. A. (2014). "Using and interpreting serum albumin and prealbumin as nutritional markers in patients on chronic dialysis." *Semin Dial* 27(6): 590–592.

Ingulli, E. G. and R. H. Mak (2014). "Growth in children with chronic kidney disease: role of nutrition, growth hormone, dialysis, and steroids." *Curr Opin Pediatr* 26(2): 187–192.

Inker, L. A., B. C. Astor, et al. (2014). "KDOQI US commentary on the 2012 KDIGO clinical practice guideline for the evaluation and management of CKD." *Am J Kidney Dis* 63(5): 713–735.

Ipema, K. J., S. Struijk, et al. (2016). "Nutritional status in nocturnal hemodialysis patients - a systematic review with meta-analysis." *PLoS One* 11(6): e0157621.

Ishii, H., H. Takahashi, et al. (2017). "The association of ankle brachial index, protein-energy wasting, and inflammation status with cardiovascular mortality in patients on chronic hemodialysis." *Nutrients* 9(4).

Jankowska, M., A. R. Qureshi, et al. (2018). "Do metabolic derangements in end-stage polycystic kidney disease differ versus other primary kidney diseases?" *Nephrology (Carlton)* 23(1): 31–36.

Jay, C. L., K. Washburn, et al. (2017). "Survival benefit in older patients associated with earlier transplant with high KDPI kidneys." *Transplantation* 101(4): 867–872.

Jebeile, H., J. Mijatovic, et al. (2016). "A systematic review and metaanalysis of energy intake and weight gain in pregnancy." *Am J Obstet Gynecol* 214(4): 465–483.

Johansson, L. (2015). "Nutrition in older adults on peritoneal dialysis." *Perit Dial Int* 35(6): 655–658.

Jonker, J. T., P. de Heer, et al. (2018). "Metabolic imaging of fatty kidney in diabesity: validation and dietary intervention." *Nephrol Dial Transplant* 33(2): 224–230.

Kaiser, L. and L. H. Allen (2008). "Position of the American Dietetic Association: nutrition and lifestyle for a healthy pregnancy outcome." *J Am Diet Assoc* 108(3): 553–561.

Kaiser, L. L. and C. G. Campbell (2014). "Practice paper of the Academy of Nutrition and Dietetics abstract: nutrition and lifestyle for a healthy pregnancy outcome." *J Acad Nutr Diet* 114(9): 1447.

Kalantar-Zadeh, K., K. C. Abbott, et al. (2005). "Survival advantages of obesity in dialysis patients." *Am J Clin Nutr* 81(3): 543–554.

Kalantar Zadeh, K., T. A. Ikizler, et al. (2003). "Malnutrition-inflammation complex syndrome in dialysis patients: causes and consequences." *Am J Kidney Dis* 42(5): 864–881.

Kalantar-Zadeh, K., C. M. Rhee, et al. (2017). "The obesity paradox in kidney disease: how to reconcile it with obesity management." *Kidney Int Rep* 2(2): 271–281.

Kalantar-Zadeh, K., P. Stenvinkel, et al. (2003). "Inflammation and nutrition in renal insufficiency." *Adv Ren Replace Ther* 10(3): 155–169.

Kalantar-Zadeh, K., A. R. Tortorici, et al. (2015). "Dietary restrictions in dialysis patients: is there anything left to eat?" *Semin Dial* 28(2): 159–168.

Kang, S. S., J. W. Chang, et al. (2017). "Nutritional status predicts 10-year mortality in patients with end-stage renal disease on hemodialysis." *Nutrients* 9(4).

Kassirer, J. P. (1978). "Clinical assessment of the kidneys." *Prog Clin Pathol* 7: 33–48.

Kent, P. S. (2007). "Issues of obesity in kidney transplantation." *J Ren Nutr* 17(2): 107–113.

Kerr, D. N., A. Robson, et al. (1967). "Diet in chronic renal failure." *Proc R Soc Med* 60(2): 115–116.

Kilbride, H. S., P. E. Stevens, et al. (2013). "Accuracy of the MDRD (Modification of Diet in Renal Disease) study and CKD-EPI (CKD Epidemiology Collaboration) equations for estimation of GFR in the elderly." *Am J Kidney Dis* 61(1): 57–66.

Kim, H., H. Lim, et al. (2014). "Compromised diet quality is associated with decreased renal function in children with chronic kidney disease." *Clin Nutr Res* 3(2): 142–149.

Kincaid-Smith, P. and K. Fairley (2005). "The investigation of hematuria." *Semin Nephrol* 25(3): 127–135.

Koetje, P. M., J. J. Spaan, et al. (2011). "Pregnancy reduces the accuracy of the estimated glomerular filtration rate based on Cockroft-Gault and MDRD formulas." *Reprod Sci* 18(5): 456–462.

Koletzko, B., C. P. Bauer, et al. (2013). "German national consensus recommendations on nutrition and lifestyle in pregnancy by the 'Healthy Start - Young Family Network'." *Ann Nutr Metab* 63(4): 311–322.

Kolpek, J. H., L. G. Ott, et al. (1989). "Comparison of urinary urea nitrogen excretion and measured energy expenditure in spinal cord injury and nonsteroid-treated severe head trauma patients." *JPEN J Parenter Enteral Nutr* 13(3): 277–280.

Komatsu, M., M. Okazaki, et al. (2015). "Geriatric nutritional risk index is a simple predictor of mortality in chronic hemodialysis patients." *Blood Purif* 39(4): 281–287.

Kontessis, P. A., I. Bossinakou, et al. (1995). "Renal, metabolic, and hormonal responses to proteins of different origin in normotensive, nonproteinuric type I diabetic patients." *Diabetes Care* 18(9): 1233.

Kopple, J. D., M. E. Swendseid, et al. (1975). "Recommendations for nutritional evaluation of patients on chronic dialysis." *Kidney Int Suppl* (2): 249–252.

Kovesdy, C. P., S. L. Furth, et al. (2017a). "Obesity and kidney disease: hidden consequences of the epidemic." *J Nephrol* 30(1): 1–10.

Kovesdy, C. P., S. L. Furth, et al. (2017b). "Obesity and kidney disease: hidden consequences of the epidemic." *Clin Kidney J* 10(1): 1–8.

Kumar, B. V. and T. Mohan (2017). "Retrospective comparison of estimated GFR using 2006 MDRD, 2009 CKD-EPI and Cockcroft-Gault with 24 hour urine creatinine clearance." *J Clin Diagn Res* 11(5): BC09–BC12.

Kurella, M., K. E. Covinsky, et al. (2007). "Octogenarians and nonagenarians starting dialysis in the United States." *Ann Intern Med* 146(3): 177–183.

Lastra, G., C. Manrique, et al. (2006). "Obesity, cardiometabolic syndrome, and chronic kidney disease: the weight of the evidence." *Adv Chronic Kidney Dis* 13(4): 365–373.

Levey, A. S., T. Greene, et al. (1999). "Dietary protein restriction and the progression of chronic renal disease: what have all of the results of the MDRD study shown? Modification of Diet in Renal Disease Study group." *J Am Soc Nephrol* 10(11): 2426–2439.

Levey, A. S., L. A. Inker, et al. (2014). "GFR estimation: from physiology to public health." *Am J Kidney Dis* 63(5): 820–834.

Lindsay, R. M. and J. Bergstrom (1994). "Membrane biocompatibility and nutrition in maintenance haemodialysis patients." *Nephrol Dial Transplant* 9(Suppl 2): 150–155.

Lindsay, R. M. and L. W. Henderson (1988). "Adequacy of dialysis." *Kidney Int Suppl* 24: S92–S99.

Locatelli, F., L. Del Vecchio, et al. (2017). "Nutritional issues with incremental dialysis: the role of low-protein diets." *Semin Dial* 30(3): 246–250.

Lovshin, J. A., M. Skrtic, et al. (2018). "Hyperfiltration, urinary albumin excretion, and ambulatory blood pressure in adolescents with Type 1 diabetes mellitus." *Am J Physiol Renal Physiol* 314(4): F667–F674.

Lozano-Kasten, F., E. Sierra-Diaz, et al. (2017). "Prevalence of albuminuria in children living in a rural agricultural and fishing subsistence community in Lake Chapala, Mexico." *Int J Environ Res Public Health* 14(12).

Lv, J., C. Yu, et al. (2017). "Adherence to healthy lifestyle and cardiovascular diseases in the Chinese population." *J Am Coll Cardiol* 69(9): 1116–1125.

Madero, M. and M. J. Sarnak (2011). "Creatinine-based formulae for estimating glomerular filtration rate: is it time to change to chronic kidney disease epidemiology collaboration equation?" *Curr Opin Nephrol Hypertens* 20(6): 622–630.

Mansy, H., T. H. Goodship, et al. (1989). "Effect of a high protein diet in patients with the nephrotic syndrome." *Clin Sci (Lond)* 77(4): 445–451.

Maroni, B. J., T. I. Steinman, et al. (1985). "A method for estimating nitrogen intake of patients with chronic renal failure." *Kidney Int* 27(1): 58–65.

Martin-Cleary, C. and A. Ortiz (2014). "CKD hotspots around the world: where, why and what the lessons are. A CKJ review series." *Clin Kidney J* 7(6): 519–523.

Masud, T., A. Manatunga, et al. (2002). "The precision of estimating protein intake of patients with chronic renal failure." *Kidney Int* 62(5): 1750–1756.

McIntyre, C. W. and S. J. Rosansky (2012). "Starting dialysis is dangerous: how do we balance the risk?" *Kidney Int* 82(4): 382–387.

Mehrotra, R., M. Rivara, et al. (2013). "Initiation of dialysis should be timely: neither early nor late." *Semin Dial* 26(6): 644–649.

Meier-Kriesche, H. U. and J. D. Schold (2005). "The impact of pretransplant dialysis on outcomes in renal transplantation." *Semin Dial* 18(6): 499–504.

Millward, D. J. (2012). "Identifying recommended dietary allowances for protein and amino acids: a critique of the 2007 WHO/FAO/UNU report." *Br J Nutr* 108(Suppl 2): S3–S21.

Mitch, W. E. and G. Remuzzi (2004). "Diets for patients with chronic kidney disease, still worth prescribing." *J Am Soc Nephrol* 15(1): 234–237.

Mitch, W. E. and G. Remuzzi (2016). "Diets for patients with chronic kidney disease, should we reconsider?" *BMC Nephrol* 17(1): 80.

Moe, S. M., M. P. Zidehsarai, et al. (2011). "Vegetarian compared with meat dietary protein source and phosphorus homeostasis in chronic kidney disease." *Clin J Am Soc Nephrol* 6(2): 257–264.

Morena, M., A. Jaussent, et al. (2017). "Treatment tolerance and patient-reported outcomes favor online hemodiafiltration compared to high-flux hemodialysis in the elderly." *Kidney Int* 91(6): 1495–1509.

Munoz-Perez, E., M. L. A. Espinosa-Cuevas, et al. (2017). "Combined assessment of nutritional status in patients with peritoneal dialysis using bioelectrical impedance vectors and malnutrition inflammation score." *Nutr Hosp* 34(5): 1125–1132.

Nagaoka, Y., R. Onda, et al. (2016). "Dietary intake in Japanese patients with kidney transplantation." *Clin Exp Nephrol* 20(6): 972–981.

Nehus, E. (2018). "Obesity and chronic kidney disease." *Curr Opin Pediatr* 30(2): 241–246.

Nogueira, J. M., M. R. Weir, et al. (2010). "A study of renal outcomes in obese living kidney donors." *Transplantation* 90(9): 993–999.

North, K. and J. Golding (2000). "A maternal vegetarian diet in pregnancy is associated with hypospadias. The ALSPAC Study Team. Avon Longitudinal Study of Pregnancy and Childhood." *BJU Int* 85(1): 107–113.

Norton, J. M., M. M. Moxey-Mims, et al. (2016). "Social determinants of racial disparities in CKD." *J Am Soc Nephrol* 27(9): 2576–2595.

Obi, Y. and K. Kalantar-Zadeh (2017). "Incremental and once- to twice-weekly hemodialysis: from experience to evidence." *Kidney Int Rep* 2(5): 781–784.

Ornish, D. (2012). "Holy Cow! What's good for you is good for our planet: comment on 'Red Meat Consumption and Mortality'." *Arch Intern Med* 172(7): 563–564.

Oscanoa, T. J., J. P. Amado, et al. (2018). "Estimation of the glomerular filtration rate in older individuals with serum creatinine-based equations: a systematic comparison between CKD-EPI and BIS1." *Arch Gerontol Geriatr* 75: 139–145.

Packham, D. K., S. E. Ivory, et al. (2011). "Proteinuria in type 2 diabetic patients with renal impairment: the changing face of diabetic nephropathy." *Nephron Clin Pract* 118(4): c331–c338.

Panichi, V., A. Cupisti, et al. (2014). "Geriatric nutritional risk index is a strong predictor of mortality in hemodialysis patients: data from the Riscavid cohort." *J Nephrol* 27(2): 193–201.

Pecoits-Filho, R., B. Lindholm, et al. (2002). "The malnutrition, inflammation, and athero-sclerosis (MIA) syndrome – the heart of the matter." *Nephrol Dial Transplant* 17(Suppl 11): 28–31.

Pedrollo, E. F., B. B. Nicoletto, et al. (2017). "Effect of an intensive nutrition intervention of a high protein and low glycemic-index diet on weight of kidney transplant recipients: study protocol for a randomized clinical trial." *Trials* 18(1): 413.

Pettigrew, S., Z. Talati, et al. (2018). "Stakeholder perceptions of a school food policy ten years on." *Public Health Nutr* 21(7): 1370–1374.

Piccoli, G. B. (2010). "Patient-based continuum of care in nephrology: why read Thomas Addis' 'Glomerular Nephritis' in 2010?" *J Nephrol* 23(2): 164–167.

Piccoli, G. B., M. Alrukhaimi, et al. (2018). "[Questions unanswered and answers unques-tioned: what we do and do not know about women and kidney diseases. Reflection on World Kidney Day and International Women's Day]." *G Ital Nefrol* 35(1).

Piccoli, G. B., I. Capizzi, et al. (2016). "Low protein diets in patients with chronic kidney dis-ease: a bridge between mainstream and complementary-alternative medicines?" *BMC Nephrol* 17(1): 76.

Piccoli, G. B., R. Clari, et al. (2015). "Vegan-vegetarian diets in pregnancy: danger or pana-cea? A systematic narrative review." *BJOG* 122(5): 623–633.

Piccoli, G. B. and A. Cupisti (2017). "'Let food be thy medicine...': lessons from low-protein diets from around the world." *BMC Nephrol* 18(1): 102.

Piccoli, G. B., A. De Pascale, et al. (2016). "Revisiting nephrocalcinosis: a single-centre per-spective. A northern Italian experience." *Nephrology (Carlton)* 21(2): 97–107.

Piccoli, G. B., M. Ferraresi, et al. (2013). "Vegetarian low-protein diets supplemented with keto analogues: a niche for the few or an option for many?" *Nephrol Dial Transplant* 28(9): 2295–2305.

Piccoli, G. B., F. Leone, et al. (2014). "Association of low-protein supplemented diets with fetal growth in pregnant women with CKD." *Clin J Am Soc Nephrol* 9(5): 864–873.

Piccoli, G. B., M. R. Moio, et al. (2017). "The diet and haemodialysis dyad: three eras, four open questions and four paradoxes. A narrative review, towards a personalized, patient-centered approach." *Nutrients* 9(4).

Piccoli, G. B., M. Nazha, et al. (2016). "Diet as a system: an observational study investigat-ing a multi-choice system of moderately restricted low-protein diets." *BMC Nephrol* 17(1): 197.

Piccoli, G. B., F. Ventrella, et al. (2016). "Low-protein diets in diabetic chronic kidney disease (CKD) patients: are they feasible and worth the effort?" *Nutrients* 8(10).

Pollock, C. A., L. S. Ibels, et al. (1995). "Total body nitrogen as a prognostic marker in maintenance dialysis." *J Am Soc Nephrol* 6(1): 82–88.

Procter, S. B. and C. G. Campbell (2014). "Position of the Academy of Nutrition and Dietetics: nutrition and lifestyle for a healthy pregnancy outcome." *J Acad Nutr Diet* 114(7): 1099–1103.

Pugh, P., P. Hemingway, et al. (2018). "Children's, parents' and other stakeholders' perspectives on early dietary self-management to delay disease progression of chronic disease in children: a protocol for a mixed studies systematic review with a narrative synthesis." *Syst Rev* 7(1): 20.

Raghavan, R. and G. Eknoyan (2014). "Acute interstitial nephritis - a reappraisal and update." *Clin Nephrol* 82(3): 149–162.

Raja, R. M., G. Ijelu, et al. (1992). "Influence of Kt/V and protein catabolic rate on hemodialysis morbidity. A long-term study." *ASAIO J* 38(3): M179–M180.

Ramakrishnan, U., F. Grant, et al. (2012). "Effect of women's nutrition before and during early pregnancy on maternal and infant outcomes: a systematic review." *Paediatr Perinat Epidemiol* 26(Suppl 1): 285–301.

Rees, L. and R. H. Mak (2011). "Nutrition and growth in children with chronic kidney disease." *Nat Rev Nephrol* 7(11): 615–623.

Reisaeter, A. V., J. Roislien, et al. (2009). "Pregnancy and birth after kidney donation: the Norwegian experience." *Am J Transplant* 9(4): 820–824.

Rippe, B. and C. M. Oberg (2016). "Albumin turnover in peritoneal and hemodialysis." *Semin Dial* 29(6): 458–462.

Ritz, E., N. Koleganova, et al. (2011). "Is there an obesity-metabolic syndrome related glomerulopathy?" *Curr Opin Nephrol Hypertens* 20(1): 44–49.

Robles, N. R., J. Villa, et al. (2015). "Non-proteinuric diabetic nephropathy." *J Clin Med* 4(9): 1761–1773.

Rodriguez-Iturbe, B. (1986). "The functional reserve capacity of the kidney." *Int J Artif Organs* 9(2): 81–84.

Rosansky, S. J. (2011). "The European Renal Best Practice Advisory Board's dialysis initiation guidelines: one size won't fit all." *Kidney Int* 80(10): 1005–1007.

Rosansky, S. J., G. Cancarini, et al. (2013). "Dialysis initiation: what's the rush?" *Semin Dial* 26(6): 650–657.

Ruchat, S. M., C. Allard, et al. (2016). "Timing of excessive weight gain during pregnancy modulates newborn anthropometry." *J Obstet Gynaecol Can* 38(2): 108–117.

Said, M. Y., P. E. Deetman, et al. (2015). "Causal path analyses of the association of protein intake with risk of mortality and graft failure in renal transplant recipients." *Clin Transplant* 29(5): 447–457.

Salahudeen, A. K. (2003). "Obesity and survival on dialysis." *Am J Kidney Dis* 41(5): 925–932.

Scales, C. D., Jr., G. E. Tasian, et al. (2016). "Urinary stone disease: advancing knowledge, patient care, and population health." *Clin J Am Soc Nephrol* 11(7): 1305–1312.

Schwingshackl, L. and G. Hoffmann (2014). "Comparison of high vs. normal/low protein diets on renal function in subjects without chronic kidney disease: a systematic review and meta-analysis." *PLoS One* 9(5): e97656.

Sherman, R. A. (2016). "Hyperphosphatemia in dialysis patients: beyond nonadherence to diet and binders." *Am J Kidney Dis* 67(2): 182–186.

Shinaberger, C. S., S. Greenland, et al. (2008). "Is controlling phosphorus by decreasing dietary protein intake beneficial or harmful in persons with chronic kidney disease?" *Am J Clin Nutr* 88(6): 1511–1518.

Silvererg, D. S. and J. C. Hunt (1966). "Dietary considerations in treating chronic renal failure." *J Am Diet Assoc* 49(5): 425–427.

Smith, M. C., P. Moran, et al. (2008). "Assessment of glomerular filtration rate during pregnancy using the MDRD formula." *BJOG* 115(1): 109–112.

So, R., S. Song, et al. (2016). "The association between renal hyperfiltration and the sources of habitual protein intake and dietary acid load in a general population with preserved renal function: the KoGES Study." *PLoS One* 11(11): e0166495.

Sood, M. M., B. Manns, et al. (2014). "Using the knowledge-to-action framework to guide the timing of dialysis initiation." *Curr Opin Nephrol Hypertens* 23(3): 321–327.

Spanner, E., R. Suri, et al. (2003). "The impact of quotidian hemodialysis on nutrition." *Am J Kidney Dis* 42(1 Suppl): 30–35.

St-Jules, D. E., D. S. Goldfarb, et al. (2016). "Nutrient non-equivalence: does restricting high-potassium plant foods help to prevent hyperkalemia in hemodialysis patients?" *J Ren Nutr* 26(5): 282–287.

Stellato, T., R. S. Rhodes, et al. (1980). "Azotemia in upper gastrointestinal hemorrhage. A review." *Am J Gastroenterol* 73(6): 486–489.

Stenvinkel, P., O. Heimburger, et al. (1999). "Strong association between malnutrition, inflammation, and atherosclerosis in chronic renal failure." *Kidney Int* 55(5): 1899–1911.

Stevens, L. A., S. Li, et al. (2010). "Prevalence of CKD and comorbid illness in elderly patients in the United States: results from the Kidney Early Evaluation Program (KEEP)." *Am J Kidney Dis* 55(3 Suppl 2): S23–S33.

Takahashi, H., K. Inoue, et al. (2017). "Comparison of nutritional risk scores for predicting mortality in Japanese chronic hemodialysis patients." *J Ren Nutr* 27(3): 201–206.

Tapson, J. S., H. Mansy, et al. (1986). "Renal functional reserve in kidney donors." *Q J Med* 60(231): 725–732.

Tattersall, J., F. Dekker, et al. (2011). "When to start dialysis: updated guidance following publication of the Initiating Dialysis Early and Late (IDEAL) study." *Nephrol Dial Transplant* 26(7): 2082–2086.

Taylor, J. M., J. M. Hamilton-Reeves, et al. (2017). "Diet and polycystic kidney disease: a pilot intervention study." *Clin Nutr* 36(2): 458–466.

Teplan, V. (2009). "Effect of keto acids on asymmetric dimethylarginine, muscle, and fat tissue in chronic kidney disease and after kidney transplantation." *J Ren Nutr* 19(5 Suppl): S27–S29.

Teplan, V., I. Valkovsky, et al. (2009). "Nutritional consequences of renal transplantation." *J Ren Nutr* 19(1): 95–100.

Tran, W. C., D. Huynh, et al. (2017). "Understanding barriers to medication, dietary, and lifestyle treatments prescribed in polycystic kidney disease." *BMC Nephrol* 18(1): 214.

van Ginhoven, T. M., R. W. de Bruin, et al. (2011). "Pre-operative dietary restriction is feasible in live-kidney donors." *Clin Transplant* 25(3): 486–494.

Vanholder, R., N. Lameire, et al. (2016). "Cost of renal replacement: how to help as many as possible while keeping expenses reasonable?" *Nephrol Dial Transplant* 31(8): 1251–1261.

Vega, A., B. Quiroga, et al. (2015). "Albumin leakage in online hemodiafiltration, more convective transport, more losses?" *Ther Apher Dial* 19(3): 267–271.

Visconti, L., V. Cernaro, et al. (2018). "The myth of water and salt: from aquaretics to tenapanor." *J Ren Nutr* 28(2): 73–82.

Walser, M., W. E. Mitch, et al. (1999). "Should protein intake be restricted in predialysis patients?" *Kidney Int* 55(3): 771–777.

Wang, A. Y., M. M. Sea, et al. (2005). "Evaluation of handgrip strength as a nutritional marker and prognostic indicator in peritoneal dialysis patients." *Am J Clin Nutr* 81(1): 79–86.

Ward, H. J. (2009). "Nutritional and metabolic issues in solid organ transplantation: targets for future research." *J Ren Nutr* 19(1): 111–122.

Watnick, S. (2013). "Financial and medico-legal implications of late-start dialysis: understanding the present policies through a window of past performance." *Semin Dial* 26(6): 702–705.

Webster, A. C., E. V. Nagler, et al. (2017). "Chronic kidney disease." *Lancet* 389(10075): 1238–1252.

Widmer, R. J., A. J. Flammer, et al. (2015). "The Mediterranean diet, its components, and cardiovascular disease." *Am J Med* 128(3): 229–238.

Wiseman, M. J., R. Hunt, et al. (1987). "Dietary composition and renal function in healthy subjects." *Nephron* 46(1): 37–42.

Wissing, K. M. and L. Pipeleers (2014). "Obesity, metabolic syndrome and diabetes mellitus after renal transplantation: prevention and treatment." *Transplant Rev (Orlando)* 28(2): 37–46.

World Health Organization (2007). "Protein and amino acid requirements in human nutrition." *World Health Organ Tech Rep Ser* (935): 1–265, back cover.

Wolfe, R. R., A. M. Cifelli, et al. (2017). "Optimizing protein intake in adults: interpretation and application of the recommended dietary allowance compared with the acceptable macronutrient distribution range." *Adv Nutr* 8(2): 266–275.

Yadin, O. (1994). "Hematuria in children." *Pediatr Ann* 23(9): 474–478, 481–475.

Zhang, J., H. Xie, et al. (2016). "Keto-supplemented low protein diet: a valid therapeutic approach for patients with steroid-resistant proteinuria during early-stage chronic kidney disease." *J Nutr Health Aging* 20(4): 420–427.

Zhang, J. Y., Y. Yin, et al. (2016). "Low-protein diet supplemented with ketoacids ameliorates proteinuria in 3/4 nephrectomised rats by directly inhibiting the intrarenal renin-angiotensin system." *Br J Nutr* 116(9): 1491–1501.

Zoccali, C., R. Vanholder, et al. (2017). "The systemic nature of CKD." *Nat Rev Nephrol* 13(6): 344–358.

10 Personalized Nutrition in Hypercholesterolemia

Aktarul Islam Siddique and Nalini Namasivayam

CONTENTS

10.1   PERSONALIZED NUTRITION IN HYPERCHOLESTEROLEMIA

Hypercholesterolemia remains the major risk factor for the development of cardiovascular diseases (CVD). CVD includes a variety of disease conditions affecting heart and blood vessels which commonly includes coronary heart disease (CHD) or ischemic heart disease or coronary artery disease and atherosclerosis. Age, gender and genetics are the most common immutable risk factors and hypercholesterolemia, hypertension, hyperlipidemia, obesity and diabetes are modifiable risk factors that play a major role in the causation and progression of the disease. High blood cholesterol is due to increased biosynthesis or reduced elimination of cholesterol from the body.

In our system, cholesterol may be generated either by *de novo* synthesis or derived from the diet. *De novo* cholesterol biosynthesis takes place in the liver, and the rate-determining reaction is catalyzed by the enzyme HMG CoA reductase. Later, the absorbed or synthesized cholesterol in the liver is distributed to the peripheral tissues through lipoproteins. The hepatic low-density lipoprotein receptor (LDLR) recaptures LDL cholesterol from the blood (Chen, Ma et al. 2011), or otherwise it accumulates, leading to atherosclerosis. Cholesterol is excreted from our body mainly through bile as bile acids.

10.2   CARDIOVASCULAR DISEASES (CVD)

CVDs persist as the major cause of morbidity and mortality around the world (Smith, Collins et al. 2012). It accounts for nearly 801,000 deaths in the US which is about one of every three deaths in the US (Benjamin, Blaha et al. 2017). Each day about 2,200 Americans die of CVD at an average of one death every 40 seconds. Globally, CVD is the foremost cause of death, accounting for more than 17.3 million deaths per year in 2013, a number that is estimated to grow to more than 23.6 million by 2030. The mortality due to CVD is higher compared to all forms of cancer and chronic lower respiratory diseases combined (Benjamin, Blaha et al. 2017).

10.3   CORONARY HEART DISEASE (CHD)

CHD is attributable to CVDs, which is the leading cause of death in the US that accounts for 45.1% of deaths, followed by stroke (16.5%), heart failure (8.5%), high blood pressure (9.1%), diseases of the arteries (3.2%) and other CVDs (Benjamin,

Blaha et al. 2017). CHD, ischemic heart disease or coronary artery disease is the term given to heart problems caused by constricted heart (coronary) arteries that supply blood to the heart muscle.

The major factor responsible for CHD mortality is the elevated plasma cholesterol concentration (Shipley, Pocock et al. 1991). Previous epidemiological and nutritional studies document that the incidence of CHD is correlated with the high concentration of total serum cholesterol and low-density lipoprotein (LDL) cholesterol. Thus, hypercholesterolemia has been reported to be the primary and an important causative factor underlying CHD (Singh, Rastogi et al. 1998). The higher incidence of CHD could be due to the higher intake of saturated fat or animal fat, which is found to be one of the major dietary factors in hypercholesterolemia (Ockene and Nicolosi 1998). Further, increased saturated fat consumption has been reported to affect cholesterol and triglyceride metabolism (Belahsen 2014). In this regard, a reduction in high serum cholesterol levels can reduce the risk of CHD.

10.4   ATHEROSCLEROSIS

Atherosclerosis, the most common underlying cause of many CVDs, is characterized by lipid and cholesterol deposition on the inner lining of blood vessels, resulting in plaque formation and eventual narrowing of the lumen (Ross 1999). It is by far the most recurrent underlying cause of coronary artery disease, carotid artery disease and peripheral arterial disease. Atherosclerosis alone is hardly fatal. When thrombosis occurs, it is superimposed on a ruptured or eroded atherosclerotic plaque, and thus precipitates the lifethreatening clinical events such as acute coronary syndromes and stroke (Naghavi, Libby et al. 2003; Spagnoli, Mauriello et al. 2004).

10.5   NUTRITIONAL THERAPY

Nutritional therapy is an evidence-based approach to changes in the diet, by adjusting quantity, quality and methods of nutrient intake to maximizing one's health potential. It mainly uses functional foods, nutraceuticals and dietary supplements to trigger the body's natural curing system. To achieve this goal, personalized nutrition should fulfill the criteria for detoxifying the body, averting vitamin and mineral deficiencies, by following healthy dietary habits. There are various factors that are involved in maintaining the overall good health of an individual. Besides a balanced nutritious diet, other factors including exposure to sunlight, exercise, fresh atmosphere and self-control also play a vital role in building healthy immunity.

Previous studies have documented an association between the risk for CVD and abdominal obesity and abnormal lipid profiles. Therefore, personalized nutrition for the prevention of CVD is focused largely on weight reduction and normalization of lipid profile through diet, exercise and medication.

A novel area of development in the field of personalized nutrition also includes the time factor. In this context, current genome-wide association studies showed a link between CLOCK (clock circadian regulator) genes and fasting glucose concentrations, obesity and metabolic syndrome, emphasizing the impact of circadian systems on human disease.

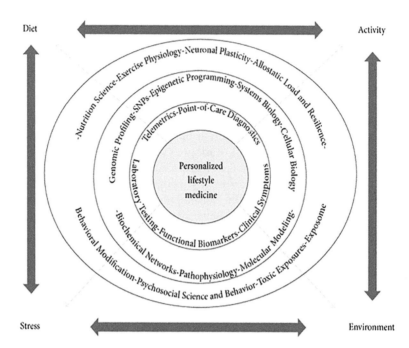

FIGURE 10.1 Personalized lifestyle medicine. (Adapted from Minich and Bland 2013.)

Personalized lifestyle medicine presents a system of medicine through the psychosocialbehavioral interface, which combines technological advances with the traditional foundation of lifestyle (Figure 10.1).

10.6 NUTRITIONAL DIETARY RESPONSE AND INDIVIDUAL VARIABILITY

The dietary response for each individual in a group varies dramatically, although the average response can be predicted for a group of individuals. It was documented that the difference in each individual may be due to the individual's genetic makeup. Epidemiologic, familial and twin studies have reported a strong genetic predisposition for CVD (Chan and Boerwinkle 1994). Studies demonstrated a link between a growing number of variants of select target genes and the risk for CVD (Ioannidis 2009; Roberts and Stewart 2012). Individuals in a group both between and within races or genders do not respond to drugs or dietary nutrients in a similar fashion. It leads to the interpretation that a reciprocal relationship might exist between the nutrients and genes.

10.7 PERSONALIZED NUTRITION AND HISTORY

From ancient times, about 3,000 years ago, the role of nutrition in health has been well documented in the writings of Ayurvedic and Chinese medicine. A prescription for personalized diet, exercise and lifestyle was always included by Ayurvedic physicians in any patient care plan during the early days. Hippocrates (460–375 BC), the Greek physician who is considered the father of modern medicine, placed great significance on nutrition in both the prevention and treatment of diseases. Only from the start of the 1900s, was there a better understanding of the metabolic pathways and also the role of macronutrients and micronutrients in health (Carpenter 2003). Moreover, a relationship between food shortages and a lower incidence of coronary artery disease in northern Europe was observed by Keys (1970). This observation was further studied in seven countries, which revealed that societies which consumed high fat (Finland and USA) encountered higher blood cholesterol levels and deaths by heart attack than those where the diets comprised of fresh fruits and vegetables. Thus, preventive efforts are warranted for the entire population rather than only the high-risk individuals (Nestle 1995; Andrade, Mohamed et al. 2009).

10.8 PERSONALIZED NUTRITION

Diet therapy continues to be the first-line treatment for patients with hypercholesterolemia. It plays a pivotal role in regulating blood cholesterol levels. A diet low in cholesterol and fat, especially saturated fats, and high in fiber is beneficial for hypercholesterolemia patients.

A collaborative study conducted by Eilat-Adar, Sinai et al. (2013) categorized nutritional information into three main sections (Figure 10.2). Firstly 'dietary patterns,' which include a low-carbohydrate diet, a low-fat diet, the Mediterranean diet, and the Dietary Approach to Stop Hypertension (DASH) diet. The second section,

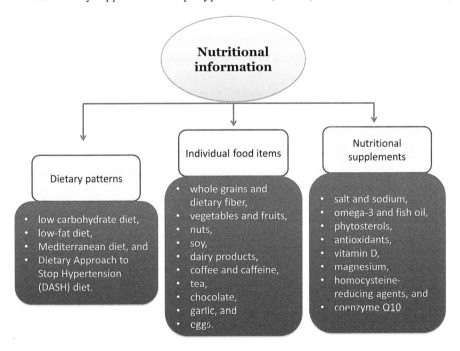

FIGURE 10.2 Nutritional information. (From Eilat-Adar et al. 2013.)

'individual food item,' includes wholegrains and dietary fiber, vegetables and fruits, nuts, soy, dairy products, coffee and caffeine, tea, chocolate, garlic and eggs. And the third section, 'nutritional supplements,' includes salt and sodium, omega-3 and fish oil, phytosterols, antioxidants, vitamin D, magnesium, homocysteine-reducing agents, and coenzyme Q10 (Eilat-Adar, Sinai et al. 2013).

10.8.1   Low-Fat Diets

The consumption of a low-fat diet is generally recommended for the prevention of CVD. The recommended low-fat diet can be achieved by choosing low-fat meats, low-fat dairy products and lowering the intake of foods containing trans-fat (Lichtenstein, Appel et al. 2006).

10.8.2   Low-Carbohydrate Diets

A low-carbohydrate diet is defined as the consumption of 30–130 g of carbohydrates per day or up to 45% of total calories (Hite, Berkowitz et al. 2011). A systematic meta-analysis study involving 1,141 obese subjects documented that low-carbohydrate diets are associated with a significant decrease in body weight, body mass index, systolic blood pressure, diastolic blood pressure, plasma triglycerides and an increase in high density lipoprotein (HDL)-cholesterol levels (Santos, Esteves et al. 2012).

10.8.3   Mediterranean Diet

The Mediterranean diet is characterized by a relatively high fat intake of about 40%–50% of total daily calories, of which saturated fat comprises ≤8% and mono-unsaturated fats 15%–25% of calories. According to a meta-analysis of seven cohort studies, adherence to the Mediterranean diet was found to be associated with a low risk of CHD with a significant decrease of overall mortality (Sofi et al. 2010).

10.8.4   Dietary Approach to Stop Hypertension (DASH) Diet

In the year 1990, the DASH diet was assembled and assessed in intervention control trials. The main objective of the DASH diet was to reduce the blood pressure and thereby the incidence of CVD by nutritional means. The recommended DASH diet includes vegetables and fruits, low-fat dairy products, wholegrains, chicken, fish and nuts. The overall DASH diet (Table 10.1) includes high calcium, potassium, magnesium and dietary fiber and lower amounts of fats, saturated fats, cholesterol and sodium than the typical Western diet (Appel, Moore et al. 1997).

It is reported that the DASH diet reduced the systolic and diastolic blood pressure by 11.4 and 5.5 mmHg respectively, and by 7.2 and 2.8 mmHg respectively in patients documented with hypertension as compared with the typical Western diets (Sacks, Obarzanek et al. 1995). When strict DASH diet was followed the theoretical decrease in the Framingham risk score for CHD was found to be 12% greater (Maruthur, Wang et al. 2009).

TABLE 10.1
Dietary Approach to Stop Hypertension (DASH) Diet Composition

Nutrient	Daily Quantity
Total fat	27% of total calories
Saturated fats	6% of total calories
Carbohydrates	55% of total calories
Protein	18% of total calories
Cholesterol	150 mg
Fiber	31 g
Potassium	4,700 mg
Magnesium	5,00 mg
Calcium	1,240 mg

Adapted from Eilat-Adar et al. (2013)

Thus, the dietary patterns explained above are proven to reduce the CVD risk factors as well as CVD mortality (Sacks, Svetkey et al. 2001).

10.8.5 DIETARY FIBER

It is reported that soluble fiber contributes to LDL reduction which is now recommended as a part of hyperlipidemia dietary regimen. Good sources of soluble fibres are fruits, vegetables, legumes, cereals, oats and whole-grains. Kelly, Summerbell et al. (2007) conducted a study in individuals with CHD or CHD risk factors supplemented with 56–85 g of fiber for the duration of four to eight weeks. The data obtained revealed that the consumption of whole-grains decreased the total cholesterol levels by 7.7 mg/dL and LDL cholesterol levels by 6.9 mg/dL (Kelly, Summerbell et al. 2007). Fiber helps in blocking of cholesterol from being absorbed into the bloodstream. Hypercholesterolemic patients are recommended to consume at least 30–40 g of fibre each day. Thus, studies reveal that diets based on plant materials are linked with a decrease in LDL cholesterol and total cholesterol of up to 15%. However, lipid levels may vary despite significant changes in the intake of fat and cholesterol, which may be due to genetic factors (Wallace, Mann et al. 2000). Another meta-analysis study by (Brown, Rosner et al. 1999) revealed that daily consumption of 2–10 g/day soluble fiber (mostly beta-glucan, psyllium, and pectin) lowered LDL cholesterol by 2.2 mg/dL, but there were no significant changes in HDL cholesterol or triglycerides.

The guidelines included in the American Heart Association, the American Dietetic Association and the National Cholesterol Education Program recommend increasing the intake of dietary soluble fiber (Marlett, McBurney et al. 2002; Lichtenstein, Appel et al. 2006). Besides, the Food and Drug Administration (2008) recommended the intake of soluble fiber from whole oats, whole-grain barley products, and barley beta fiber.

10.8.6 FRUITS AND VEGETABLES

High consumption of fruits and vegetables are considered to be part of a healthy diet. Fruits and vegetables consist of high amounts of protective ingredients including potassium, folate, vitamins, fiber and other phenolic compounds (Van Duyn and Pivonka 2000). These nutrients play a vital role in lowering blood pressure and antioxidant stress and improving lipoprotein profile, insulin sensitivity and hemostasis regulation (Appel, Moore et al. 1997; Van Duyn and Pivonka 2000; Bazzano, Serdula et al. 2003). Various cohort studies have addressed the relationship between intake of vegetables and fruits and CHD. In a pooled analysis of cohort studies, Pereira, O'Reilly et al. (2004) reported an inverse association between intake of fiber from fruit and vegetables and CHD risk. Fruit and vegetable consumption is also reported to lower blood pressure (Maruthur, Wang et al. 2009). Besides, other meta-analysis studies documented an inverse link between consumption of fruits and vegetables and the incidence of stroke (Dauchet, Amouyel et al. 2005; He, Nowson et al. 2006).

10.8.7 NUTS

Nuts are nutrient-dense foods, rich in unsaturated fatty acids and other bioactive compounds including minerals, fibers, tocopherols, phytosterols and phenolic compounds (Brufau, Boatella et al. 2006). Epidemiological data shows consumption of nuts is associated with reduction in CHD (Sabate and Ang 2009). A study conducted by Kris-Etherton, Hu et al. (2008) reported that high intake of nuts is associated with 35% reduction in CVD. Further, a pooled analysis was performed using data from 25 intervention trials of nut consumption among normolipidemic and hypercholesterolemia subjects who were not on lipid-lowering drugs. Daily consumption of 67 g of nuts reduced the LDL cholesterol concentration by a mean value of 10.2 mg/dL, but with no significant change in HDL cholesterol levels. Moreover, mean triglyceride levels were found to be reduced by 20.6 mg/dL (Sabate, Oda et al. 2010). This effect could be attributed to the high polyunsaturated fatty acids and low saturated fatty acids content of nuts.

10.8.8 SOY

Soy is a type of protein that is found in soybeans and is considered to be an individual's diet as an alternative to animal protein. Soybean contains low amounts of saturated fat and no cholesterol. Soy proteins are a good source of fiber, iron, calcium, zinc and B vitamins (Montgomery 2003). In the year 1999, FDA accepted the type of foods containing soy proteins as protective agents against CHD (FDA 1999).

10.8.9 DAIRY PRODUCTS

Dairy products consist of large amounts of minerals (calcium, potassium and magnesium), proteins (casein and whey) and vitamins (riboflavin and vitamin B-12) which can offer potent beneficial effects against CVD. A meta-analysis study performed on individuals who consumed increased amounts of dairy products revealed reduced risk for ischemic heart disease as well as stroke (Elwood, Pickering et al. 2010).

10.8.10 Coffee

Consumption of coffee has long been suspected to trigger the development of CVD. But recently there is accumulating evidence that suggest a protective association between moderate consumption of coffee and CHD morbidity and CVD mortality (de Koning Gans, Uiterwaal et al. 2010; Sugiyama, Kuriyama et al. 2010).

10.8.11 Tea

Tea has been used as the most popular beverage for 4,000 years. Green tea is documented to reduce LDL cholesterol levels (Hooper, Kroon et al. 2008). Tea consumption is associated with anti-inflammatory, antioxidant and antiproliferative effects which reduced CVD risk (Deka and Vita 2011).

10.8.12 Omega-3 Fatty Acids

Omega-3 fatty acids are reported to possess beneficial effects against hypercholesterolemia and overall beneficial effects on health and general well-being (Siddiqui, Shaikh et al. 2004). In the year 2002, the American Heart Association recommended a revised form for dietary intake of omega-3 fatty acids (Kris-Etherton, Harris et al. 2002) as follows:

- Individuals reported without CHD were advised to consume fish (preferably oily) twice a week including foods and oils rich in alpha lipoic acid (ALA) (flaxseed, canola, soy, walnuts). The overall intake should be ≈500 mg/das compared to the current intake of <100 mg/d of omega-3 fatty acids.
- Individuals reported with CHD were suggested to consume 1 g/d of eicosapentaenoic acid (EPA) plus DHA, preferably from oily fish. They can also consume EPA plus DHA supplements, but only after consultation with a physician.
- Individuals documented with hypertriglyceridemia were advised to consume 2–4 g/d of EPA plus DHA under a physician's care.

Based on the scientific studies documented, the FDA declared a qualified health claim asserting a connection between omega-3 fatty acids and reduced risk of CHD (Meyer, Mann et al. 2003).

10.8.13 Plant Sterols and Stanols

Plant sterols and stanols are reported to inhibit cholesterol absorption and are considered to be active agents in lowering circulating cholesterol levels (Escurriol, Cofan et al. 2009). These compounds are commonly found in margarines. Consuming up to 9 g/day of products enriched with plant stanol/sterol esters is reported to reduce serum LDL cholesterol levels by 17.4% (Mensink, de Jong et al. 2010).

10.8.14 Flavonoids

Accumulating evidence from several epidemiological studies emphasizes an inverse association between dietary flavonoids and the incidence of CHD (Rimm, Katan et al. 1996; Maron 2004). A meta-analysis of various studies demonstrated that high dietary intake of flavonols from various sources like fruits, vegetables and tea are associated with a reduced risk of CHD mortality (Huxley and Neil 2003). In addition, several studies of fatal CHD were evaluated, with intake of flavonoid classes, flavone and flavonol which showed that combined consumption of both flavone and flavonol lowered the risk of fatal CHD (Hertog, Feskens et al. 1993; Hertog, Feskens et al. 1997). A recent 30 days' study conducted among hypercholesterolemia patients when administered with bergamot (*C. bergamia* Risso & Poiteau) fruit extract, which is rich in naringin, neoeriocitrin, neohesperidin, melitidin and brutieridin revealed a dose-dependent reduction of total and LDL cholesterol levels with subsequent increase in HDL-cholesterol levels (Mollace, Sacco et al. 2011). Thus, flavonoids by virtue of their hypocholesterolemic property may be relevant in the prevention of various heart diseases including CHD.

10.8.15 GARLIC

For a long time, garlic has been the subject of clinical trials in lowering cholesterol levels in humans. Earlier meta-analysis trials suggested a significant reduction in serum total cholesterol levels in garlic supplemented participants (Warshafsky, Kamer et al. 1993; Silagy and Neil 1994; Stevinson, Pittler et al. 2000). Recently a study conducted by Sobenin, Andrianova et al. (2008) showed that garlic tablets (Allicor—comprising 600 mg of garlic) moderately altered the lipid levels such as: 7.6% decrease in cholesterol; 11.7% decrease in LDL cholesterol and a substantial increase in 11.5% HDL cholesterol. Thus, garlic displays potent beneficial effects due to its inherent hypocholesterolemic properties.

10.9   HEART-HEALTHY DIETS

The National Heart, Lung, and Blood Institute (NHLBI, NIH (US)) includes the following under a heart-healthy diet:

- Fat-free or low-fat dairy products.
- Fish rich in omega-3 fatty acids, such as salmon, tuna and trout.
- Fruit, such as apples, bananas, oranges, pears and prunes.
- Legumes, such as kidney beans, lentils, chickpeas, black-eyed peas and lima beans.
- Vegetables, mainly broccoli, cabbage and carrots.
- Whole-grains, such as oatmeal, brown rice and corn tortillas.
- Monounsaturated and polyunsaturated fats-containing products:
  - Avocados.
  - Corn, sunflower, and soybean oils.
  - Nuts and seeds, such as walnuts.
  - Olive, canola, peanut, safflower and sesame oils.
  - Peanut butter.
  - Salmon and trout.
  - Tofu.

10.10   HEART-UNHEALTHY DIETS

The following are heart-unhealthy diets that are to be avoided (NHLBI, NIH (US)).

- Excessive red meat.
- Palm and coconut oils.
- Sugary foods and beverages.
- Saturated fat.
- Trans-fat (trans-fatty acids).

10.11   PUBLIC HEALTH RECOMMENDATIONS FOR NUTRITION AND LIFESTYLE

For the purpose of prevention or as a part of treatment for diseases, several opinion leader organizations have recommended and published lifestyle medicine, consisting primarily of diet and physical activity. The US Department of Agriculture and the US Department of Health and Human Services published an extensive guideline on the healthy dietary pattern for the American population in the year 2010 (2010). The lifestyle medicine recommendations for general populations are illustrated in Figure 10.3.

10.12   THERAPEUTIC MEASURES

Despite outstanding advancement in the area of the management of CVDs, significant challenges are warranted for the better outcome of the treatment of CVD.

10.12.1   Antioxidant Therapy

A variety of antioxidants available from several biological and chemical origin can be involved in the therapeutic management of CVD. Dietary antioxidants that help in the management of CVD are mainly comprised of polyphenols, which include the flavonoids and phenolic acid. Additionally, other antioxidants that are present in the diet include vitamin E, carotenes, vitamin C, lycopene, etc.

10.12.1.1   Coenzyme Q 10 (CoQ10)

CoQ10, an antioxidant, plays a vital role in preventing the oxidation of cell membranes, LDL cholesterol, HDL cholesterol and various other blood components (Crane 2001). CoQ10 acts as an inhibitor of lipid peroxidation at the initial as well as intermediate steps, thus validating it as an essential antioxidant (Ernster and Forsmark-Andree 1993). FDA has approved CoQ10 as a supplement for the management of CVD. LiQ10® (syrup) and Q-Gel® and Q-Nol® (soft gelatin capsules) are among the foremost commercially available formulations of CoQ10 (Swarnakar, Jain et al. 2011). Daily consumption of CoQ10 (150 mg/d) is documented to decrease oxidative stress, inflammatory markers (IL-6) and improve resistance against oxidation of LDL cholesterol (Witting, Pettersson et al. 2000; Lee, Huang et al. 2012).

Lifestyle medicine recommendations

Balancing Calories to manage weight	Foods and food components to be reduced	Foods and nutrients to increase
• Prevent and/or reduce overweight through improved eating and physical activity. • Control total calorie intake to manage body weight. • To increase physical activity. • Maintain appropriate calorie balance during each stage of life.	• Reduce daily sodium intake to less than 2,300 mg and person who are 51 or older are recommended to further reduce the intake to 1,500 mg. • Consume less than 10 percent of saturated fatty acids. • Consume less than 300 mg per day of dietary cholesterol. • Consume as low as possible of trans fatty acid. • Reduce the intake of calories from solid fats and added sugars. • Limit the consumption of foods that contain refined grains, especially refined grain foods.	• Increase vegetable and fruit intake, especially dark-green, red and orange vegetables and beans and peas. • Increase whole-grain intake. • Increase intake of fat-free or low-fat milk and milk products. • Prefer a variety of protein foods, including seafood, lean meat and poultry, eggs, beans and peas, soy products, and unsalted nuts and seeds. • Consume protein foods that are lower in solid fats and calories. • Use of oils against solid fats is recommended. • Choose foods that are rich in potassium, dietary fiber, calcium, and vitamin D.

FIGURE 10.3   Lifestyle medicine recommendations. (US Department of Agriculture and US Department of Health and Human Services 2010.)

10.12.1.2   Carotenoids

Carotenoids are dietary antioxidants that play a major role in controlling CVD. Since their carotenoids are not synthesized in the human body they are obtained from diet or other supplements. Commercially there are various carotenoid supplements available for controlling CVD which include AstaSana™ (Astaxanthin), CaroCare® (beta carotene) and Redivivo® (lycopene). Dietary intake of beta carotene has also been documented to be of benefit in different CVD including atherosclerosis and hypertension (Gammone, Riccioni et al. 2015). Beta carotene acts through a different mechanism and prevents lipid peroxidation enhances lymphocyte proliferation and modulates vascular nitric oxide bioavailability (Jialal, Norkus et al. 1991; Robbins and Topol 2002). Lycopene, a red pigment present in fruits and vegetables also possess antioxidant properties and plays a role in the management of various CVD (Muller, Caris-Veyrat et al. 2016).

10.12.1.3 Polyphenols

Polyphenols act through diverse mechanisms and prevent CVDs. It is reported to inhibit ROS and reactive nitrogen species by directly scavenging peroxynitrite, hydroxyl radicals, superoxide anion and nitric oxide (Tangney and Rasmussen 2013). Among the various polyphenols available, resveratrol has shown a substantial role in the management of various CVDs including atherosclerosis, hypertension, ischemic heart disease, stroke, arrhythmia and chemotherapy-induced cardiotoxicity, heart failure and diabetic cardiomyopathy.

10.12.2 MEDICINES CURRENTLY USED TO TREAT CVDs

- Anticoagulants.
- Aspirin.
- Angiotensin-converting enzyme (ACE) inhibitors.
- Beta-blockers.
- Calcium channel blockers.
- Nitroglycerin.
- Glycoprotein IIb-IIIa.
- Statins.
- Fish oil and other supplements high in omega-3 fatty acids.

10.12.3 CHOLESTEROL-LOWERING DRUGS

Medicines can help lower cholesterol levels but it can't guarantee to cure it, so it is recommended to continue taking medicine to keep cholesterol levels in the recommended stage. Five major types of cholesterol-lowering medicines commercially available include:

- Statins: reported to lower LDL cholesterol levels.
- Bile acid sequestrants which also help lower LDL cholesterol.
- Nicotinic acid documented to lower LDL cholesterol and triglycerides and raise HDL cholesterol levels.
- Fibrates lower triglycerides, and also can raise HDL cholesterol levels.
- Ezetimibe lowers LDL cholesterol levels.

10.12.3.1 Antiplatelets

Antiplatelets help to inhibit the functions of platelets in the blood by reducing the bloodclotting process thus preventing the platelets from sticking to blood vessel walls. They are also known as anti-aggregants.

At present two types of antiplatelet drugs have been approved:

- **Acetylsalicylic acid (ASA):** People with CVD are usually recommended to take one 100 mg tablet per day.
- **Clopidogrel:** Frequently used by patients who can't tolerate ASA. One tablet of 75 mg is taken per day.

Studies report that ASA prevents heart attacks or strokes in about five to ten out of 100 people who have CHD. Thus, people with CHD, with regular ASA intake can be prevented from major complications like heart attacks or strokes. In addition, the drug clopidogrel is also known to function equally with ASA in preventing complications.

10.12.3.2   Statins

Statins are one of the most common medicines used to lower cholesterol levels in the blood. Recently they have been thought to protect the blood vessel walls by inhibiting inflammation. The approved statins are simvastatin, atorvastatin, fluvastatin, lovastatin and pravastatin. Among them, simvastatin and atorvastatin are most commonly used, taken at the dose of 40 mg per day and 10 mg per day respectively (Mills, Wu et al. 2011). Studies show that people with coronary artery disease who take statins regularly over a period of time have reduced the risk of coronary artery disease complications and statins are generally welltolerated by the patients (Naci, Brugts et al. 2013).

10.12.3.3   Beta-Blockers

Beta-blockers are mainly recommended for people who have coronary artery disease as well as high blood pressure or heart failure (Ko, Hebert et al. 2004). Besides, beta-blockers also help in relieving angina. Various types of beta-blockers are available. Among them, bisoprolol and metoprolol are the ones most commonly used. Bisoprolol and metoprolol are usually used at the dose of 2.5 to 5 mg per day and 50 to 100 mg per day respectively.

10.12.3.4   ACE Inhibitors and Angiotensin II Antagonists

ACE inhibitors and angiotensin II antagonists, also known as sartans are considered to be blood pressure-lowering drugs. ACE inhibitors and angiotensin II antagonists function by dilating (widening) the blood vessels, thereby lowering the blood pressure.

10.12.4   Conventional Therapeutic Strategies

A varied range of conventional therapeutic strategies has been developed for the treatment of CVDs including gene delivery or therapy, cell transplantation and stem cell therapy, tissue factor inhibitors, micro RNSs and or other small molecules.

10.12.4.1   Gene Delivery or Therapy

The evolution of vector technology has emerged as a significant technique for understanding the advanced molecular mechanisms underlying CVDs. Gene therapy appears to be an attractive alternative to current pharmacological therapies which may be considered beneficial in the treatment of CVDs (Katz, Fargnoli et al. 2013). Moreover, the recent advancement in recombinant technology such as gene transfer has stimulated some hope that this technique can be used to improve the practice of cardiovascular medicine. In animal models, gene therapy coding for antioxidants, heat shock proteins (HSPs), mitogen-activated protein kinase (MAPK) and numerous other proteins have revealed significant cardioprotection (Guellich, Mehel et al. 2014).

10.12.4.2   Cell Transplantation and Stem Cell Therapy

Stem cell-based therapies have emerged as a novel promising therapeutic approach for patients with different heart diseases. Studies on cell-based therapies suggest an improvement in cardiac function and regeneration of the damaged tissue (Segers and Lee 2008).

Further, the therapeutic strategies underlying tissue factor inhibitors, micro RNAs and other small molecules are warranted to be beneficial for the treatment of CVDs. In addition, the potential technology utilizing nanoscale dimension materials can play a role in bridging the gap between molecular and cellular interactions, and has the potential to revolutionize medicine. Moreover nanomedicine, when compared to traditional medicine offers treatment as effectively as possible with negligible side effects, and an enhanced clinical result (Singh, Garg et al. 2016). Among the conventional therapeutic measures, gene therapy plays a promising role in the treatment and prevention of CVDs (Perk, De Backer et al. 2012).

10.13   CONCLUSION

CHD persists as the leading cause of death worldwide and hypercholesterolemia remains the major risk factor for the development of CHD. Early detection of hypercholesterolemia can help prevent premature CHD as well as CVD. Diet therapy plays a vital role in the treatment of hypercholesterolemia. Maintenance of a personalized healthy diet containing diverse foods is recommended for lowering the risk of CVD. The diet should include a variety of fruits, vegetables, whole-grains, whole wheat bread, legumes, and high-fiber low-salt food items. Vegetable oils, mainly olive and canola oils, excluding palm and coconut oils, are preferable to animal fats. Moreover, there are certain other compounds that may confer benefits for CVD such as nuts, almonds, avocado and tahini, low-fat dairy products, green tea and two to three servings of fatty fish per week. The Mediterranean diet has been documented to reduce cardiovascular morbidity and mortality in both primary and secondary prevention. Additionally, other dietary patterns that could be suggested are a low-fat diet for people reported with cardiovascular risk, the DASH diet for people with hypertension, and lowcarbohydrate diets for overweight people or those suffering from metabolic syndrome.

ABBREVIATIONS

ACE: Angiotensin-converting enzyme
ALA: Alpha Lipoic acid
CHD: Coronary heart disease
CoQ10:   Coenzyme Q10
CVD: Cardiovascular disease
DASH: Dietary Approach to Stop Hypertension
EPA: Eicosapentaenoic acid

FDA: Food and Drug Administration
HDL: High-density lipoprotein
HSP: Heat shock protein
LDL: Low-density lipoprotein
NHLBI: National Heart, Lung, and Blood Institute

REFERENCES

Andrade, J., A. Mohamed, et al. (2009). "Ancel keys and the lipid hypothesis: from early breakthroughs to current management of dyslipidemia." *BC Med J* 51(2): 66–72.

Appel, L.J., T. J. Moore, et al. (1997). "A clinical trial of the effects of dietary patterns on blood pressure. DASH collaborative research group." *N Engl J Med* 336(16): 1117–1124.

Bazzano, L. A., M. K. Serdula, et al. (2003). "Dietary intake of fruits and vegetables and risk of cardiovascular disease." *Curr Atheroscler Rep* 5(6): 492–499.

Belahsen, R. (2014). "Nutrition transition and food sustainability." *Proc Nutr Soc* 73(3): 385–388.

Benjamin, E. J., M. J. Blaha, et al. (2017). "Heart disease and stroke statistics-2017 update: a report from the American Heart Association." *Circulation* 135(10): e146–e603.

Brown, L., B. Rosner, et al. (1999). "Cholesterol-lowering effects of dietary fiber: a meta-analysis." *Am J Clin Nutr* 69(1): 30–42.

Brufau, G., J. Boatella, et al. (2006). "Nuts: source of energy and macronutrients." *Br J Nutr* 96(2 Suppl): S24–S28.

Carpenter, K. J. (2003). "A short history of nutritional science: part 1 (1785–1885)." *J Nutr* 133(3): 638–645.

Chan, L. and E. Boerwinkle (1994). "Gene-environment interactions and gene therapy in atherosclerosis." *Cardiol Rev* 2(3): 130–137.

Chen, Z.-Y., K. Y. Ma, et al. (2011). "Role and classification of cholesterol-lowering functional foods." *J Funct Foods* 3(2): 61–69.

Crane, F. L. (2001). "Biochemical functions of coenzyme Q10." *J Am Coll Nutr* 20(6): 591–598.

Dauchet, L., P. Amouyel, et al. (2005). "Fruit and vegetable consumption and risk of stroke: a meta-analysis of cohort studies." *Neurology* 65(8): 1193–1197.

de Koning Gans, J. M., C. S. Uiterwaal, et al. (2010). "Tea and coffee consumption and cardiovascular morbidity and mortality." *Arterioscler Thromb Vasc Biol* 30(8): 1665–1671.

Deka, A. and J. A. Vita (2011). "Tea and cardiovascular disease." *Pharmacol Res* 64(2): 136–145.

Eilat-Adar, S., T. Sinai, et al. (2013). "Nutritional recommendations for cardiovascular disease prevention." *Nutrients* 5(9): 3646–3683.

Elwood, P. C., J. E. Pickering, et al. (2010). "The consumption of milk and dairy foods and the incidence of vascular disease and diabetes: an overview of the evidence." *Lipids* 45(10): 925–939.

Ernster, L. and P. Forsmark-Andree (1993). "Ubiquinol: an endogenous antioxidant in aerobic organisms." *Clin Investig* 71(8 Suppl): S60–S65.

Escurriol, V., M. Cofan, et al. (2009). "Serum sterol responses to increasing plant sterol intake from natural foods in the Mediterranean diet." *Eur J Nutr* 48(6): 373–382.

Food and Drug Administration, HHS (2008). "Food labeling: health claims; soluble fiber from certain foods and risk of coronary heart disease. Interim final rule." *Fed Regist* 73(37): 9938–9947.

Gammone, M. A., G. Riccioni, et al. (2015). "Carotenoids: potential allies of cardiovascular health?" *Food Nutr Res* 59: 26762.

Guellich, A., H. Mehel, et al. (2014). "Cyclic AMP synthesis and hydrolysis in the normal and failing heart." *Pflugers Arch* 466(6): 1163–1175.

He, F. J., C. A. Nowson, et al. (2006). "Fruit and vegetable consumption and stroke: meta-analysis of cohort studies." *Lancet* 367(9507): 320–326.

Hertog, M. G., E. J. Feskens, et al. (1993). "Dietary antioxidant flavonoids and risk of coronary heart disease: the Zutphen Elderly Study." *Lancet* 342(8878): 1007–1011.

Hertog, M. G., E. J. Feskens, et al. (1997). "Antioxidant flavonols and coronary heart disease risk." *Lancet* 349(9053): 699.

Hite, A. H., V. G. Berkowitz, et al. (2011). "Low-carbohydrate diet review: shifting the paradigm." *Nutr Clin Pract* 26(3): 300–308.

Hooper, L., P. A. Kroon, et al. (2008). "Flavonoids, flavonoid-rich foods, and cardiovascular risk: a meta-analysis of randomized controlled trials." *Am J Clin Nutr* 88(1): 38–50.

Huxley, R. R. and H. A. Neil (2003). "The relation between dietary flavonol intake and coronary heart disease mortality: a meta-analysis of prospective cohort studies." *Eur J Clin Nutr* 57(8): 904–908.

Ioannidis, J. P. (2009). "Prediction of cardiovascular disease outcomes and established cardiovascular risk factors by genome-wide association markers." *Circ Cardiovasc Genet* 2(1): 7–15.

Jialal, I., E. P. Norkus, et al. (1991). "β-Carotene inhibits the oxidative modification of low-density lipoprotein." *Biochim Biophys Acta* 1086(1): 134–138.

Katz, M. G., A. S. Fargnoli, et al. (2013). "Gene therapy delivery systems for enhancing viral and nonviral vectors for cardiac diseases: current concepts and future applications." *Hum Gene Ther* 24(11): 914–927.

Kelly, S. A., C. D. Summerbell, et al. (2007). "Wholegrain cereals for coronary heart disease." *Cochrane Database Syst Rev* (2): CD005051.

Keys, A. (1970). "Coronary heart disease in seven countries." *Circulation* 41(1): 186–195.

Ko, D. T., P. R. Hebert, et al. (2004). "Adverse effects of beta-blocker therapy for patients with heart failure: a quantitative overview of randomized trials." *Arch Intern Med* 164(13): 1389–1394.

Kris-Etherton, P. M., W. S. Harris, et al. (2002). "Fish consumption, fish oil, omega-3 fatty acids, and cardiovascular disease." *Circulation* 106(21): 2747–2757.

Kris-Etherton, P. M., F. B. Hu, et al. (2008). "The role of tree nuts and peanuts in the prevention of coronary heart disease: multiple potential mechanisms." *J Nutr* 138(9): 1746S–1751S.

Lee, B. J., Y. C. Huang, et al. (2012). "Coenzyme Q10 supplementation reduces oxidative stress and increases antioxidant enzyme activity in patients with coronary artery disease." *Nutrition* 28(3): 250–255.

Lichtenstein, A. H., L. J. Appel, et al. (2006). "Diet and lifestyle recommendations revision 2006: a scientific statement from the American Heart Association Nutrition Committee." *Circulation* 114(1): 82–96.

Marlett, J. A., M. I. McBurney, et al. (2002). "Position of the American Dietetic Association: health implications of dietary fiber." *J Am Diet Assoc* 102(7): 993–1000.

Maron, D. J. (2004). "Flavonoids for reduction of atherosclerotic risk." *Curr Atheroscler Rep* 6(1): 73–78.

Maruthur, N. M., N. Y. Wang, et al. (2009). "Lifestyle interventions reduce coronary heart disease risk: results from the PREMIER Trial." *Circulation* 119(15): 2026–2031.

Mensink, R. P., A. de Jong, et al. (2010). "Plant stanols dose-dependently decrease LDL-cholesterol concentrations, but not cholesterol-standardized fat-soluble antioxidant concentrations, at intakes up to 9 g/d." *Am J Clin Nutr* 92(1): 24–33.

Meyer, B. J., N. J. Mann, et al. (2003). "Dietary intakes and food sources of omega-6 and omega-3 polyunsaturated fatty acids." *Lipids* 38(4): 391–398.

Mills, E. J., P. Wu, et al. (2011). "Efficacy and safety of statin treatment for cardiovascular disease: a network meta-analysis of 170,255 patients from 76 randomized trials." *QJM* 104(2): 109–124.

Minich, D. M. and J. S. Bland (2013). "Personalized lifestyle medicine: relevance for nutrition and lifestyle recommendations." *Sci World J* 26: 129841. doi: 10.1155/2013/129841.

Mollace, V., I. Sacco, et al. (2011). "Hypolipemic and hypoglycaemic activity of bergamot polyphenols: from animal models to human studies." *Fitoterapia* 82(3): 309–316.

Montgomery, K. S. (2003). "Soy protein." *J Perinat Educ* 12(3): 42–45.

Muller, L., C. Caris-Veyrat, et al. (2016). "Lycopene and its antioxidant role in the prevention of cardiovascular diseases-a critical review." *Crit Rev Food Sci Nutr* 56(11): 1868–1879.

Naci, H., J. Brugts, et al. (2013). "Comparative tolerability and harms of individual statins: a study-level network meta-analysis of 246 955 participants from 135 randomized, controlled trials." *Circ Cardiovasc Qual Outcomes* 6(4): 390–399.

Naghavi, M., P. Libby, et al. (2003). "From vulnerable plaque to vulnerable patient: a call for new definitions and risk assessment strategies: part I." *Circulation* 108(14): 1664–1672.

Nestle, M. (1995). "Mediterranean diets: historical and research overview." *Am J Clin Nutr* 61(6 Suppl): 1313S–1320S.

Ockene, I. S. and R. Nicolosi (1998). "Dietary fat intake and the risk of coronary heart disease in women." *N Engl J Med* 338(13): 917; author reply 918–919.

Pereira, M. A., E. O'Reilly, et al. (2004). "Dietary fiber and risk of coronary heart disease: a pooled analysis of cohort studies." *Arch Intern Med* 164(4): 370–376.

Perk, J., G. De Backer, et al. (2012). "European guidelines on cardiovascular disease prevention in clinical practice (version 2012). The fifth joint task force of the European Society of Cardiology and other societies on cardiovascular disease prevention in clinical practice (constituted by representatives of nine societies and by invited experts)." *Eur Heart J* 33(13): 1635–1701.

Rimm, E. B., M. B. Katan, et al. (1996). "Relation between intake of flavonoids and risk for coronary heart disease in male health professionals." *Ann Intern Med* 125(5): 384–389.

Robbins, M. and E. J. Topol (2002). "Inflammation in acute coronary syndromes." *Cleve Clin J Med* 69(2 Suppl): SII130–SII142.

Roberts, R. and A. F. Stewart (2012). "Genetics of coronary artery disease in the 21st century." *Clin Cardiol* 35(9): 536–540.

Ross, R. (1999). "Atherosclerosis—an inflammatory disease." *N Engl J Med* 340(2): 115–126.

Sabate, J. and Y. Ang (2009). "Nuts and health outcomes: new epidemiologic evidence." *Am J Clin Nutr* 89(5): 1643S–1648S.

Sabate, J., K. Oda, et al. (2010). "Nut consumption and blood lipid levels: a pooled analysis of 25 intervention trials." *Arch Intern Med* 170(9): 821–827.

Sacks, F. M., E. Obarzanek, et al. (1995). "Rationale and design of the Dietary Approaches to Stop Hypertension trial (DASH): a multicenter controlled-feeding study of dietary patterns to lower blood pressure." *Ann Epidemiol* 5(2): 108–118.

Sacks, F. M., L. P. Svetkey, et al. (2001). "Effects on blood pressure of reduced dietary sodium and the Dietary Approaches to Stop Hypertension (DASH) diet. DASH-Sodium Collaborative Research Group." *N Engl J Med* 344(1): 3–10.

Santos, F. L., S. S. Esteves, et al. (2012). "Systematic review and meta-analysis of clinical trials of the effects of low carbohydrate diets on cardiovascular risk factors." *Obes Rev* 13(11): 1048–1066.

Segers, V. F. and R. T. Lee (2008). "Stem-cell therapy for cardiac disease." *Nature* 451(7181): 937–942.

Shipley, M. J., S. J. Pocock, et al. (1991). "Does plasma cholesterol concentration predict mortality from coronary heart disease in elderly people? 18 year follow up in Whitehall study." *BMJ* 303(6794): 89–92.

Siddiqui, R. A., S. R. Shaikh, et al. (2004). "Omega 3-fatty acids: health benefits and cellular mechanisms of action." *Mini Rev Med Chem* 4(8): 859–871.

Silagy, C. and A. Neil (1994). "Garlic as a lipid lowering agent—a meta-analysis." *J R Coll Physicians Lond* 28(1): 39–45.

Singh, B., T. Garg, et al. (2016). "Recent advancements in the cardiovascular drug carriers." *Artif Cells Nanomed Biotechnol* 44(1): 216–225.

Singh, R. B., V. Rastogi, et al. (1998). "Serum cholesterol and coronary artery disease in populations with low cholesterol levels: the Indian paradox." *Int J Cardiol* 65(1): 81–90.

Smith, S. C., Jr., A. Collins, et al. (2012). "Our time: a call to save preventable death from cardiovascular disease (heart disease and stroke)." *Eur Heart J* 33(23): 2910–2916.

Sobenin, I. A., I. V. Andrianova, et al. (2008). "Lipid-lowering effects of time-released garlic powder tablets in double-blinded placebo-controlled randomized study." *J Atheroscler Thromb* 15(6): 334–338.

Sofi, F., R. Abbate, et al. (2010). "Accruing evidence on benefits of adherence to the Mediterranean diet on health: An updated systematic review and meta-analysis." *Am J Clin Nutr* 92(5): 1189–1196.

Spagnoli, L. G., A. Mauriello, et al. (2004). "Extracranial thrombotically active carotid plaque as a risk factor for ischemic stroke." *JAMA* 292(15): 1845–1852.

Stevinson, C., M. H. Pittler, et al. (2000). "Garlic for treating hypercholesterolemia. A meta-analysis of randomized clinical trials." *Ann Intern Med* 133(6): 420–429.

Sugiyama, K., S. Kuriyama, et al. (2010). "Coffee consumption and mortality due to all causes, cardiovascular disease, and cancer in Japanese women." *J Nutr* 140(5): 1007–1013.

Swarnakar, N. K., A. K. Jain, et al. (2011). "Oral bioavailability, therapeutic efficacy and reactive oxygen species scavenging properties of coenzyme Q10-loaded polymeric nanoparticles." *Biomaterials* 32(28): 6860–6874.

Tangney, C. C. and H. E. Rasmussen (2013). "Polyphenols, inflammation, and cardiovascular disease." *Curr Atheroscler Rep* 15(5): 324.

U.S. Department of Agriculture and U.S. Department of Health and Human Services (2010). *Dietary Guidelines for Americans*, Washington (DC), U.S. Government Printing Office. 7th Edition. www.dietaryguidelines.gov

Van Duyn, M. A. and E. Pivonka (2000). "Overview of the health benefits of fruit and vegetable consumption for the dietetics professional: selected literature." *J Am Diet Assoc* 100(12): 1511–1521.

Wallace, A. J., J. I. Mann, et al. (2000). "Variants in the cholesterol ester transfer protein and lipoprotein lipase genes are predictors of plasma cholesterol response to dietary change." *Atherosclerosis* 152(2): 327–336.

Warshafsky, S., R. S. Kamer, et al. (1993). "Effect of garlic on total serum cholesterol. A meta-analysis." *Ann Intern Med* 119(7 Pt 1): 599–605.

Witting, P. K., K. Pettersson, et al. (2000). "Anti-atherogenic effect of coenzyme Q10 in apolipoprotein E gene knockout mice." *Free Radic Biol Med* 29(3–4): 295–305.

11 Omega-3 Fatty Acids in the Prevention of Maternal and Offspring Metabolic Disorders

Olatunji Anthony Akerele and
Sukhinder Kaur Cheema

CONTENTS

11.1 INTRODUCTION

Maternal nutritional status during pregnancy is a major factor in healthy prenatal development and programming for adult diseases (Laker, Wlodek et al. 2013). Maternal diet is also critical for a successful pregnancy (Grieger and Clifton 2014; Marangoni, Cetin et al. 2016). The development of several chronic diseases has been clearly associated with early life insults *in utero* (Barker, Hales et al. 1993; Laker, Wlodek et al. 2013). Conditions characterized by severe undernutrition during pregnancy, as typified by the Biafran famine (Hult, Tornhammar et al. 2010) and the Dutch Hunger Winter (Schulz 2010) have the potential to impact fetal health negatively. Extreme nutritional deficiency at critical periods of pregnancy increases the risk of cardio-metabolic diseases in the offspring, which manifest at childhood or later in life (Voortman, van den Hooven et al. 2015). Several studies have now established that the quantity, as well as the quality, of dietary fats consumed during pregnancy have profound health implication on both maternal and fetal health during and after pregnancy (Coletta, Bell et al. 2010; Schwab, Lauritzen et al. 2014). In this respect, the type and the amount of essential fatty acids consumed during pregnancy is very crucial. Omega (n)-6 and n-3 polyunsaturated fatty acids (PUFA), the essential fatty acids (Abedi and Sahari 2014), are vital for fetal growth and development (Birch, Garfield et al. 2007; Gomez Candela, Bermejo Lopez et al. 2011). Dietary shifts over the years to a Western diet has caused a drastic change in the ratio of n-6 to n-3 fatty acids from about 1–2 : 1, as consumed by the hunter gatherers (Paleolithic diet) to current intake of about 20–30 : 1 (Gomez Candela, Bermejo Lopez et al. 2011). This transition has been found to promote the pathogenesis of chronic diseases such as cardiovascular disease (CVD), diabetes and obesity (Simopoulos 2006; Simopoulos 2016). Metabolism of n-3 PUFA produce antiinflammatory lipids mediators, known to reduce the risk of these chronic diseases (Mozaffarian and Wu 2011; Mori 2014), while n-6 PUFA are generally considered as proinflammatory (Calder 2009). As such, a diet with a balanced intake of n-6 and n-3 PUFAs produces fewer inflammatory and fewer immunosuppressive eicosanoids (Abedi and Sahari 2014), thereby protecting maternal health, and improving fetal growth and development. The marine-derived n-3 PUFAs such as eicosapentaenoic acid (EPA) and docosahexaenoic acid (DHA) are important in the overall fetal growth, as well as the development of vital organs such as the brain and eyes (retina) during pregnancy (Uauy, Birch et al. 1992; Singh 2005). The brain has the largest amount of lipids (60% dry weight), compared to other organs in the body (Chang, Ke et al. 2009). DHA constitute about 10–15% of total fatty acids in the brain, and this represents more than 97% of total n-3 PUFA (O'Brien, Fillerup et al. 1964; Makrides, Neumann et al. 1994). It has been shown that there is acceleration of fetal brain growth during the second trimester (Coletta, Bell et al. 2010); the accretion of DHA in the brain is most rapid during the third trimester of pregnancy and the first year after birth (Clandinin, Chappell et al. 1980; Martinez and Mougan 1998). As such, inadequate intake of DHA during pregnancy has been associated with impaired cognitive functions and lower visual acuity during development (Cheatham, Colombo et al. 2006). Mean n-3 PUFA intake of 90% of Canadian women is only 82 mg per day, which is far below the recommendation of the International Society for the Study of Fatty Acids and Lipids for North Americans (300 mg/day) (Denomme, Stark et al. 2005).

We have previously described the role of n-3 PUFAs as moderators and mediators of positive fetal outcome (Akerele and Cheema 2016), revealing that maternal diet enriched in n-3 PUFA increased fetal sustainability during pregnancy. Evidence from other studies have also shown that intake of marine-derived n-3 PUFA during pregnancy reduced the risk of adverse pregnancy conditions such as gestational diabetes, pre-eclampsia, maternal obesity and pre-term birth (Makrides, Duley et al. 2006; Redman and Sargent 2009; Rylander, Sandanger et al. 2014; Haghiac, Yang et al. 2015). As such, low intake of marine-derived n-3 PUFA or impaired endogenous synthesis plays a key role in the pathophysiology of these adverse pregnancy conditions. Thus, maternal n-3 PUFA intake is important in the prevention of adverse pregnancy outcomes and prevents cardio-metabolic disorders in offspring.

11.2 ADVERSE PREGNANCY CONDITIONS AND OFFSPRING HEALTH: ROLES OF N-3 PUFA

11.2.1 GESTATIONAL *DIABETES MELLITUS* (GDM)

GDM is a condition characterized by glucose intolerance during pregnancy (Coustan 2013). GDM exposes the fetus to uterine environment with excess glucose, thus resulting in fetal overnutrition. The developmental overnutrition hypothesis was first proposed by Pedersen in 1954; at birth, infants born to diabetic women were bigger, fatter and more edematous, compared to those born to normal mothers (Pedersen 1954). The relationship between gestational diabetes and fetal overgrowth was subsequently associated with hyperglycemia (Ryan 2013). Hyperglycemia during pregnancy increases the transfer of glucose across the placental to the fetus, and the fetal pancreas responds to increased glucose concentrations by producing and releasing more insulin (Pedersen 1954). Fetal hyperinsulinemia leads to insulin-mediated fetal growth (macrosomia) (Figure 11.1), which is the underlying cause of most of the fetal problems, collectively known as diabetic fetopathy (Schwartz, Gruppuso et al. 1994). Macrosomic babies are also putatively programmed for glucose intolerance, obesity and diabetes during pregnancy and these manifest either at childhood or at adulthood (Catalano, Kirwan et al. 2003).

A retrospective study on 84 women with GDM has revealed that diabetes during pregnancy induces dyslipidemia, which was characterized by elevated triglycerides and total cholesterol in maternal circulation (McGrowder, Grant et al. 2009). A similar retrospective study carried out in India on maternal and neonatal outcomes of gestational diabetes revealed that in pregnancy terminated by cesarean delivery, long-term progression to type 2 diabetes, in-born nursery admissions and increased

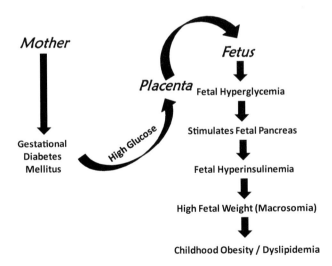

FIGURE 11.1 Gestational diabetes and fetal health outcomes. Gestational diabetes can program the fetus for hyperinsulinemia, which will eventually lead to macrosomia and childhood obesity.

neonatal birth weight are the major outcomes of GDM (Sreelakshmi, Nair et al. 2015). Outcomes from 4,873 women attending a university hospital antenatal diabetes clinic in Sydney, Australia revealed that pre-eclampsia (PE), pre-term delivery, cesarean section, and neonatal jaundice were more prevalent in women with preexisting diabetes and early GDM (Sweeting, Ross et al. 2016). The incidence of babies large for their gestational age was confirmed in a recent study to be 28.2% higher in mothers affected by GDM, compared to control (Kumari, Dalal et al. 2018). According to WHO criteria, GDM has been consistently associated with perinatal mortality, PE, macrosomia and cesarean delivery (Wendland, Torloni et al. 2012).

Intake of n-3 PUFA, especially DHA was observed to be significantly lower in women with gestational diabetes (Chen, Scholl et al. 2010). More so, women with GDM have different plasma fatty acids profile compared to nondiabetic women (Chen, Scholl et al. 2010), suggesting change in maternal fatty acids metabolism in diabetic mothers. GDM has also been shown to alter lipids metabolism in the offspring (Kilby, Neary et al. 1998). Several studies have established the beneficial health effects of n-3 PUFA supplementation during pregnancy on maternal and fetal health. A prospective population-based cohort study showed that fish consumption (75–100 g/d) among Norwegian women caused 30% reduction in the risk of developing type 2 diabetes, compared to those who did not consume fish (Rylander, Sandanger et al. 2014). Beneficial effects of n-3 PUFA on GDM is also well documented in animal studies. Supplementing maternal diet with longer chain n-3 PUFA from fish oil increased utilization of peripheral glucose in animals (Lardinois and Starich 1991). Dietary fish oil has also been shown to elicit beneficial effects on insulin-stimulated glucose metabolism in insulin-resistant and slightly diabetic rats (Luo, Rizkalla et al. 1996). Interestingly, consumption of a diet enriched in n-3 PUFA, especially EPA, diminished the incidence of macrosomia from 64% to 48% in pregnant diabetic rats (Yessoufou, Soulaimann et al. 2006). Animal studies also revealed that n-3 PUFA improved hyperlipidemia and restored antioxidant status of diabetic dams and their offspring (Soulimane-Mokhtari, Guermouche et al. 2005; Yessoufou, Soulaimann et al. 2006).

11.2.2 PRE-ECLAMPSIA (PE)

PE is a pregnancy specific disorder that complicates pregnancy at mid-gestation, usually after the 20th week of gestation. PE is characterized by elevated blood pressure, an abnormal amount of protein in the urine and other systemic disturbances (Lain and Roberts 2002). PE is a major obstetrical problem worldwide affecting about 8% of all pregnancies (Lain and Roberts 2002; Duley 2009), thus increasing both maternal and neonatal morbidity and mortality. About 10–15% of direct maternal mortality has been associated with PE (Duley 2009). Infants born to women with PE have about a five-fold higher risk of perinatal mortality, compared to those from mothers without PE (Roberts, Pearson et al. 2003). Of keen interest is the fact that PE exhibits a bi-modal distribution pattern on fetal growth; early onset increases the risk of intra-uterine growth restriction and delivering pre-term newborns, while late onset increases the risk of large for gestational age/macrosomic babies (Verlohren, Melchiorre et al. 2014).

Many of the mortalities attributed to PE may be preventable through advance understanding of the pathophysiology of PE. Altered n-3 PUFA levels have been documented in women with PE; maternal plasma DHA and total n-3 PUFA levels were lower in PE, which coincided with increased oxidative stress, compared to the control (Mehendale, Kilari et al. 2008). Placental oxidative stress plays a pivotal role in the pathogenesis of PE. Strong evidence exists that placental oxidative stress is a major intermediary event inducing placental inflammatory response and endothelial dysfunction in PE (Redman and Sargent 2009; Hansson, Naav et al. 2014). As such, addition of antioxidants or blocking inflammatory pathways with n-3 PUFA could provide a therapeutic platform to attenuate these adverse effects. Supplementing maternal diet with n-3 PUFA during pregnancy increased the levels of inflamma-tionresolving mediators, thereby reducing placental oxidative damage (Jones, Mark et al. 2013). Similarly, maternal n-3 PUFA supplementation reduced placental F_2-Isoprostanes levels, which is a reliable marker of placental oxidative damage (Jones, Mark et al. 2013). These studies highlight the therapeutic potential of n-3 PUFA in eliciting inflammation resolution in the placental interface.

A number of studies have examined the effect of maternal diets enriched in n-3 PUFA on the risk of PE. A prospective cohort study revealed that high intake of EPA and DHA (100 mg/day), or fish consumption during the first trimester of pregnancy reduced the risk of PE significantly (Oken, Ning et al. 2007). Data from other obser-vational studies showed that women with higher levels of n-6 PUFA in the erythro-cytes and platelet were 7.6 times more likely to have their pregnancies complicated by PE (Williams, Zingheim et al. 1995; Velzing-Aarts, van der Klis et al. 1999). As such, a 15% increase in the ratio of n-3 to n-6 fatty acids has been associated with a 46% reduction in risk of PE during pregnancy (Williams, Zingheim et al. 1995). However, despite best efforts to understand the nature of PE in clinical studies, it remains elusive. Methods for prevention and management PE are also limited. Thus, more robust intervention studies are required to confirm the beneficial effects of n-3 PUFA on PE during pregnancy.

11.2.3 MATERNAL OBESITY

Obesity is a major public health concern and is estimated to trend upwards for the foreseeable future (Kelly, Yang et al. 2008). Lifestyle changes such as high caloric intake and drastic reduction in required physical activity have contributed substan-tially to the prevalence of obesity across the globe. The increasing rate of mater-nal obesity poses a principal challenge to obstetric practice (Heslehurst, Ells et al. 2007). According to the report from the Confidential Enquiry into Maternal and Child Health (CEMACH), 35% of all women who died in the United Kingdom between year 2000 and 2002 were obese (Lewis and Drife 2004). Obesity in preg-nancy has been associated with several adverse maternal health outcomes and preg-nancy complications like gestational diabetes, large for gestational age babies and PE (Guelinckx, Devlieger et al. 2008). Maternal obesity also increases fetal risk for stillbirth and congenital anomalies (Leddy, Power et al. 2008). Obesity in pregnancy can also affect fetal health later in life; excessive availability of nutrients *in utero* has been shown to program for adverse cardiovascular function in the offspring

(Poston 2011). Besides, maternal obesity is associated with systemic increase in the production of pro-inflammatory cytokines during pregnancy (Madan, Davis et al. 2009). It also stimulates macrophage accumulation and inflammation in the placenta (Challier, Basu et al. 2008), thus exposing the fetus to an inflammatory environment during development.

Supplementation of maternal diet with longer chain n-3 PUFA has been well documented to reduce inflammation in obese pregnant women; DHA and EPA (1200 mg/day) from week 16 gestation to delivery exert potent anti-inflammatory properties by lowering the expression of inflammatory cytokines in both adipose and placental tissue (Haghiac, Yang et al. 2015). Since n-3 and n-6 PUFAs are metabolized by same group of enzymes, an established mechanism through which n-3 PUFA inhibits the production of pro-inflammatory cytokines is by preventing the metabolism of n-6 PUFA into downstream pro-inflammatory eicosanoids (Schmitz and Ecker 2008) (Figure 11.2). It is therefore possible that excess intake of n-6 PUFA as typified by Western diets play a role in adverse pregnancy outcomes in obese women. A study using Fat-1 transgenic mice, which converts n-6 to n-3 PUFA endogenously, revealed that reduction in n-6 PUFA reduced maternal obesity-associated inflammation and limits adverse developmental programming in mice offspring (Heerwagen, Stewart et al. 2013). Fat-1 mice are thus protected from adverse effects of high n-6 PUFA diet, such as systemic inflammation and macrophage accumulation in adipose tissue. More so, male offspring from Fat-1 mothers fed a high fat diet displayed less hepatic lipid accumulation and adipose tissue inflammation, compared with offspring from wild type mothers (Heerwagen, Stewart et al. 2013). N-3 PUFA also modulates the secretion of adipokines and decreased the production of tumor necrosis factor alpha and interleukin-6 in the adipose tissue of rats with high-fat diet-induced obesity (Perez-Echarri, PerezMatute et al. 2008).

Other line of evidence suggests that n-3 PUFA could elicit antiobesity effects during pregnancy, especially towards parturition (Muir, Liu et al. 2018). Although there is limited knowledge on the mechanism through which n-3 PUFA prevents maternal

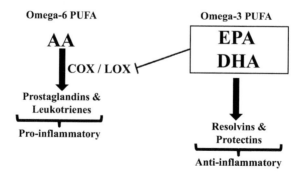

FIGURE 11.2 Anti-inflammatory property of omega-3 PUFA. Omega-3 PUFA inhibits the production of pro-inflammatory cytokines by preventing the metabolism of omega-6 PUFA into downstream pro-inflammatory eicosanoids. AA, Arachidonic acid; COX, Cyclooxygenase; EPA, Eicosapentaenoic acid; DHA, Docosahexaenoic acid; LOX, Lipoxygenase; PUFA, Polyunsaturated fatty acids.

obesity, evidence suggests that n-3 PUFA induces fatty acid re-esterification and creates substrate cycle in the adipose tissue, which results in energy expenditure and prevents fat accumulation (Janovska, Flachs et al. 2013). Fish oil has also been shown to increase resting energy expenditure, fat oxidation and reduced body fat, compared to the control in a single blind intervention study (Couet, Delarue et al. 1997). Animal studies also suggest that increased consumption of marine-derived n-3 PUFA can protect against the development of obesity in animals exposed to an obesogenic diet and reduce body fat when already obese (Buckley and Howe 2009). For instance, intake of the longer chain n-3 PUFA, specifically EPA and DHA from krill oil has been shown to preferentially increase fatty acid oxidation and prevent lipid accumulation in rat liver (Ferramosca, Conte et al. 2012). Incorporation of fish oil (representing 15% of energy) into a high fat diet caused 20% and 30% reduction in subcutaneous (inguinal) and visceral (retroperitoneal and epididymal) adipose tissues, respectively in rats (Hainault, Carolotti et al. 1993). Reduction in fat accumulation has been attributed to the ability of n-3 PUFA to elicit reduction in adipocyte hypertrophy (Belzung, Raclot et al. 1993). However, a recent study from our lab showed for the first time that n-3 PUFA prevents adipocyte hypertrophy in mice by downregulating the expressions of fatty acid binding protein (FABP)-4 and diacylglycerol acetyl transferase (DGAT)-2 in a sex-specific manner (Balogun and Cheema 2016). Attenuation of adipose hypertrophy by longer chain n-3 PUFA has also been associated with increased plasma concentration of leptin and reduction in food intake, suggesting that some of the effect of n-3 PUFA on body fat may be mediated by an appetite-suppressing effect (Perez-Matute, Perez-Echarri et al. 2007). Furthermore, enriching pre-term infant formula with longer chain n-3 PUFA has also been demonstrated to reduce body fat accumulation in growing infants (Groh-Wargo, Jacobs et al. 2005). A number of studies have also investigated the effects of supplementing maternal diet with n-3 PUFA during pregnancy on growth and body composition of infants (Lauritzen, Hoppe et al. 2005; Lucia Bergmann, Bergmann et al. 2007; Helland, Smith et al. 2008). Bergman et al. (2007) indicated that DHA intake by pregnant and lactating mothers may reduce body mass index in late infancy (Lucia Bergmann, Bergmann et al. 2007). However, longer-term human studies are required to confirm whether increasing intake of n-3 PUFA during pregnancy might be an effective strategy to combat maternal and fetal obesity.

11.2.4 PRE-TERM BIRTH (PTB)

PTB is a leading cause of neonatal morbidity and mortality across the globe (Johnston, Gooch et al. 2014). It is a recurrent problem in obstetrics relating to delivery before 37 gestational weeks; about 15 million babies are estimated to be born prematurely every year (WHO 2018). The economic burden caused by PTB cannot be over-emphasized; for instance, Canada spends about $587.1 million per year on conditions related to PTB (Johnston, Gooch et al. 2014). Although the etiology of PTB is complex and multifactorial, a handful of studies have implicated inflammation in the etiopathogenesis of PTB (Koucky, Germanova et al. 2009; Hudic, StrayPedersen et al. 2015). Studies have reported an inverse association between intake of n-3 PUFA during pregnancy and PTB. Earliest observational studies revealed that women consuming high amounts of marine foods (rich source of n-3 PUFA) in the Faroe Islands had a very low risk of birth before the 37th week of pregnancy (Olsen and Joensen 1985). A similar study conducted in Denmark showed that the risk of PTB was higher in women who never ate fish compared to those who consumed fish at least once per week (Olsen and Secher 2002). A fish supplementation trial conducted in London revealed a 20.4% reduction in PTB in the supplemented group, compared to the control (Olsen and Secher 1990).

Fish oil supplementation (2.7 g/d of EPA and DHA) at 20 weeks gestation reduced the recurrence of PTB from about 33% to 21% (Olsen, Secher et al. 2000). Cohort studies have provided evidence showing a dose dependent effect of marine foods consumption during pregnancy and the prevention of PTB (Olsen, Grandjean et al. 1993; Olsen and Secher 2002). Others have also found that an increased n-3 PUFA intake during pregnancy reduced the risk of PTB, especially in high-risk pregnancies (Olsen and Secher 2002; Makrides, Duley et al. 2006; Horvath, Koletzko et al. 2007). As such, improving maternal n-3 PUFA status during pregnancy could be harnessed as a promising prophylactic intervention strategy to prevent PTB. The mechanism of action of n-3 PUFA in relation to reduction in the risk of PTB during pregnancy has been attributed to its involvement in pregnancy duration modulation (Olsen, Hansen et al. 1991).

Data from intervention trials, as well as observational studies, have suggested a positive association between intake of n-3 PUFA during pregnancy and gestation length (Olsen and Joensen 1985; Olsen and Secher 1990; Olsen, Hansen et al. 1991). Of interest is the observation that about a 1% relative increase in cord serum phospholipid DHA was associated with approximately 1.5 days increase in gestation length (Grandjean, Bjerve et al. 2001). DOMInO (DHA to Optimize Mother and Infant Outcome) study, which is the largest single randomized controlled trial of n-3 PUFA supplementation during pregnancy in five Australian prenatal centers, also provided evidence showing that DHA-rich fish oil capsules (providing 800 mg/d of DHA) supplementation during pregnancy increased mean duration of gestation (Makrides, Gibson et al. 2010). Women who received longer chain n-3 PUFA supplements from fish oil during pregnancy had longer gestation period, an average of four days for most pregnancy and higher in high-risk pregnancies (Olsen, Sorensen et al. 1992; Makrides, Duley et al. 2006). Intake of eggs enriched with 133 mg DHA/egg/day during the last trimester was also shown to increase gestation length by approximately six days (Smuts, Huang et al. 2003). Also, pregnant women who received fish oil (approximately 100 mg DHA/day), from 15 weeks gestation until delivery, showed increased gestation length in infants with higher umbilical cord plasma DHA in the fish oil group (Malcolm, McCulloch et al. 2003).

It is likely that n-3 PUFA supplementation during pregnancy influences gestation length by regulating pregnancy establishment activities as well as labor induction. A randomized controlled trial suggested that fish oil supplementation may alter pregnancy duration by inhibiting the production of prostaglandins, which play key roles in labor induction at term, thereby increasing the gestation period (Olsen, Soorensen et al. 1994). Altogether, it appears that maternal diet enriched in n-3 PUFA may influence various activities involved in timely pregnancy establishment and labor (Akerele and Cheema 2016), requiring a more robust and mechanistic investigation. However,

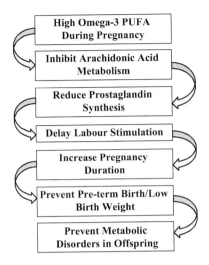

FIGURE 11.3 Omega-3 PUFA and pregnancy outcomes. Omega-3 PUFA regulates pregnancy duration and prevents metabolic disorders in the offspring. PUFA, Polyunsaturated fatty acid.

the observed variations in the gestation length can clearly be attributed to varying doses of n-3 PUFA during intervention, timing of the intervention, as well as the duration of the treatment. Aside from these factors, the availability of longer chain n3 PUFA during pregnancy is also dependent on endogenous synthesis from its precursor (alpha-linolenic acid (ALA)). Alteration in this metabolic pathway could have a significant impact on maternal n-3 PUFA status, especially EPA and DHA during pregnancy. A proposed mechanism by which n-3 PUFA prevents PTB and low birth weight to improve metabolic function of offspring is depicted in Figure 11.3.

11.3 N-3 PUFA METABOLISM DURING PREGNANCY

Humans lack the enzyme required for the insertion of a cis double bond at 3rd carbon of n-3, hence, the simplest form of n-3PUFA (ALA) must be obtained from diet (Lee, Lee et al. 2016). Once consumed, longer chain n-3 PUFA, such as EPA and DHA are produced endogenously from ALA through a series of desaturation and elongation processes (Leonard, Pereira et al. 2004) (Figure 11.4). A study by Burdge and Wootton (Burdge and Wootton 2002) revealed that the conversion of ALA to EPA and DHA was substantially higher in women of reproductive age, compared to men of similar age group (Burdge and Wootton 2002). This finding was supported by a kinetic study revealing that the rate of DHA synthesis was about four-fold higher in females, compared to males (Pawlosky, Hibbeln et al. 2003).

One plausible explanation for greater DHA synthesis in females is that the sex hormone (estrogen) may influence the enzymatic synthesis of longer chain fatty acids. Synthesis of DHA was observed to be three-fold higher in women using contraceptives containing 17α-ethynylestradiol, compared to those who did not use synthetic estrogen (Burdge and Wootton 2002). Administration of oral ethynylestradiol

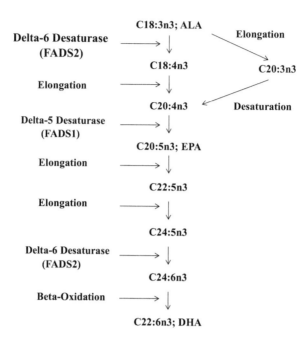

FIGURE 11.4 Synthesis of longer chain omega-3 PUFA. Longer chain omega-3 PUFAs are produced endogenously from ALA through a series of desaturation and elongation processes. (From Pathway modified from Bokor et al. (2010). *Journal of Lipid Research* **51**(8): 2325–2333); Permission obtained from publisher. ALA, Alpha linolenic acid; EPA, Eicosapentaenoic acid; FADS, Fatty acid desaturase; DHA, Docosahexaenoic acid; PUFA, Polyunsaturated fatty acid.

also increased the concentration of DHA in plasma cholesteryl esters by 42%, compared to control (Giltay, Gooren et al. 2004). Apparently, estrogen plays a key role in upregulating the elongation and desaturation pathway in women, which may contribute to the physiological increase in plasma DHA concentration during pregnancy. Greater fractional conversion of ALA to DHA in women could also in part be due to a significantly lower rate of dietary ALA utilization for beta-oxidation (Williams and Burdge 2006; Abedi and Sahari 2014).

Another possible biological significance of greater DHA synthesis capacity in women is to meet the demands of the fetus and neonate for this essential fatty acid. Since desaturase activity in the developing human liver of neonates appear to be lower than in adults (Poisson, Dupuy et al. 1993), the fetus depends on DHA from maternal circulation in order to satisfy their DHA requirement. As such, maternal plasma phosphatidylcholine DHA increases by approximately 33% between mid- and late-gestation (Postle, Al et al. 1995). The increase in maternal blood volume (hemodilution) (Gregersen and Rawson 1959) could also be a significant adaptation to an increase in maternal DHA during pregnancy. The maternal circulating estrogen level increases during pregnancy due to increased synthesis by the placenta, and this could also contribute substantially to increased conversion of ALA to DHA during pregnancy. Alteration in this conversion pathway could affect the proportion of EPA and DHA in circulation (Lemaitre, Tanaka et al. 2011), and perhaps impact pregnancy progression and fetal development negatively. Maternal genetic profile is also a major factor regulating the efficiency of longer chain n-3 PUFA synthesis endogenously. Single nucleotide polymorphisms (SNPs) have been identified to limit/ameliorate the synthesis of longer chain n-3 PUFA, with concomitant impact on cognitive function in the offspring (Malinowska, Wiśniewski et al. 2017).

11.4 SINGLE NUCLEOTIDE POLYMORPHISMS (SNPs) AFFECTING N-3 PUFA METABOLISM

The elongation step is catalyzed by elongases (elongation of very long chain fatty acids (ELOVL)), while the desaturation step is catalyzed by fatty acid desaturases (FADS) such as Δ-5 and Δ-6 desaturases, which are regulated by FADS1 and FADS2 gene, respectively (Drag, Gozdzialska et al. 2017). Both desaturases are expressed in most human tissues; however, the highest expression levels were observed in the liver (Cho, Nakamura et al. 1999). Desaturation is the rate limiting step in the pathway for the synthesis of longer chain PUFA. Deletion of the FADS2 gene in mice has been shown to abolish the initial step in the pathway for enzymatic synthesis of longer chain PUFA, thus resulting in the longer chain PUFA deficiency, with resultant alterations in reproductive function (Stoffel, Holz et al. 2008).

Several studies have also shown that SNPs in the FADS gene affect the proportion of longer chain PUFA in human tissues (Schaeffer, Gohlke et al. 2006; Bokor, Dumont et al. 2010). For instance, polymorphism in FADS1 and FADS2 gene has been associated with an increased proportion of ALA, with a drastic reduction in the corresponding longer chain n-3 PUFAs such as EPA, docosapentenoic acid (DPA) and DHA (Table 11.1). Meta-analysis of genome-wide association studies in five population-based cohorts comprising 8,866 subjects of European ancestry also revealed that SNPs in FADS1 and FADS2 were associated with higher levels of ALA, and lower levels of EPA and DPA (Lemaitre, Tanaka et al. 2011).

Other lines of evidence have also revealed a consistent association between SNPs in FADS gene cluster and hypertriglyceridemia (Aulchenko, Ripatti et al. 2009; Kathiresan, Willer et al. 2009), suggesting that alterations in desaturase activities affect plasma triglyceride metabolism. Recent genome-wide studies have also associated polymorphism in ELOVL2 gene with increased proportions of n-3 PUFA elongation substrates, with a drastic reduction in the proportion of their elongation products (Lemaitre, Tanaka et al. 2011; Tintle, Pottala et al. 2015).

11.4.1 SNPs AND PREGNANCY OUTCOMES

Emerging evidence has demonstrated that variations in FADS are associated with alterations in the composition of essential fatty acids, which may subsequently modify the propensity of susceptibility to diseases (Merino, Ma et al. 2010). A recent study has revealed that a minor allele carrier of FADS is associated with increased risk of obesity in pregnant women (de la Garza Puentes, Montes Goyanes et al. 2017). Of interest is the fact that children of mothers carrying variant gene associated with higher FADS2 activity have been shown to have a significant advantage in

TABLE 11.1

SNPs in FADS Affecting Synthesis of Longer Chain Omega-3 Polyunsaturated Fatty Acids

SNPs	Outcomes	References
FADS1 (rs174537)	↑ALA ↓EPA in serum	Tanaka et al. (2009)
FADS1 (rs174544)	↑ALA ↓EPA and DPA in serum	Schaeffer et al. (2006)
FADS1 (rs968567)	↑ALA ↓EPA and DHA in serum	Bokor et al. (2010)
FADS2 (rs174572)	↑ALA ↓EPA in serum	Bokor et al. (2010)
FADS2 (rs174583)	↑ALA ↓EPA and DPA in serum	Schaeffer et al. (2006)
FADS2 (rs174583)	↑ALA ↓EPA/ALA in red blood cells	Martinelli et al. (2008)
FADS2 (rs174589)	↑ALA ↓EPA in serum	Bokor et al. (2010)
FADS2 (rs174611)	↑ALA ↓EPA/ALA in red blood cells	Martinelli et al. (2008)

ALA, Alpha linolenic acid; DPA, docosapentaenoic acid; EPA, Eicosapentaenoic acid; FADS, Fatty acid desaturase; DHA, Docosahexaenoic acid; SNPs, Single nucleotide polymorphisms

cognition at 14 months, perhaps due to higher levels of DHA in colostrum (Morales, Bustamante et al. 2011). This suggests that the expression of FADS during pregnancy and breastfeeding is crucial for child cognition. Minor allele carriers of rs174448 and rs174575 were found to be responsible for a decrease in DHA synthesis in the offspring (Jensen, Harslof et al. 2014). This study also documented a sex-specific effect of FADS SNPs in the offspring. Certain SNPs in the FADS gene have been linked to diseases such as coronary artery disease (Li, Lin et al. 2013) and type 2 diabetes (Huang, Sun et al. 2014).

The effects of FADS SNPs on PUFA metabolism has also been shown to be race-specific (Mathias, Sergeant et al. 2011). A larger proportion of African-Americans carry the FADS genetic variant that increases the production of AA, compared to European Americans (Mathias, Sergeant et al. 2011); this could increase the risk of inflammatory diseases in both the mother and the fetus as AA could trigger inflammatory response. Minor allele carriers of rs2236212 and rs3798713 (ELOVL2) has also been nominally associated with a lower DHA : DPA index in pregnant overweight/obese mothers (de la Garza Puentes, Montes Goyanes et al. 2017). The Framingham heart offspring study, a genome-wide association study, also found that SNPs in the ELOVL2 gene contributed to reduced efficiency of the n-3 conversion pathway, thus resulting in more DPA and less DHA (Tintle, Pottala et al. 2015). However, the effect of ELOVL polymorphisms on the risk of allergy development in the offspring is yet to be established.

Unequivocally, genetic variants within the FADS cluster and ELOVL are determinants of longer chain n-3 PUFA levels in circulation and tissues. Evidence clearly supports that the associations are robust to individuals, as well as ethnicity (Mathias, Sergeant et al. 2011). Thus, it is important to consider that gene-PUFA interactions could be differentially driving the risk of inflammatory diseases during pregnancy in diverse populations, thereby leading to disparities in pregnancy outcomes. This calls for developing a personalized nutrition model to meet individual requirement for longer chain n-3 PUFA.

11.5 PERSONALIZED NUTRITION AND RECOMMENDATIONS FOR N-3 PUFA

An understanding of genetic variations has posed a big question on the appropriateness of the one-size-fits-all recommended dietary allowance. Recent studies on SNPs revealed that the differences from one person to another may be greater than previously thought. SNPs in the genes encoding the fatty acid desaturase and elongase enzymes affect longer chain PUFA production. As such, what is adequate for some may be deficient for others; it is therefore pertinent to consider individual genetic variations and move towards personalized nutrition, especially during pregnancy. Nutritional intervention has been suggested as a tool for improving virtually any condition. Information about person's SNPs status would proffer a platform for designing a nutritional intervention strategy that can improve the maternal health status. For instance, DHA intake during pregnancy may prevent inflammation-inclined adverse pregnancy outcomes in mothers with low ALA to DHA conversion efficiency.

Not only will personalized nutrition during pregnancy optimize maternal health and prevent adverse pregnancy conditions related to longer chain n-3 PUFA deficiency, it will also reduce the risk of chronic diseases such as obesity and diabetes in the offspring. A crosssectional study on Danish infants revealed that maternal FADS gene variants modify the association between maternal longer chain n-3 PUFA intake in pregnancy and birth weight of the offspring (Molto-Puigmarti, van Dongen et al. 2014). Among the women homozygous for the minor allele, those at the 75th percentile of DHA intake had lower risk of low birth weight, compared to those at the 25th percentile of intake. This indicates that supplementing mothers carrying homozygous minor allele of FADS during pregnancy with DHA is important in preventing adverse pregnancy outcomes, as well as meeting maternal and fetal requirements for DHA (Molto-Puigmarti, van Dongen et al. 2014). Genetic variants of FADS1 and FADS2 (rs174575) have also been shown to reduce blood and breast milk DHA in pregnancy and lactation (Xie and Innis 2008), which suggests the importance of personalized DHA intake in these mothers during pregnancy to prevent the adverse effect of DHA deficiency on the cognitive capability of the offspring.

11.6 CONCLUSIONS

Evidence suggest that longer chain n-3 PUFA proffers beneficial effects on maternal obesity and associated metabolic disturbances, such as gestational diabetes, PE and PTB. Low maternal n-3 PUFA status during pregnancy has also been associated with the development of CVD, diabetes and obesity in the offspring at early life or at adulthood. Moreover, SNPs in FADs as well as in ELOVL2 alter maternal longer chain n-3 PUFA status during pregnancy, which could negatively impact pregnancy progression, as well as fetal health outcomes. The concept of personalized nutrition considers the complex interaction between nutrients and genes in order to create an individual-based dietary recommendation which complements a person's unique genetic profile. Thus, highlighting individual genetic architecture when designing dietary recommendations, especially the intake of longer chain n-3 PUFA, could serve as a promising approach for the prevention and management of cardio-metabolic disorders in the mother and the offspring.

ABBREVIATIONS

ALA:	Alpha linolenic acid
CEMACH:	Confidential Enquiry into Maternal and Child Health
COX:	Cyclooxygenase
CVD:	Cardiovascular disease
DGAT-2:	Diacylglycerol acetyltransferase-2
DHA:	Docosahexaenoic acid
DOMInO:	DHA to Optimize Mother and Infant Outcome
DPA:	Docosapentaenoic acid
ELOVL:	Elongation of very long fatty acids
EPA:	Eicosapentaenoic acid
FABP-4:	Fatty acid binding protein-4
FADS:	Fatty acid desaturases
GDM:	Gestational *diabetes mellitus*
LOX:	Lipoxygenase
PE:	Pre-eclampsia
PTB:	Pre-term birth
PUFA:	Polyunsaturated fatty acids
SNPs:	Single nucleotide polymorphisms
WHO:	World Health Organization

REFERENCES

Abedi, E. and M. A. Sahari (2014). "Long-chain polyunsaturated fatty acid sources and evaluation of their nutritional and functional properties." *Food Sci Nutr* 2(5): 443–463.

Akerele, O. A. and S. K. Cheema (2016). "A balance of omega-3 and omega-6 polyunsaturated fatty acids is important in pregnancy." *J Nutr Intermed Metab* 5: 23–33.

Aulchenko, Y. S., S. Ripatti, et al. (2009). "Loci influencing lipid levels and coronary heart disease risk in 16 European population cohorts." *Nat Genet* 41(1): 47–55.

Balogun, K. A. and S. K. Cheema (2016). "Dietary omega-3 fatty acids prevented adipocyte hypertrophy by downregulating DGAT-2 and FABP-4 in a sex-dependent fashion." *Lipids* 51(1): 25–38.

Barker, D. J., C. N. Hales, et al. (1993). "Type 2 (non-insulin-dependent) diabetes mellitus, hypertension and hyperlipidaemia (syndrome X): relation to reduced fetal growth." *Diabetologia* 36(1): 62–67.

Belzung, F., T. Raclot, et al. (1993). "Fish oil n-3 fatty acids selectively limit the hypertrophy of abdominal fat depots in growing rats fed high-fat diets." *Am J Physiol* **264**(6 Pt 2): R1111–R1118.

Birch, E. E., S. Garfield, et al. (2007). "Visual acuity and cognitive outcomes at 4 years of age in a double-blind, randomized trial of long-chain polyunsaturated fatty acid-supplemented infant formula." *Early Hum Dev* **83**(5): 279–284.

Bokor, S., J. Dumont, et al. (2010). "Single nucleotide polymorphisms in the FADS gene cluster are associated with delta-5 and delta-6 desaturase activities estimated by serum fatty acid ratios." *J Lipid Res* **51**(8): 2325–2333.

Buckley, J. D. and P. R. Howe (2009). "Anti-obesity effects of long-chain omega-3 polyunsaturated fatty acids." *Obes Rev* **10**(6): 648–659.

Burdge, G. C. and S. A. Wootton (2002). "Conversion of alpha-linolenic acid to eicosapentaenoic, docosapentaenoic and docosahexaenoic acids in young women." *Br J Nutr* **88**(4): 411–420.

Calder, P. C. (2009). "Polyunsaturated fatty acids and inflammatory processes: new twists in an old tale." *Biochimie* **91**(6): 791–795.

Catalano, P. M., J. P. Kirwan, et al. (2003). "Gestational diabetes and insulin resistance: role in short- and long-term implications for mother and fetus." *J Nutr* **133**(5 Suppl 2): 1674S–1683S.

Challier, J. C., S. Basu, et al. (2008). "Obesity in pregnancy stimulates macrophage accumulation and inflammation in the placenta." *Placenta* **29**(3): 274–281.

Chang, C. Y., D. S. Ke, et al. (2009). "Essential fatty acids and human brain." *Acta Neurol Taiwan* **18**(4): 231–241.

Cheatham, C. L., J. Colombo, et al. (2006). "N-3 fatty acids and cognitive and visual acuity development: methodologic and conceptual considerations." *Am J Clin Nutr* **83**(6 Suppl): 1458S–1466S.

Chen, X., T. O. Scholl, et al. (2010). "Differences in maternal circulating fatty acid composition and dietary fat intake in women with gestational diabetes mellitus or mild gestational hyperglycemia." *Diabetes Care* **33**(9): 2049–2054.

Cho, H. P., M. Nakamura, et al. (1999). "Cloning, expression, and fatty acid regulation of the human delta-5 desaturase." *J Biol Chem* **274**(52): 37335–37339.

Clandinin, M. T., J. E. Chappell, et al. (1980). "Intrauterine fatty acid accretion rates in human brain: implications for fatty acid requirements." *Early Hum Dev* **4**(2): 121–129.

Coletta, J. M., S. J. Bell, et al. (2010). "Omega-3 fatty acids and pregnancy." *Rev Obstet Gynecol* **3**(4): 163.

Couet, C., J. Delarue, et al. (1997). "Effect of dietary fish oil on body fat mass and basal fat oxidation in healthy adults." *Int J Obes Relat Metab Disord* **21**(8): 637–643.

Coustan, D. R. (2013). "Gestational diabetes mellitus." *Clin Chem* **59**(9): 1310–1321.

de la Garza Puentes, A., R. Montes Goyanes, et al. (2017). "Association of maternal weight with FADS and ELOVL genetic variants and fatty acid levels- the PREOBE follow-up." *PLoS One* **12**(6): e0179135.

Denomme, J., K. D. Stark, et al. (2005). "Directly quantitated dietary (n-3) fatty acid intakes of pregnant Canadian women are lower than current dietary recommendations." *J Nutr* **135**(2): 206–211.

Drag, J., A. Gozdzialska, et al. (2017). "Effect of high carbohydrate diet on elongase and desaturase activity and accompanying gene expression in rat's liver." *Genes Nutr* **12**: 2.

Duley, L. (2009). *The Global Impact of Pre-eclampsia and Eclampsia. Seminars in Perinatology*, Elsevier.

Ferramosca, A., A. Conte, et al. (2012). "A krill oil supplemented diet suppresses hepatic steatosis in high-fat fed rats." *PLoS One* **7**(6): e38797.

Giltay, E. J., L. J. Gooren, et al. (2004). "Docosahexaenoic acid concentrations are higher in women than in men because of estrogenic effects." *Am J Clin Nutr* **80**(5): 1167–1174.

Gomez Candela, C., L. M. Bermejo Lopez, et al. (2011). "Importance of a balanced omega 6/omega 3 ratio for the maintenance of health: nutritional recommendations." *Nutr Hosp* **26**(2): 323–329.

Grandjean, P., K. S. Bjerve, et al. (2001). "Birthweight in a fishing community: significance of essential fatty acids and marine food contaminants." *Int J Epidemiol* **30**(6): 1272–1278.

Gregersen, M. I. and R. A. Rawson (1959). "Blood volume." *Physiol Rev* **39**(2): 307–342.

Grieger, J. A. and V. L. Clifton (2014). "A review of the impact of dietary intakes in human pregnancy on infant birthweight." *Nutrients* **7**(1): 153–178.

Groh-Wargo, S., J. Jacobs, et al. (2005). "Body composition in preterm infants who are fed long-chain polyunsaturated fatty acids: a prospective, randomized, controlled trial." *Pediatr Res* **57**(5 Pt 1): 712–718.

Guelinckx, I., R. Devlieger, et al. (2008). "Maternal obesity: pregnancy complications, gestational weight gain and nutrition." *Obes Rev* **9**(2): 140–150.

Haghiac, M., X. H. Yang, et al. (2015). "Dietary omega-3 fatty acid supplementation reduces inflammation in obese pregnant women: a randomized double-blind controlled clinical trial." *PLoS One* **10**(9): e0137309.

Hainault, I., M. Carolotti, et al. (1993). "Fish oil in a high lard diet prevents obesity, hyperlipemia, and adipocyte insulin resistance in rats." *Ann NY Acad Sci* **683**: 98–101.

Hansson, S. R., A. Naav, et al. (2014). "Oxidative stress in preeclampsia and the role of free fetal hemoglobin." *Front Physiol* **5**: 516.

Heerwagen, M. J., M. S. Stewart, et al. (2013). "Transgenic increase in N-3/n-6 fatty acid ratio reduces maternal obesity-associated inflammation and limits adverse developmental programming in mice." *PLoS One* **8**(6): e67791.

Helland, I. B., L. Smith, et al. (2008). "Effect of supplementing pregnant and lactating mothers with n-3 very-long-chain fatty acids on children's IQ and body mass index at 7 years of age." *Pediatrics* **122**(2): e472–e479.

Heslehurst, N., L. J. Ells, et al. (2007). "Trends in maternal obesity incidence rates, demographic predictors, and health inequalities in 36,821 women over a 15-year period." *BJOG* **114**(2): 187–194.

Horvath, A., B. Koletzko, et al. (2007). "Effect of supplementation of women in high-risk pregnancies with long-chain polyunsaturated fatty acids on pregnancy outcomes and growth measures at birth: a meta-analysis of randomized controlled trials." *Br J Nutr* **98**(2): 253–259.

Huang, T., J. Sun, et al. (2014). "Genetic variants in desaturase gene, erythrocyte fatty acids, and risk for type 2 diabetes in Chinese Hans." *Nutrition* **30**(7–8): 897–902.

Hudic, I., B. Stray-Pedersen, et al. (2015). "Preterm birth: pathophysiology, prevention, diagnosis, and treatment." *Biomed Res Int* **2015**: 417965.

Hult, M., P. Tornhammar, et al. (2010). "Hypertension, diabetes and overweight: looming legacies of the Biafran famine." *PLoS One* **5**(10): e13582.

Janovska, P., P. Flachs, et al. (2013). "Anti-obesity effect of n-3 polyunsaturated fatty acids in mice fed high-fat diet is independent of cold-induced thermogenesis." *Physiol Res* **62**(2): 153–161.

Jensen, H. A., L. B. Harslof, et al. (2014). "FADS single-nucleotide polymorphisms are associated with behavioral outcomes in children, and the effect varies between sexes and is dependent on PPAR genotype." *Am J Clin Nutr* **100**(3): 826–832.

Johnston, K. M., K. Gooch, et al. (2014). "The economic burden of prematurity in Canada." *BMC Pediatr* **14**: 93.

Jones, M. L., P. J. Mark, et al. (2013). "Maternal dietary omega-3 fatty acid intake increases resolvin and protectin levels in the rat placenta." *J Lipid Res* **54**(8): 2247–2254.

Kathiresan, S., C. J. Willer, et al. (2009). "Common variants at 30 loci contribute to polygenic dyslipidemia." *Nat Genet* **41**(1): 56–65.

Kelly, T., W. Yang, et al. (2008). "Global burden of obesity in 2005 and projections to 2030." *Int J Obes (Lond)* **32**(9): 1431–1437.

Kilby, M. D., R. H. Neary, et al. (1998). "Fetal and maternal lipoprotein metabolism in human pregnancy complicated by type I diabetes mellitus." *J Clin Endocrinol Metab* **83**(5): 1736–1741.

Koucky, M., A. Germanova, et al. (2009). "Pathophysiology of preterm labour." *Prague Med Rep* **110**(1): 13–24.

Kumari, R., V. Dalal, et al. (2018). "Maternal and perinatal outcome in gestational diabetes mellitus in a tertiary care hospital in Delhi." *Indian J Endocrinol Metab* **22**(1): 116–120.

Lain, K. Y. and J. M. Roberts (2002). "Contemporary concepts of the pathogenesis and management of preeclampsia." *JAMA* **287**(24): 3183–3186.

Laker, R. C., M. E. Wlodek, et al. (2013). "Epigenetic origins of metabolic disease: the impact of the maternal condition to the offspring epigenome and later health consequences." *Food Sci Hum Wellness* **2**(1): 1–11.

Lardinois, C. K. and G. H. Starich (1991). "Polyunsaturated fats enhance peripheral glucose utilization in rats." *J Am Coll Nutr* **10**(4): 340–345.

Lauritzen, L., C. Hoppe, et al. (2005). "Maternal fish oil supplementation in lactation and growth during the first 2.5 years of life." *Pediatr Res* **58**(2): 235–242.

Leddy, M. A., M. L. Power, et al. (2008). "The impact of maternal obesity on maternal and fetal health." *Rev Obstet Gynecol* **1**(4): 170.

Lee, J. M., H. Lee, et al. (2016). "Fatty acid desaturases, polyunsaturated fatty acid regulation, and biotechnological advances." *Nutrients* **8**(1): 23–36.

Lemaitre, R. N., T. Tanaka, et al. (2011). "Genetic loci associated with plasma phospholipid n-3 fatty acids: a meta-analysis of genome-wide association studies from the CHARGE Consortium." *PLoS Genet* **7**(7): e1002193.

Leonard, A. E., S. L. Pereira, et al. (2004). "Elongation of long-chain fatty acids." *Prog Lipid Res* **43**(1): 36–54.

Lewis, G. and J. Drife (2004). "Confidential enquiry into maternal and child health. Why mothers die 2000–2002." *Sixth report of the Confidential Enquiries into Maternal Deaths in the United Kingdom.*

Li, S. W., K. Lin, et al. (2013). "FADS gene polymorphisms confer the risk of coronary artery disease in a Chinese Han population through the altered desaturase activities: based on high-resolution melting analysis." *PLoS One* **8**(1): e55869.

Lucia Bergmann, R., K. E. Bergmann, et al. (2007). "Does maternal docosahexaenoic acid supplementation during pregnancy and lactation lower BMI in late infancy?" *J Perinat Med* **35**(4): 295–300.

Luo, J., S. W. Rizkalla, et al. (1996). "Dietary (n-3) polyunsaturated fatty acids improve adipocyte insulin action and glucose metabolism in insulin-resistant rats: relation to membrane fatty acids." *J Nutr* **126**(8): 1951–1958.

Madan, J. C., J. M. Davis, et al. (2009). "Maternal obesity and markers of inflammation in pregnancy." *Cytokine* **47**(1): 61–64.

Makrides, M., L. Duley, et al. (2006). "Marine oil, and other prostaglandin precursor, supplementation for pregnancy uncomplicated by pre-eclampsia or intrauterine growth restriction." *Cochrane Database Syst Rev* (3): CD003402.

Makrides, M., R. A. Gibson, et al. (2010). "Effect of DHA supplementation during pregnancy on maternal depression and neurodevelopment of young children: a randomized controlled trial." *JAMA* **304**(15): 1675–1683.

Makrides, M., M. A. Neumann, et al. (1994). "Fatty acid composition of brain, retina, and erythrocytes in breast- and formula-fed infants." *Am J Clin Nutr* **60**(2): 189–194.

Malcolm, C. A., D. L. McCulloch, et al. (2003). "Maternal docosahexaenoic acid supplementation during pregnancy and visual evoked potential development in term infants: a double blind, prospective, randomised trial." *Arch Dis Child Fetal Neonatal Ed* **88**(5): F383–F390.

Malinowska, M. A., O. W. Wiśniewski, et al. (2017). "Single nucleotide polymorphisms in desaturases genes–effect on docosahexaenoic acid levels in maternal and fetal tissues and early development of the child." *J Med Sci* **86**(2): 177–185.

Marangoni, F., I. Cetin, et al. (2016). "Maternal diet and nutrient requirements in pregnancy and breastfeeding. An italian consensus document." *Nutrients* **8**(10): E629–646.

Martinez, M. and I. Mougan (1998). "Fatty acid composition of human brain phospholipids during normal development." *J Neurochem* **71**(6): 2528–2533.

Martinelli, N., Girelli, D. et al. 2008. "FADS genotypes and desaturase activity estimated by the ratio of arachidonic acid to linoleic acid are Associated with inflammation and coronary artery disease." *Am J Clin Nutrit***88**(4). Oxford University Press: 941–49.

Mathias, R. A., S. Sergeant, et al. (2011). "The impact of FADS genetic variants on omega6 polyunsaturated fatty acid metabolism in African Americans." *BMC Genet* **12**: 50.

McGrowder, D., K. Grant, et al. (2009). "Lipid profile and clinical characteristics of women with gestational diabetes mellitus and preeclampsia." *J Med Biochem* **28**(2): 72–81.

Mehendale, S., A. Kilari, et al. (2008). "Fatty acids, antioxidants, and oxidative stress in pre-eclampsia." *Int J Gynaecol Obstet* **100**(3): 234–238.

Merino, D. M., D. W. Ma, et al. (2010). "Genetic variation in lipid desaturases and its impact on the development of human disease." *Lipids Health Dis* **9**: 63.

Molto-Puigmarti, C., M. C. van Dongen, et al. (2014). "Maternal but not fetal FADS gene variants modify the association between maternal long-chain PUFA intake in pregnancy and birth weight." *J Nutr* **144**(9): 1430–1437.

Morales, E., M. Bustamante, et al. (2011). "Genetic variants of the FADS gene cluster and ELOVL gene family, colostrums LC-PUFA levels, breastfeeding, and child cognition." *PLoS One* **6**(2): e17181.

Mori, T. A. (2014). "Dietary n-3 PUFA and CVD: a review of the evidence." *Proc Nutr Soc* **73**(1): 57–64.

Mozaffarian, D. and J. H. Wu (2011). "Omega-3 fatty acids and cardiovascular disease: effects on risk factors, molecular pathways, and clinical events." *J Am Coll Cardiol* **58**(20): 2047–2067.

Muir, R., G. Liu, et al. (2018). "Maternal obesity-induced decreases in plasma, hepatic and uterine polyunsaturated fatty acids during labour is reversed through improved nutrition at conception." *Sci Rep* **8**(1): 3389.

O'Brien, J. S., D. L. Fillerup, et al. (1964). "Quantification and fatty acid and fatty aldehyde composition of ethanolamine, choline, and serine glycerophosphatides in human cerebral grey and white matter." *J Lipid Res* **5**(3): 329–338.

Oken, E., Y. Ning, et al. (2007). "Diet during pregnancy and risk of preeclampsia or gestational hypertension." *Ann Epidemiol* **17**(9): 663–668.

Olsen, S. F., P. Grandjean, et al. (1993). "Frequency of seafood intake in pregnancy as a determinant of birth weight: evidence for a dose dependent relationship." *J Epidemiol Community Health* **47**(6): 436–440.

Olsen, S. F., H. S. Hansen, et al. (1991). "Gestational age in relation to marine n-3 fatty acids in maternal erythrocytes: a study of women in the Faroe Islands and Denmark." *Am J Obstet Gynecol* **164**(5 Pt 1): 1203–1209.

Olsen, S. F. and H. D. Joensen (1985). "High liveborn birth weights in the Faroes: a comparison between birth weights in the Faroes and in Denmark." *J Epidemiol Community Health* **39**(1): 27–32.

Olsen, S. F. and N. J. Secher (1990). "A possible preventive effect of low-dose fish oil on early delivery and pre-eclampsia: indications from a 50-year-old controlled trial." *Br J Nutr* **64**(3): 599–609.

Olsen, S. F., N. J. Secher, et al. (2000). "Randomised clinical trials of fish oil supplementation in high risk pregnancies. Fish Oil Trials In Pregnancy (FOTIP) Team." *BJOG* **107**(3): 382–395.

Olsen, S. F. and N. J. Secher (2002). "Low consumption of seafood in early pregnancy as a risk factor for preterm delivery: prospective cohort study." *BMJ* **324**(7335): 447.

Olsen, S. F., J. D. Sorensen, et al. (1992). "Randomised controlled trial of effect of fish-oil supplementation on pregnancy duration." *Lancet* **339**(8800): 1003–1007.

Olsen, S. F., J. D. Soorensen, et al. (1994). "[Fish oil supplementation and duration of pregnancy. A randomized controlled trial]." *Ugeskr Laeger* **156**(9): 1302–1307.

Pawlosky, R., J. Hibbeln, et al. (2003). "n-3 Fatty acid metabolism in women." *Br J Nutr* **90**(5): 993–994; discussion 994–995.

Pedersen, J. (1954). "Weight and length at birth of infants of diabetic mothers." *Acta Endocrinol (Copenh)* **16**(4): 330–342.

Perez-Echarri, N., P. Perez-Matute, et al. (2008). "Differential inflammatory status in rats susceptible or resistant to diet-induced obesity: effects of EPA ethyl ester treatment." *Eur J Nutr* **47**(7): 380–386.

Perez-Matute, P., N. Perez-Echarri, et al. (2007). "Eicosapentaenoic acid actions on adiposity and insulin resistance in control and high-fat-fed rats: role of apoptosis, adiponectin and tumour necrosis factor-alpha." *Br J Nutr* **97**(2): 389–398.

Poisson, J. P., R. P. Dupuy, et al. (1993). "Evidence that liver microsomes of human neonates desaturate essential fatty acids." *Biochim Biophys Acta* **1167**(2): 109–113.

Postle, A. D., M. D. Al, et al. (1995). "The composition of individual molecular species of plasma phosphatidylcholine in human pregnancy." *Early Hum Dev* **43**(1): 47–58.

Poston, L. (2011). "Influence of maternal nutritional status on vascular function in the offspring." *Microcirculation* **18**(4): 256–262.

Redman, C. W. and I. L. Sargent (2009). "Placental stress and pre-eclampsia: a revised view." *Placenta* **30**(Suppl A): S38–S42.

Roberts, J. M., G. Pearson, et al. (2003). "Summary of the NHLBI working group on research on hypertension during pregnancy." *Hypertension* **41**(3): 437–445.

Ryan, E. A. (2013). "Balancing weight and glucose in gestational diabetes mellitus." *Diabetes Care* **36**(1): 6–7.

Rylander, C., T. M. Sandanger, et al. (2014). "Consumption of lean fish reduces the risk of type 2 diabetes mellitus: a prospective population based cohort study of Norwegian women." *PLoS One* **9**(2): e89845.

Schaeffer, L., H. Gohlke, et al. (2006). "Common genetic variants of the FADS1 FADS2 gene cluster and their reconstructed haplotypes are associated with the fatty acid composition in phospholipids." *Hum Mol Genet* **15**(11): 1745–1756.

Schmitz, G. and J. Ecker (2008). "The opposing effects of n-3 and n-6 fatty acids." *Prog Lipid Res* **47**(2): 147–155.

Schulz, L. C. (2010). "The Dutch Hunger Winter and the developmental origins of health and disease." *Proc Natl Acad Sci USA* **107**(39): 16757–16758.

Schwab, U., L. Lauritzen, et al. (2014). "Effect of the amount and type of dietary fat on cardiometabolic risk factors and risk of developing type 2 diabetes, cardiovascular diseases, and cancer: a systematic review." *Food Nutr Res* **58**: 25145–25171.

Schwartz, R., P. A. Gruppuso, et al. (1994). "Hyperinsulinemia and macrosomia in the fetus of the diabetic mother." *Diabetes Care* **17**(7): 640–648.

Simopoulos, A. P. (2006). "Evolutionary aspects of diet, the omega-6/omega-3 ratio and genetic variation: nutritional implications for chronic diseases." *Biomed Pharmacother* **60**(9): 502–507.

Simopoulos, A. P. (2016). "An increase in the omega-6/omega-3 fatty acid ratio increases the risk for obesity." *Nutrients* **8**(3): 128.

Singh, M. (2005). "Essential fatty acids, DHA and human brain." *Indian J Pediatr* **72**(3): 239–242.

Smuts, C. M., M. Huang, et al. (2003). "A randomized trial of docosahexaenoic acid supplementation during the third trimester of pregnancy." *Obstet Gynecol* **101**(3): 469–479.

Soulimane-Mokhtari, N. A., B. Guermouche, et al. (2005). "Modulation of lipid metabolism by n-3 polyunsaturated fatty acids in gestational diabetic rats and their macrosomic offspring." *Clin Sci (Lond)* **109**(3): 287–295.

Sreelakshmi, P. R., S. Nair, et al. (2015). "Maternal and neonatal outcomes of gestational diabetes: a retrospective cohort study from Southern India." *J Fam Med Prim Care* **4**(3): 395–398.

Stoffel, W., B. Holz, et al. (2008). "Delta6-desaturase (FADS2) deficiency unveils the role of omega3- and omega6-polyunsaturated fatty acids." *EMBO J* **27**(17): 2281–2292.

Sweeting, A. N., G. P. Ross, et al. (2016). "Gestational diabetes mellitus in early pregnancy: evidence for poor pregnancy outcomes despite treatment." *Diabetes Care* **39**(1): 75–81.

Tanaka, T., Shen, J. et al. (2009). "Genome-wide association study of plasma polyunsaturated fatty acids in the InCHIANTI study." *PLoS Genetics* **5**(1): e1000338.

Tintle, N. L., J. V. Pottala, et al. (2015). "A genome-wide association study of saturated, mono- and polyunsaturated red blood cell fatty acids in the Framingham Heart Offspring Study." *Prostaglandins Leukot Essent Fatty Acids* **94**: 65–72.

Uauy, R., E. Birch, et al. (1992). "Visual and brain function measurements in studies of n-3 fatty acid requirements of infants." *J Pediatr* **120**(4 Pt 2): S168–S180.

Velzing-Aarts, F. V., F. R. van der Klis, et al. (1999). "Umbilical vessels of preeclamptic women have low contents of both n-3 and n-6 long-chain polyunsaturated fatty acids." *Am J Clin Nutr* **69**(2): 293–298.

Verlohren, S., K. Melchiorre, et al. (2014). "Uterine artery Doppler, birth weight and timing of onset of pre-eclampsia: providing insights into the dual etiology of late-onset pre-eclampsia." *Ultrasound Obstet Gynecol* **44**(3): 293–298.

Voortman, T., E. H. van den Hooven, et al. (2015). "Effects of polyunsaturated fatty acid intake and status during pregnancy, lactation, and early childhood on cardiometabolic health: a systematic review." *Prog Lipid Res* **59**: 67–87.

Wendland, E. M., M. R. Torloni, et al. (2012). "Gestational diabetes and pregnancy outcomes— a systematic review of the World Health Organization (WHO) and the International Association of Diabetes in Pregnancy Study Groups (IADPSG) diagnostic criteria." *BMC Pregnancy Childbirth* **12**: 23.

WHO (2018). "Preterm birth." Retrieved from http://www.Who.Int/Mediacentre/Factsheets/ Fs363/En/. Accessed on July, 2018.

Williams, C. M. and G. Burdge (2006). "Long-chain n-3 PUFA: plant v. marine sources." *Proc Nutr Soc* **65**(1): 42–50.

Williams, M. A., R. W. Zingheim, et al. (1995). "Omega-3 fatty acids in maternal erythro- cytes and risk of preeclampsia." *Epidemiology* **6**(3): 232–237.

Xie, L. and S. M. Innis (2008). "Genetic variants of the FADS1 FADS2 gene cluster are associated with altered (n-6) and (n-3) essential fatty acids in plasma and erythrocyte phospholipids in women during pregnancy and in breast milk during lactation." *J Nutr* **138**(11): 2222–2228.

Yessoufou, A., N. Soulaimann, et al. (2006). "N-3 fatty acids modulate antioxidant status in diabetic rats and their macrosomic offspring." *Int J Obes (Lond)* **30**(5): 739–750.

12 Challenges to the Clinical Implementation of Personalized Nutrition

Diego Accorsi and Nilanjana Maulik

CONTENTS

12.1 INTRODUCTION

Adequate nutrition is understood not only to play a key role in maintaining health, but also in the management of many conditions such as obesity (Arterburn, Maciejewski et al. 2005), diabetes and prediabetes (Evert, Dennison et al. 2019), cardiovascular disease, liver disease, kidney disease (Kalantar-Zadeh and Fouque 2017), cancer (Shiao, Grayson et al. 2018), malnourishment and/or nutritional deficiencies and gastrointestinal (GI) conditions such as inflammatory bowel disease (IBD)/irritable bowel syndrome (IBS) to mention a few. However, thus far our holistic understanding of what constitutes nutritional 'adequacy' is lacking, especially when it comes to the intricate and multifactorial mechanisms underlying its role in disease prevention. Personalized nutrition refers to individually tailored nutritional planning based on patient-specific parameters to promote good health, and prevent or manage disease states. This contrasts with 'stratified' nutrition, which attempts to categorize individuals into discrete risk groups, each group receiving its own sets of nutritional recommendations (Ordovas, Ferguson et al. 2018). Furthermore, current medical nutritional advice by clinicians tends to be somewhat generalized, and even in those fields where substantial nutritional research has led to modest improvement in nutritional recommendations (for example, diabetes), nutritional education is often lacking (Evert, Dennison et al. 2019). Personalized nutrition applies to both healthy individuals as well as those with underlying genetic and physiologic susceptibilities. Parameters considered for a personalized approach include internal ones such as patient age, genetic makeup (nutrigenetics—genes affecting how an individual processes nutrients—and nutrigenomics—how food affects gene expression—are both often used synonymously with personalized nutrition but

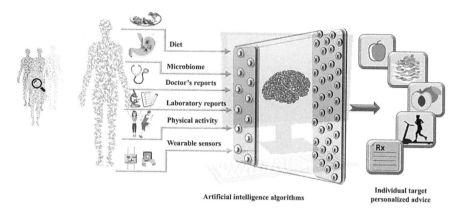

FIGURE 12.1 Proposed pathway for the generation of personalized nutritional advice encompassing data acquisition from an individual, data-processing through artificial intelligence algorithms and prediction interpretation and delivery. (Reprinted under a creative commons license, original work by Verma, M. et al. (2018) "Challenges in Personalized Nutrition and Health." *Frontiers in Nutrition* **5:** 117.)

represent only a fraction of it), epigenomics (changes to the expression of an organism's genes), microbiomics (the study of organisms living within the human body, especially gut microbes), metabolomics (an individual's specific metabolite makeup which translates into cellular activity and physiology), as well as external factors like the environment (including exposures), food processing/availability and physical activity, etc. (Verma, Hontecillas et al. 2018). Current advances in science and -omics, biomolecular markers, computational analysis, patient data collection and storage, data-tracking and artificial intelligence algorithms have made it possible to gather and process nutritional data on a patient-to-patient, nearly real-time basis (Figure 12.1). However, while we have the tools necessary to acheive this, there are many challenges to the testing of personalized nutritional approaches, let alone their implementation in real-life clinical scenarios. This chapter will seek to identify the barriers to the practical and clinical implementation of personalized nutrition at several time points through the pipeline.

12.2   CHALLENGES TO DATA ACQUISITION AND INTERPRETATION

The more patient variables that are analyzed, the more tailored nutritional recommendations can become important. If every single variable ranging from genetic makeup to daily life exposures that makes an individual unique could be recorded and tracked for progress over time, it would create a large amount of information. Attempt to do so for an entire population and it can quickly become an overwhelming task. This concept is known as 'big data,' which is defined as a collection of complex and quantitatively large data which are too difficult to manage through conventional means (Ross, Wei et al. 2014). In a context where 'everything matters,' it becomes evident that the implementation of personalized nutrition and the monitoring of its effects over time on an individual requires a system that allows for the storage, access to and capacity for real-time updates/monitoring of big data. Luckily, the healthcare field is well acquainted with this concept, as electronic health records (EHRs) are perfect examples of it. As EHRs become more widely used, it is estimated that the amount of electronic data in the United States would soon reach the zettabyte range (10^{21} gigabytes) (Raghupathi and Raghupathi 2014). However, there are still barriers to data acquisition.

Just because EHRs can contain massive amounts of information, it does not mean that at present they contain large amounts of nutritional data; nutritional data is both sparse and heterogeneous. By existing within cyberspace, EHRs have the potential to be linked to other databases containing crucial information such as food content information, food handling and transportation, and calorie counting mobile applications. Even more missing data can be harnessed from wearable sensor devices such as portable glucometers and activity trackers. For example, in a study by Zeevi et al., personalized nutrition was provided to patients based on measurements of postprandial blood glucose obtained through glucometers, as well as real-time nutritional intake and activity data inputted by patients via their smartphones (Zeevi, Korem et al. 2015). Furthermore, significant amounts of nutritional information are available in alternative formats including, for example, handwritten doctors' notes and food diaries. While it is possible to digitalize this information, the process is not standardized and often fraught with potential for error. Lastly, it is important to remember that there always exists the risk that data is missing altogether because of error, intentional omission or patient non-compliance. If all else fails, statistical analysis could help fill in the blanks but current methods may introduce bias and are computationally intensive (Verma, Hontecillas et al. 2018).

Once all necessary patient variables have been acquired and stored, they must be analyzed. However, part of the definition of big data is a complexity and depth that makes conventional analysis difficult (Ross, Wei et al. 2014). To that end it may be necessary to incorporate less conventional methods, such as artificial intelligence. With these technologies, it becomes possible to analyze large amounts of complex and interacting data with pre-crafted computational parameters and algorithms based on current understanding of underlying genetic, molecular and pathological mechanisms. For example, Leber et al. used one such algorithm to predict outcomes using an '*in silico*' pipeline, for multiple artificial subjects undergoing alternative treatments for *Clostridium difficile* (anti-toxin antibodies, fecal transplant, immunomodulators, etc.), using data collected regarding immune response (e.g. T-cell ratios), gut microbiome and time of clearance, while adhering to many known biological and molecular pathways (Leber, Hontecillas et al. 2017). Using this pipeline, the authors managed to establish the therapeutic potential of bis-(benzimidazolyl)-terephthalanilide (BTTs) NSC61610, an immunomodulator involved in the lanthionine synthetase C-like 2 (LANCL2) pathway, in the management of *C. difficile* colitis, an effect which was then validated through an *in vivo* mice study.

Employing similar technologies, based on a patient's real time dataset and understanding of underlying mechanisms, we are now potentially poised to predict the effect of a nutritional intervention on one or many potential outcome measures. However, the final interpretation and validation of these predictions is in the hands of professionally trained individuals. For one, it becomes imperative that training be standardized and frequently regulated. Training would be multifaceted, with scientific, computational and mathematical components; indeed, practitioners with a solid understanding of personalized nutrition may become a precious commodity. In addition, there is the issue of scientific validation; the animal model employed by Leber et al. was sufficient to establish a correlation between the use of NSC61610 and decreased colonic inflammation and immune infiltration (Leber, Hontecillas et al. 2017), but how does one go about establishing correlation between much more complex parameter such as nutrition and health?

12.3   CHALLENGES TO THE VALIDATION OF PERSONALIZED NUTRITION

To ascertain the efficacy of personalized nutritional recommendations, it becomes necessary to validate and test the data derived from artificial computational models. Ideally this would be achieved through the highest level of evidence available currently: a statistically significant correlation in a clinical, randomized controlled trial (RCT). However, this is factually difficult to accomplish in the context of personalized nutrition. Because of this, there is a sparsity of RCTs in the field and most present studies have been retrospective or observational in nature.

Taking one such existing RCT as an example, Corella et al investigated the effect of Mediterranean diet (MedDiet with extra virgin olive oil, MedDiet with mixed nuts and a control low-fat diet) on major cardiovascular events tracked over a period of 4.8 years in diabetics with and without the TCF7L2-rs7903146 polymorphism (TT, CT and CC). They found that increased adherence to a Mediterranean diet (a score out of 14 as measured by a self-administered questionnaire) led to improved fasting glucose level and baseline lipids, as well as lower incidence of stroke, with the most improvement observed in homozygous mutated individuals (Corella, Carrasco et al. 2013). This study highlights several challenges to the scientific testing of dietary implementations.

Firstly, the issue of study controls. In disease subgroups, such as diabetics, it may be possible to compare individualized nutritional recommendations to standard disease-specific dietary recommendations such as a low-fat diet; however, this method fails to consider otherwise healthy individuals who are also viable targets for personalized nutritional feedback; what constitutes the standard 'regular' diet? It also fails to consider geographical, behavioral and socio-cultural variables that influence the very makeup of the food patients have access to; indeed, even how food is cooked will alter its nutritional makeup. Thus, there's no optimal control for diet, and for similar reasons it also becomes immediately apparent that qualitative and quantitative manipulation of diet in a controlled manner would be difficult to accomplish.

Secondly, the current design of RCTs does not lend itself to the analysis of gene-food interactions. Current studies are vastly nutrigenetic-focused, and are further limited to tiny segments of the genome. They also tend to limit outcome data to changes in few, specific biomarkers, such as blood glucose levels or lipid levels (Ordovas, Ferguson et al. 2018). This is because there is a pervasive need in modern research to identify one-to-one, statistically significant, reproducible and reversible correlations between two variables. Based on Corella's results, it would be easy to infer that if a patient is homozygous for the TCF7L2-rs7903146 polymorphism, then consuming a Mediterranean diet will always and predictably lower your lipid levels. However, the truth is that one cannot apply these standard principles to the study of personalized nutrition because gene-environment interactions are not straightforward. The reason is that the evidence acquired from RCTs is probabilistic, meaning that RCTs excel at predicting mean outcomes from a given intervention or genotype, but cannot predict outcomes for an individual; there will always be those individuals who will stray from the mean due to other variables not tested. Thus, experts have concluded that the study of personalized nutrition models will rely on a different form of validation; either one that will look at multiple variables at the same time, or one that will look at individuals rather than groups, or so-called n-of-1 trials.

12.4   CHALLENGES TO THE CLINICAL IMPLEMENTATION OF PERSONALIZED NUTRITION

Evidence suggests that it is possible to facilitate a change in behavior using personalized advice as the catalyst. Horne et al. for example reviewed the results of 26 studies meeting inclusion criteria to assess whether genetic testing interventions had an impact on behavioral change, namely smoking, nutrition and physical activity (Horne, Madill et al. 2018). What was noted was that the provision of genetic testing facilitated behavioral change, and even more interestingly was the fact that these results were more evident in the context of nutrition. These results are echoed in the Food4Me trial, what is to date one of the few, most recognized and largest RCTs investigating the effect of personalized nutritional feedback which included everything from dietary recommendations to genetic information, on patient behavior and, subsequently, on clinical endpoints (Table 12.1). For example, it was noticed that patients who were informed they were risk carriers of the FTO genotype were more likely to have larger changes in adiposity biomarkers and weight loss compared to those patients who were not informed (Celis-Morales, Marsaux et al. 2017). What's more, it also seems that increased frequency of personalized feedback is only weakly associated with improved outcomes, albeit limited only to the first three months after initiation of intervention, and coming at the cost of slightly higher dropout rates at three and six months (Celis-Morales, Livingstone et al. 2019). It seems at first glance therefore that there's something about the content of personalized nutritional feedback that patients respond to. In a study by Vallèe Marcotte et al., French Canadian public perceptions specifically to genetic testing for nutritional advice were evaluated through online surveys (Vallee Marcotte, Cormier et al. 2018). In general, this population had overall positive views on nutrigenetic testing, citing 'good health' and 'disease prevention' as the primary advantages over routine nutritional advice.

But what are patients' concerns regarding personalized nutrition? Vallèe Marcotte at al. reported most patients concerned themselves with public access to their private health information (Vallee Marcotte, Cormier et al. 2018). Indeed, personalized nutrition involves the digitalization of large amounts of personal health data over multiple databases and devices, and therefore the issue of health information protection becomes a necessity. Data gathering, interpretation and dissemination must

TABLE.12.1
The Food4Me Trial

	Level 1 (L1): dietary group Personalized advice was provided for weight, PA and dietary intake.	Level 2 (L2): dietary + phenotypic group Personalized advice was provided for weight, WC, PA, dietary intake and blood markers	Level 3 (L3): Dietary + phenotypic + genomic group Personalized advice was provided for weight, WC, PA, dietary intake, and blood and genomic markers
* Feedback on how food group intakes compare with guidelines (to optimize the consumption of fruits and vegetables, whole-grain products, fish, dairy products, and meat)	✓	✓	✓
* Participant anthropometric profile (weight, BMI, WC)	✓	✓	✓
* Participant PA profile (Baecke Questionnaire and Accelerometry)	✓	✓	✓
* Participant nutritional profile based on the online-FFQ (protein, carbohydrates, total fat, monounsaturated fat, polyunsaturated fat, saturated fat, salt, omega-3, fiber, calcium, iron, vitamin A, folate, thiamine, riboflavin, vitamin B12, vitamin C)	✓	✓	✓
* Participant blood profile related to nutrition (glucose, total cholesterol, carotenes, n3 index)	✗	✓	✓
* Participant genetic profile related to nutrition (MTHFR, FTO, TCF7L2, APOE ε4 and FADS1 genes)	✗	✗	✓

Participants were enrolled randomly into four different groups including a control (standard nutritional advice) and three others as outlined above, each with increasing amounts of information provided at feedback. These were subsequently subdivided into low (month 0, 3, 6) and high (0, 1, 2, 3 and 6) frequency feedback. Adapted from CelisMorales, C. et al. (2019). "Frequent Nutritional Feedback, Personalized Advice, and Behavioral Changes: Findings from the European Food4Me Internet-Based RCT N - Appendix 1: Methods and Study Design." *Am J Prev Med* **57**(2): 209–219.*

necessarily comply with the Health Insurance Portability and Accountability Act. Another issue of debate is the cost of personalized nutrition. The implementation of technologies such as monitoring devices and n-of-1 trials may make individualized health recommendations prohibitively costly when deployed at the level of a population, which is counter-intuitive in the context of an initiative that is meant to diminish health disparities. To this end it may become a necessity to limit the use of these technologies only to what is absolutely necessary in order to ensure health equity (Ordovas, Ferguson et al. 2018). Another solution is to rely on digitalization to minimize costs; Celis-Morales et al. stated that digitalization of health feedback as opposed to face-to-face feedback (as was done in the Food4Me trial, Figure 12.2) could be more economically sustainable by 1) being employed across multiple devices such as computers and phones which are widely available, thus reaching

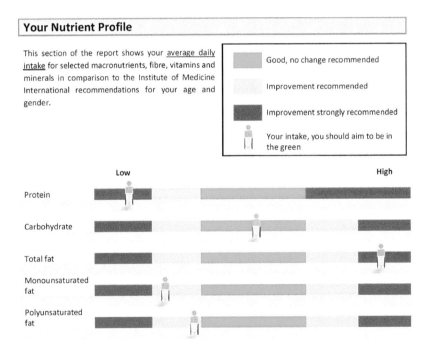

FIGURE 12.2 Example of personalized nutritional feedback computer interface as used in the Food4Me trial. Note that pictorial representation makes it easier for users to understand how to modify their food consumption. These reports should be available and scalable to target as many individuals as possible. (Reprinted with permission from Celis-Morales, C. et al. (2019). "Frequent Nutritional Feedback, Personalized Advice, and Behavioral Changes: Findings from the European Food4Me Internet-Based RCT N - Appendix 1: Methods and Study Design." *Am J Prev Med* **57**(2): 209–219.)

wide geographical areas across diverse socioeconomic backgrounds; 2) being scalable, meaning that content can be provided in multiple forms (such as inclusion and exclusion of certain content, etc.); and 3) because while digital platform development may have greater upfront cost, applied over a population it would actually lower costs, including maintenance and updating (Celis-Morales, Lara et al. 2015). But what about health coverage? As it stands, medical nutritional therapy is covered in most of Europe and the US only in the context of certain comorbidities (examples include obesity, diabetes and kidney disease) as well as certain high risk individuals (Fischer, Berezowska et al. 2016). For otherwise 'healthy' individuals for whom advice would be more preventative in nature, it is largely up to them to determine whether to pay for more personalized nutritional advice. Fischer et al. reported that, from a survey of 8,233 individuals across eight different European countries, about 30% were willing to pay more for personalized nutritional advice, and those individuals were willing to pay on average about 150% what they would for generalized advice. This is important because for the foreseeable future the advancement of personalized nutrition will also depend on these consumers' willingness to pay, at least until the time changes can be made in the field of health coverage.

12.5   SUMMARY AND FUTURE DIRECTIONS

This chapter has reviewed the challenges inherent to the clinical implementation of personalized nutrition. Acquiring all the necessary patient information to derive effective nutritional feedback will result in 'big data,' which is complex. EHRs lend themselves as good starting tools for the acquisition and storage of big data, while missing information could potentially be gathered realtime through monitoring devices and portable applications. Data analysis would be complex through regular means and will likely necessitate the implementation of advanced artificial intelligence tools to make predictions based on data input and known underlying biomolecular and genetic principles. Once predictions can be made, it will be necessary to validate these analytical tools before clinical implementation; RCTs are, however, difficult to execute in the context of personalized nutrition, and therefore its study will most likely necessitate new scientific approaches that either involve multiple variables or a single subject (n-of-1 trials for example). Once validated, predictions from analytical technologies will need to be translated into nutritional feedback for patients, which will not only require rigorous, multifaceted training of clinicians, but will also need to consider patient concerns such as privacy and cost. It is hoped that once these hurdles are overcome, personalized nutrition can help achieve better health outcomes, including both disease management and prevention, across all populations.

ABBREVIATIONS

BTTs: Bis-(benzimidazolyl)-terephthalanilide
***C. difficile*:**   *Clostridium difficile*
EHRs: Electronic health records
FTO: Fat-mass & Obesity
GI: Gastrointestinal

IBD:	Inflammatory bowel disease
IBS:	Irritable bowel syndrome
LANCL2:	Lanthionine synthetase C-like 2
MedDiet:	Mediterranean diet
RCT:	Randomized controlled trial

REFERENCES

(2019). Cancer Research UK"Current opportunities to catalyze research in nutrition and cancer prevention - an interdisciplinary perspective." *BMC Med* 17(1): 148.

Arterburn, D. E., M. L. Maciejewski, et al. (2005). "Impact of morbid obesity on medical expenditures in adults." *Int J Obes (Lond)* 29(3): 334–339.

Celis-Morales, C., J. Lara, et al. (2015). "Personalising nutritional guidance for more effective behaviour change." *Proc Nutr Soc* 74(2): 130–138.

Celis-Morales, C., K. M. Livingstone, et al. (2019). "Frequent nutritional feedback, personalized advice, and behavioral changes: findings from the European Food4Me internet-based RCT." *Am J Prev Med* 57(2): 209–219.

Celis-Morales, C., C. F. Marsaux, et al. (2017). "Can genetic-based advice help you lose weight? Findings from the Food4Me European randomized controlled trial." *Am J Clin Nutr* 105(5): 1204–1213.

Corella, D., P. Carrasco, et al. (2013). "Mediterranean diet reduces the adverse effect of the TCF7L2-rs7903146 polymorphism on cardiovascular risk factors and stroke incidence: a randomized controlled trial in a high-cardiovascular-risk population." *Diabetes Care* 36(11): 3803–3811.

Evert, A. B., M. Dennison, et al. (2019). "Nutrition therapy for adults with diabetes or prediabetes: a consensus report." *Diabetes Care* 42(5): 731–754.

Fischer, A. R., A. Berezowska, et al. (2016). "Willingness to pay for personalised nutrition across Europe." *Eur J Public Health* 26(4): 640–644.

Horne, J., J. Madill, et al. (2018). "A systematic review of genetic testing and lifestyle behaviour change: are we using high-quality genetic interventions and considering behaviour change theory?" *Lifestyle Genom* 11(1): 49–63.

Kalantar-Zadeh, K. and D. Fouque (2017). "Nutritional management of chronic kidney disease." *N Engl J Med* 377(18): 1765–1776.

Leber, A., R. Hontecillas, et al. (2017). "Modeling new immunoregulatory therapeutics as antimicrobial alternatives for treating Clostridium difficile infection." *Artif Intell Med* 78: 1–13.

Ordovas, J. M., L. R. Ferguson, et al. (2018). "Personalised nutrition and health." *BMJ* 361: bmj k2173.

Raghupathi, W. and V. Raghupathi (2014). "Big data analytics in healthcare: promise and potential." *Health Inf Sci Syst* 2: 3.

Ross, M. K., W. Wei, et al. (2014). "'Big data' and the electronic health record." *Yearb Med Inform* 9: 97–104.

Shiao, S. P. K., J. Grayson, et al. (2018). "Personalized nutrition—genes, diet, and related interactive parameters as predictors of cancer in multiethnic colorectal cancer families." *Nutrients* 10(6), 795.

Vallee Marcotte, B., H. Cormier, et al. (2018). "Nutrigenetic testing for personalized nutrition: an evaluation of public perceptions, attitudes, and concerns in a population of French Canadians." *Lifestyle Genom* 11(3–6): 155–162.

Verma, M., R. Hontecillas, et al. (2018). "Challenges in personalized nutrition and health." *Front Nutr* 5: 117.

Zeevi, D., T. Korem, et al. (2015). "Personalized nutrition by prediction of glycemic responses." *Cell* 163(5): 1079–1094.

13 The Importance of Personalized Nutrition in Psychological Disorders

Gulsah Kaner Tohtak

CONTENTS

Psychological disorders should be considered as a major problem of public health worldwide and should not be neglected or ignored.

Personalized dietary counseling from a registered dietitian should be considered as a part of the treatment.

13.1 INTRODUCTION

According to the World Health Organization (WHO), mental health is 'a state of well-being in which every individual realizes his or her own potential, can cope with the normal stresses of life, can work productively and fruitfully and is able to make a contribution to her or his community' (WHO 2014a). Thanks to positive mental health, which is a fundamental element of democratic society, social cohesion is strengthened, peace and stability are improved and economic development is enhanced in societies (Lesage, Vasiliadis et al. 2006).

Since mental health problems are encountered in all stages of life, cultures and populations, provision of mental health should be adjusted to every individuals' needs (Lesage, Vasiliadis et al. 2006). Mental health conditions refer to changes in the functioning of the brain and nervous system, which lead to alterations in perception of the environment and responses given to it. It is thought that there is a relationship between long-lasting disability and significant mortality through suicide, medical illnesses and accidental death (Davison, Ng et al. 2012). Since they may appear in different forms, individuals may have different experiences. The WHO states that 350 million people suffer from depression, 50 million from epilepsy and 21 million from schizophrenia. Furthermore, 900,000 suicides occur each year. Although 76–85% of these individuals have severe mental health conditions, unfortunately those coming from low- and middle-income countries do not receive any treatment (WHO 2014b). Published by the American Psychiatric Association, the Diagnostic and Statistical Manual of Mental Disorders (DSM) forms a common language to group mental health conditions (Davison, Ng et al. 2012).

Nutrition is involved in the behavior and mood, as well as in the pathology and treatment of mental illnesses. Nutrition sustains the structure and function of neurons and neurotransmitters in the nervous system, which is one of the most crucial contributions to mental health. Neurotransmitter production depends on whether there are sufficient amounts of nutrients or not. These nutrients include amino acids (tryptophan, tyrosine and glutamine), minerals (zinc, copper, iron, selenium, magnesium) and B vitamins (B_1, B_2, B_3, B_6, B_9, B_{12}). Thanks to these neurotransmitters, the body is able to communicate within itself and also with the environment (Teitelbaum, Weiss et al. 2017).

The number of epidemiological studies which analyze the correlations between dietary patterns and mental states has increased steadily during the past ten years. The common conclusion reached by both cross-sectional and longitudinal studies is that the Western diet (a low-quality diet in which one consumes mostly processed foods including fast food, processed meat, refined grains, soft beverages and sweets/ sugars) increases the risk of developing psychiatric symptoms such as depression and anxiety. On the other hand, the Mediterranean diet (in which one consumes mostly fruits, vegetables, legumes, whole-grain products, fish and a low/moderate consumption of meat, dairy products and alcohol) and the Dietary Approaches to Stop Hypertension (DASH) diet (high consumption of whole-grains, fruits, vegetables, moderate consumption of legumes, nuts and low-fat dairy products and low consumption red and processed meats and sodium) are identified with a lower risk of depression (Jacka, Pasco et al. 2013; Khayyatzadeh, Mehramiz et al. 2018).

This chapter is discussed in two sections. The first one studies the prevention of illnesses and the role of macro and micronutrients in mental function, and the second one, on the other hand, analyzes the nutrition problems seen in four selected different mental illnesses (schizophrenia, autism spectrum disorders, anxiety, dementia and Alzheimer's disease) and includes nutrition advice for such problems.

13.2 THE ROLE OF NUTRIENTS IN MENTAL FUNCTION

13.2.1 CARBOHYDRATES

Carbohydrates are found in nature as polysaccharides and are important elements of organism structure and function. They have a positive effect on the mood and behavior of human beings. Consumption of meals containing high levels of carbohydrates leads to insulin release in the body. Insulin not only helps blood sugar infuse into cells to be used for energy, but also enables tryptophan to enter the brain. Neurotransmitters' levels are affected by the levels of tryptophan in the brain. Following a diet which contains low amounts of carbohydrate can provoke depression since what prompts the production of serotonin and tryptophan contributing to well-being is carbohydrate-rich foods (Rao, Asha et al. 2008).

Known for their high levels of fat and refined sugar content, Western diets are observed to reduce brain-derived neurotrophic factor (BDNF), which is thought to have a crucial impact on the pathophysiology of depressive illness in animals. Depressive patients have lower BDNF levels. There are few reports investigating the relationship between sugar dietary patterns and depression. In a case series of diabetic Jordanian female patients, a link was found between uncontrolled glycemic level and the prevalence of undiagnosed depression. Moreover, an observational study conducted in eight countries associated the sugar intake with depression, on the basis of the changes it caused in endorphin levels and oxidative stress (Westover and Marangell 2002; Al-Amer, Sobeh et al. 2011).

Cognition seems to be positively affected by chronic sugar consumption. According to Hoer et al. (Hoerr, Fogel et al. 2017), sugar and sweetener intake is positively and significantly related to anxiety disorders, mood disorders, impulse control disorders and substance use disorders.

Four studies on soft beverages, including two which analyzed adolescents, one young adults and the other general adult population, have established a relationship between increased consumption of soft beverages and increased mental health burden (Lien, Lien et al. 2006; Estherlydia and John 2009; Shi, Taylor et al. 2010; Pan, Zhang et al. 2011). Moreover, it was found that Australian adolescents following the Western diet had higher rates of attention deficit hyperactivity disorder (ADHD) compared to those following a 'healthy' diet (Howard, Robinson et al. 2011).

13.2.2 PROTEINS

Proteins consist of amino acids and are the essential elements of life. The human body is able to produce as many as 12 of the amino acids in the body's proteins. Receiving the other eight, which are called the essential amino acids, is only possible through diet. All essential amino acids exist in a high-quality protein diet. Meats, eggs and dairy products, such as milk, are among the foods which contain a great deal of high-quality protein. Proteins found in plants such as beans, peas and grains may not contain one or two of the essential amino acids in sufficient amounts. It is possible that protein intake and in turn individual amino acids have an impact on brain functions and mental health (Rao, Asha et al. 2008).

13.2.2.1 Tryptophan: A Serotonin Precursor

A number of protein-based foods and dietary proteins such as meats, dairy products, fruits and seeds include tryptophan, which is one of the essential amino acids. Availability of tryptophan is also increased by high glycemic content and glycemic-loaded meals. Apart from its contribution to protein formation, tryptophan is also a precursor for the neurotransmitter serotonin (Jenkins, Nguyen et al. 2016).

Being one of the crucial neurotransmitters, serotonin has an effect on mental health. Most of the serotonin goes out of the central nervous system (CNS); therefore, it plays a role in various physiological processes in different organs. Nevertheless, 2% of the serotonin that remains in the CNS has a crucial impact on the etiology of a number of mental disorders. Synapses are affected by both receptors and transporters. Serotonin neurotransmitter activates the 5-hydroxytryptamine (5-HT) receptors while it is reuptaken from the synaptic cleft by the 5-HT transporter. There may be a relationship between the altered receptor and/or serotonin transporter and mental disorders (Lin, Lee et al. 2014). There is little evidence that suggests that the tryptophan levels in depressed patients and particularly those suffering from a melancholic depression are decreased (Parker and Brotchie 2011).

13.2.2.2 Tyrosine and Phenylalanine: Catecholamine Precursors

Catecholamines (norepinephrine, dopamine and epinephrine) are synthesized from tyrosine by neurons in the brain. Being one of the essential amino acids, L-phenylalanine is the direct precursor of tyrosine; however, the body receives the two through a normal diet. After being formed by tyrosine, dihydroxyphenylalanine (DOPA) is transformed into dopamine, which in return forms norepinephrine and then epinephrine. Neurotransmitters norepinephrine and dopamine are involved in depression. Few studies have investigated whether there is a relationship between tyrosine deficiency and depression (Parker and Brotchie 2011). Nevertheless, plasma concentrations of tryptophan, tyrosine and phenylalanine levels are observed to be lower among individuals with ADHD (Bornstein, Baker et al. 1990).

13.2.3 FATS

13.2.3.1 Saturated Fatty Acids

The numerous types of fatty acids are classified into three basic groups: saturated (SFA), monounsaturated (MUFA) and polyunsaturated fatty acids (Mizunoya,

Ohnuki et al. 2013). Animal products such as butter, cows' milk, meat, salmon and egg yolks, and some plant products such as chocolate, cocoa butter, coconut and palm kernel oils are the main sources of saturated fatty acids (de Souza, Mente et al. 2015).

According to a recent review analyzing the effect of SFA on dementia, a positive correlation between SFA intake and Alzheimer's disease (AD) has been found in three of the four studies reviewed. In the fourth one, on the other hand, an inverse relationship was observed. SFA consumption was found to have a positive relationship with total dementia in one of two studies, with mild cognitive impairment in one of four studies and with cognitive decline in two of four studies (Barnard, Bunner et al. 2014). Previous studies have also detected correlations between MUFA/SFA changes and behavioral and cognitive results. Sartorius et al. (Sartorius, Ketterer et al. 2012) proposed that a high-SFA diet and acute intraventricular injection of palmitic acid in mice inhibited insulin signaling in the brain and locomotor activity in response to acute intraventricular injection of insulin. Similarly, normal wakefulness and sleep behavior were disrupted compared to a high-MUFA diet.

13.2.3.2 Polyunsaturated Fatty Acids: Omega-3 and Omega-6

The effects of polyunsaturated fatty acids (PUFAs) on health are well known. α-linolenic acid (ALA), eicosapentaenoic acid (EPA) and docosahexaenoic acid (DHA) are the components of Omega-3 (ω-3) PUFAs, which can be chiefly found in fatty fish, other kinds of seafood, certain nuts and seeds. According to certain randomized controlled studies, ω-3 PUFA intake improves or even prevents some physical illnesses such as inflammatory and cardiovascular diseases; however, other studies have not reported such data (Giudetti and Cagnazzo 2012; Hoogeveen, Geleijnse et al. 2015; La Rovere and Christensen 2015).

Some of the potential mechanisms for the neuroprotective effect and therapeutic value of ω-3 fatty acids on depression are their modulation of BDNF and their cAMP-response element binding protein expression, which appears through suppressing pro-inflammatory cytokines. On one hand, ω-3 fatty acids increase the release of serotonergic and dopaminergic neurotransmitter; on the other hand, they decrease the production of inflammatory eicosanoids (Bae and Kim 2018). Compared to rats which were put on a control-based diet, rats which were supplemented with the ω-3 PUFA of 10% cod liver oil for 20 days demonstrated an increase in the cortex levels of serotonin and dopamine, which are the two main target neurotransmitters in psychiatric illnesses (Varghese, Shameena et al. 2001). According to a meta-analysis of randomized controlled trials in humans, a diet with ω-3 fatty acids contributes to the treatment of depressive symptoms (Grosso, Pajak et al. 2014). Moreover, many studies suggest that ω-3 fatty acids can be beneficial for depression treatment in elderly individuals. These studies also report that ω-3 PUFAs supplementation decreases the emergence of depressive symptoms in elderly female patients (Kiecolt-Glaser, Belury et al. 2007; Rondanelli, Giacosa et al. 2010). Despite the fact that specific mechanisms are still unclear, clinical research has demonstrated that adequate amount of EPA intake is important for an individual's general mental health and that it works as an adjunctive treatment for depression. Generally, EPA's performance is better when it is consumed with DHA. These acids naturally coexist in foods.

Preferred and selectively stored in the brain and nerve cells, DHA constitutes a large part of the brain tissue mass. The brain's growth, development and maturation—all need DHA. DHA functions in neurotransmission, lipid messaging, genetic expression and cell membrane synthesis. Additionally, it provides vital structural contributions; DHA is concentrated in the phospholipids of brain cell membranes. Fatty fish such as salmon, sardines and tuna have high levels of EPA and DHA. The fresher the fish is, the higher level of the EPA and DHA it has (Teitelbaum, Weiss et al. 2017).

Linoleic and arachidonic acids, which exist in plant and vegetable seeds and oils, margarines and numerous processed foods, are the components of Omega-6 (ω -6) PUFAs. A correlation has been established between high ω-6 PUFA intake and chronic inflammatory diseases, cardiovascular diseases, obesity, rheumatoid arthritis and AD. The ratio of ω-6 to ω-3 PUFAs has shown a dramatic increase in modern diets. Since people consume much more SFAs, this may have led to an increase in the prevalence of depressive disorders. Some studies suggested that depressed patients had a significant reduction in ω-3 PUFAs (EPA, DHA or total ω-3 PUFA), but not in ω-6 levels. Increased ω-6/ω-3 ratio (arachidonic acid/eicosapentaenoic acid ratio) may also accompany depression (Owen, Rees et al. 2008; Su 2008; Riemer, Maes et al. 2010).

13.2.3.3 Trans-Fatty Acids

Being a type of unsaturated fatty acid, trans-fatty acids are constituted by the hydrogenation of vegetable oils and are chiefly found in industrial products. The Federal Drug Administration (FDA) demanded in 2006 that trans-fats must be listed on food labels. In 2013, the FDA decided that the use of hydrogenated oils in foods was unsafe. Moreover, in June 2015, trans-fat was recommended to be removed from food products within the next three years (Ford, Jaceldo-Siegl et al. 2016).

Consumption of trans-fatty acids is pro-inflammatory, as experimental evidence has shown. The brain and the nervous system are adversely affected by trans-fat. Received through diet, trans-fat is incorporated into the membranes of the brain cells and alters the ability of neurons to communicate, which can reduce mental performance. Increasing evidence confirms that trans-fat may play a role in the development of AD and cognitive decline with age (Ginter and Simko 2016). In a prospective cohort study that included 12,059 Spanish university graduates without depression, a detrimental relationship between the intake of trans-fat and the risk for depression was found (Sanchez-Villegas, Verberne et al. 2011). Moreover, the Whitehall II Study suggested that there is a relationship between the trans-fat intake and recurrent depressive symptoms in women (Akbaraly, Sabia et al. 2013). Similarly, in the Adventist Health Study 2, lower dietary trans-fat intake was found to have beneficial impacts on emotional behavior (Ford, Jaceldo-Siegl et al. 2016).

13.2.4 Vitamins

In order for the brain to function optimally and for neurotransmitters to be produced, vitamin B complex is needed. The B vitamins consist of eight water-soluble essential nutrients which are: B_1 (thiamine), B_2 (riboflavin), B_3 (niacin), B_5 (pantothenic acid), B_6 (pyridoxine), B_7 (biotin), B_9 (folate/folic acid) and B_{12} (cobalamin). In adults, there is a relationship between low vitamin B status (notably vitamins B_6, B_{12} and folate) and internalizing behavioral problems like depression (Skarupski, Tangney et al. 2010).

From a neurotransmitter production perspective, B vitamins have a high biological plausibility in influencing energy, brain function and mood modulation. In terms of externalizing behaviors such as delinquency and aggression, it is possible for B vitamin supplementation to reduce aggressive and antisocial behavior. However, there are only few population studies in this area (Zaalberg, Nijman et al. 2010).

13.2.4.1 Vitamin B_1 (Thiamine)

Being the most bioactive form of thiamine, thiamine diphosphate (TDP) is the key coenzyme participating in glucose metabolism and biosynthesis of neurotransmitters such as acetylcholine, gamma-aminobutyric acid (GABA), glutamate, aspartate and serotonin. Antidepression therapies potentially aim at norepinephrine, serotonin, glutamate and their receptors. The brains of patients with AD, Parkinson's disease, dementia and other neurodegenerative diseases have revealed low TDP concentrations and thiamin-dependent enzymatic activities (Zhang, Ding et al. 2013). High doses of thiamin supplement have shown ameliorating effects on learning disorders and behavioral problems in young children (Fattal, Friedmann et al. 2011). Serum thiamin levels have been observed to be lower in patients with AD compared to those with other kinds of dementia (Lu'o'ng and Nguyễn 2011).

13.2.4.2 Vitamin B_2 (Riboflavin)

Riboflavin participates in determining the circulating concentrations of homocysteine. It may exert some influence by reducing the metabolism of other B vitamins, especially that of folate and B_6 (Markley 2012). Riboflavin, which plays an antioxidative role in the metabolism of vitamins B_3 and B_6, can be associated with flavin-glutathione reductase (Batchelor, Kwandou et al. 2017). After six weeks' use of daily supplementation of 400 mg riboflavin, it was noticed that migraine frequency had decreased by 50% to 69% in individuals experiencing frequent migraines (Markley 2012).

13.2.4.3 Vitamin B_3 (Niacin)

Niacin is the essential component of the nicotinamide adenine dinucleotide hydrate (NADH) molecule, which plays an important role in neurotransmitter dopamine production. Dementia is among the signs of pellagra, which is also called 'the niacin-deficiency disease.' Pellagra can lead to delirium during alcohol withdrawal (Oldham and Ivkovic 2012).

Although niacin's psychiatric manifestations, such as irritability, poor concentration, anxiety, fatigue, restlessness, apathy, as well as depression, are quite common (Prakash, Gandotra et al. 2008), it is easy to overlook them because of their non-specific nature. Niacin response abnormality is counted as one of the most widely replicated peripheral biomarkers of schizophrenia. Many comprehensive reviews have focused on its relationship to schizophrenia (Buretic-Tomljanovic, Giacometti et al. 2008; Nadalin, Buretic-Tomljanovic et al. 2010). Moreover, it seems that niacin response abnormality has a high specificity for schizophrenia among those who have schizophrenia when compared to those who have bipolar disorder (Yao, Dougherty et al. 2016).

13.2.4.4 Vitamin B$_6$ (Pyridoxine), Vitamin B$_9$ (Folic Acid) and Vitamin B$_{12}$ (Cobalamin)

The role of B vitamins, especially those of B$_6$, B$_9$ and B$_{12}$ in affecting homocysteine levels in the body and in cognition and brain function has drawn remarkable interest. Vitamin B$_6$ has three forms which occur naturally; pyridoxine, pyridoxal and pyridoxamine. These three forms may also be encountered as phosphorylated compounds. Pyridoxal-5'-phosphate is the primary form of B6 which exists in food and in the body. It acts as a carbonyl-reactive coenzyme for a number of reactions which participate in amino acid metabolism including the metabolism of sulfur-containing amino acids such as homocysteine, formation of neurotransmitters epinephrine, norepinephrine, serotonin and γ- amino butyric acid, taurine synthesis and conversion of tryptophan to niacin (Parletta, Milte et al. 2013). According to Herbison et al. (Herbison, Hickling et al. 2012), there is a relationship between low intakes of vitamin B$_6$ and folate and higher incidence of internalizing disorders. Supporting these findings, adolescents in Japan who follow a non-Western diet and consume more B$_6$ and folate were observed to have reduced the depression symptoms (Murakami, Miyake et al. 2010).

There is a relationship between folate deficiency and depression and cognitive decline and dementia. According to epidemiological, biochemical and gene association studies, folate deficiency is a risk factor for schizophrenia. It is possible for folate plus vitamin B$_{12}$ supplementation to ameliorate the negative symptoms of schizophrenia. Furthermore, low levels of folate are correlated with AD, whereas higher levels of folate reduce the risk of developing AD by 50% (Luchsinger, Tang et al. 2007).

Stupor, apathy, negativism, memory and judgment disorders or even psychoses, depression and dementia may be counted among the mental symptoms of vitamin B$_{12}$ deficiency (Grober, Kisters et al. 2013). Depression, bipolar disorder, panic disorder, psychosis, phobias and dementia are some psychiatric disorders which patients suffering from B$_{12}$ deficiency are diagnosed with. A number of cross-sectional and prospective studies relate moderately elevated concentrations of homocysteine (10% μmol/L) to a higher risk of dementia and especially AD. Both in people with AD and healthy elderly people, raised plasma concentrations of homocysteine are correlated with regional and whole-brain apathy (Teitelbaum, Weiss et al. 2017).

13.2.4.5 Dietary Antioxidants (Vitamins A, C and E)

Being efficient dietary antioxidants, vitamins A, C and E can avert cytotoxicity stemming from free radicals. They also serve the function of direct scavengers of reactive oxygen species (ROS) and for upregulating antioxidant enzyme activity (Parletta, Milte et al. 2013). Antioxidants can act as a protection against aging and chronic diseases. Additionally, they can play a biological role in the etiology of depression. Individuals who have depression have been observed to have higher levels of oxidative stress (Payne, Steck et al. 2012).

It has been discovered that consuming more fruits and vegetables brings better cognitive test results, fewer depressive symptoms and lower risk of developing depression. Individuals with depression have been observed to have a lower beta carotene, vitamin C and fiber intake compared to those without depression (Park, You et al. 2010). With their hydrophilic and hydrophobic characteristics, Vitamins C and E cooperate by providing complete antioxidant defense (Mahadik, Evans et al. 2001).

As demonstrated by studies, flow mediated dilation in elderly people is augmented by the combination of antioxidants vitamin C, vitamin E and alpha-lipoic acid, probably through decreasing plasma-free radicals and restoring the endothelial function (Wray, Nishiyama et al. 2012).

13.2.4.6 Vitamin D

Vitamin D has long been known to participate in protection of the bones. Nonetheless, a great number of extra-bone effects have been found recently. According to studies, it is a possibility that vitamin D efficiency plays a role in the etiology of mood disorders. The fact that vitamin D receptors (VDR) are found in numerous parts of the brain provides support for this relationship (Bellikci Koyu and Buyuktuncer 2015).

Clinical research has shown that vitamin D is correlated with the presence of mood disorders, aspects of cognitive disorders and also with an increased risk of major and minor depression in the elderly. Many studies affirming that vitamin D deficiency is associated with depression in adult populations have led to significant interest in vitamin D's effect on major depression (Anglin, Samaan et al. 2013; Ju, Lee et al. 2013). Having inconsistent results, some of the recently published meta-analyses have investigated the role of vitamin D supplementation in depression symptoms in adults (Li, Mbuagbaw et al. 2014; Shaffer, Edmondson et al. 2014; Spedding 2014).

Assessment of circulating levels of 25(OH)D serum and the combined product of skin synthesis from sun exposure and dietary sources are generally used to test the serum levels of vitamin D. No 'official' agreement has been reached yet concerning the blood levels of 25(OH)D which signal for deficiency, adequacy and inadequacy of vitamin D, those particularly associated with the brain's health. Despite the fact that levels of 60 to 80 ng/mL can be useful in the treatment of cancer, diabetes, autoimmune illness or depression, a number of practitioners prefer 25(OH)D serum levels of at least 30 ng/mL or 75 nmol/L (Teitelbaum, Weiss et al. 2017).

13.2.5 MINERALS

13.2.5.1 Zinc

Elements of zinc, iron, copper and selenium play a significant role in growth and development of individuals. Neuromodulation and cell functions need zinc and iron elements to operate properly (Momčilović, Prejac et al. 2010). Also, antioxidant protection takes place with the help of enzymes, which contain copper and selenium elements (Roman, Jitaru et al. 2014; Li, Wang et al. 2018).

Psychogeriatric patients are usually observed to suffer from zinc deficiency (Gronli, Kvamme et al. 2013). Some authors suggested that there is a correlation between the dietary zinc intake and depression; however, this topic is still controversial. According to some studies, zinc intake has a negative correlation with depression, whereas others have shown no correlation at all (Li, Wang et al. 2018). There are some dynamics suggested to convey the association between depression and zinc intake. To begin with, zinc intake regulates the brain's zinc homeostasis (Takeda, Minami et al. 2001). In addition, neural transmission might be affected by the existence of zinc, such as the activation of N-methyl-D-aspartic acid (NMDA) receptors and the BDNF activity as a common element of depression (Takeda and Tamano 2009; Szewczyk, Kubera et al. 2011; Toth 2011). Moreover, the pathophysiology of depression might be affected by the antioxidant properties of zinc (Swardfager, Herrmann et al. 2013).

Studies have also exhibited that zinc supplementation as an adjunctive treatment to antidepressant drugs yields better results. In an experiment conducted by Nowac et al. (Nowak, Siwek et al. 2003), patients were given either zinc supplements or placebo with antidepressants. In this group of 14 patients, it was observed that the ones who had received zinc supplementation significantly reduced their scores on the Beck's Depression Inventory and the Hamilton Depression Rating Scale only after six and 12 weeks of the experiment compared to the placebo receivers.

13.2.5.2 Iron

Existence of the iron element is crucial for many major functions in our body, such as: brain growth, development, cell differentiation, protein synthesis, hormone production and cellular energy metabolism (Khedr, Hamed et al. 2008). In addition, neurotransmitters like dopamine and serotonin cannot be synthesized without iron, which is also crucial for brain parenchyma (Beard, Connor et al. 1993). Studies also have shown that iron is responsible for reducing the NMDA levels in brain tissues and its toxic effects (Cheah, Kim et al. 2006; Yu, Feng et al. 2011).

Anemia, a very common health issue, is mainly caused by iron deficiency, which more than the 25% of people are suffering from (Kassebaum, Jasrasaria et al. 2014). Other effects of iron deficiency are low levels of brain myelination and failure in monoamine metabolism. Glutamate and GABA homeostasis is also related to the iron levels in the brain. Low levels might bring up various conditions like memory deficiency, lower learning capacity, reduced motor skills and emotional and psychological disorders (Kim and Wessling-Resnick 2014). In addition, apathy, depression and fatigue are known to be related to iron deficiency. Childhood anemia, which is also associated with iron deficiency, may cause very serious complications like psychiatric disorders, mood disorders, autism spectrum disorder, ADHD and developmental disorders (Chen, Su et al. 2013). Some studies suggested that increasing iron consumption can reduce depression symptoms, whereas others suggested no apparent correlation. Fulkerson et al.'s (Fulkerson, Sherwood et al. 2004), Tseng et al.'s (Tseng, Cheng et al. 2018) and Wang et al.'s (Wang, Huang et al. 2017) studies demonstrated the correlation between ADHD and lower serum ferritin levels and iron deficiency. In addition, in the studies of Konofal et al. (Konofal, Lecendreux et al. 2008) and Sever et al. (Sever, Ashkenazi et al. 1997), it was revealed that after increasing the serum ferritin of ADHD patients by iron supplementation, ADHD symptoms can be reduced. Although further support with future studies and experiments is needed, one can suggest that increased iron intake may ease the symptoms of ADHD.

13.2.5.3 Copper

Copper has some very beneficial effects like erythropoiesis, myelin formation, synthesis of hormones, antioxidant protection and immune system modulation. Many biochemical processes such as the cholesterol and glucose metabolism are regulated with the help of copper. Copper is also a main ingredient for many proteins and enzymes. Hence, it is no surprise to find abundance of copper in the liver, heart and brain, which have the tissues with the highest metabolic activity (Młyniec, Gaweł et al. 2015).

Although copper is crucial for a healthy diet, it can be toxic at high levels of consumption and may lead to Fenton-type redox reactions, resulting in oxidative cell damage and cell death. Fortunately, it is not very common to be poisoned by excessive copper, thanks to the homeostatic mechanisms that are responsible for copper absorption and excretion (Turnlund, Keyes et al. 2005).

Copper is not only a major component of enzymes like monoamine oxidase, dopamine beta-hydroxylase and tyrosine hydroxylase, it also takes place in the catecholaminergic pathways that are associated with the pathophysiology of depression (Heninger, Delgado et al. 1996). Some studies suggest that copper deficiency can lead to some neuropsychiatric disorders such as depression (Crayton and Walsh 2007; Russo 2011). The study on the long recognized age-related metal-transporting proteins and compounds in important parts of the attentional circuits suggests that copper intake may lead to cognitive decline (Zatta, Drago et al. 2008). In a study conducted on a group of healthy, elderly women, Lam et al. (Lam, Kritz-Silverstein et al. 2008) reported an inverse correlation between serum copper concentrations and cognitive performance. Also, according to studies, women with serum copper over 2.15 mg/L perform significantly worse than women with serum copper below 0.9 mg/L in the long-term and short-term memory tests. In addition to that, high copper concentration is observed in AD, Parkinson's disease and Wilson's disease patients and might also cause lower intelligence in young adolescents (Osredkar and Sustar 2011; Bandmann, Weiss et al. 2015).

13.2.5.4 Selenium

Selenium is another nutrient of vital importance to the human health. It is found in very low amounts in the body. Selenium performs as an antioxidant and as an immunomodulator in both humans and animals, and is also involved in the operation of particular endocrine pathways (Ullah, Liu et al. 2018). Dysfunction of the immune systems, mental failure and higher mortality risk are associated with low levels of selenium (Ullah, Liu et al. 2018). Also, cognitive decline and AD were found to be related to reduced levels of selenium (Cardoso, Ong et al. 2010).

Dopamine, which plays a role in the pathophysiology of depression and other psychiatric illnesses, has been shown to be affected by selenium's modulatory effects. Some studies reported lower glutathione peroxidase antioxidant enzyme activity among individuals with major depression than those in the control group; thus, the results suggest that lack of selenium can lead to depression through antioxidant pathways (Maes, Galecki et al. 2011). Another nested case control study conducted in Australia reports that lower dietary selenium consumption is associated with an increased risk of *de novo* depression (Pasco, Jacka et al. 2012). There aren't many studies that concentrate on the association between selenium and depressive symptoms in young adults. A study conducted by Conner et al. (Conner, Richardson et al. 2015) showed that an optimal range of serum selenium, between 82 and 85 mcg/L to be exact, was associated with reduced risk of depressive symptoms in young adults. In this range of selenium, the glutathione peroxidase level is maximal.

There is also a relationship between selenium supplementation and postpartum depression (Mokhber, Namjoo et al. 2011). Another double-blind crossover experiment, conducted in Britain on 50 people, demonstrated that selenium intake of 100 g/day for five weeks reduces the symptoms of depression, and the effect was more significant in those with lower selenium intake levels (Benton and Cook 1991). On the other hand, an experiment with a larger randomized-controlled trial among 501 elderly volunteers in the UK showed no such effect (Rayman, Thompson et al. 2006).

13.2.5.5 Magnesium

Another crucial element for the metabolism is magnesium, which is an important coenzyme for many enzymes that play a role in phosphate transfer and energy metabolism. Stabilizing genes, DNA replication, synthesis of protein and nucleic acids and metabolism of macronutrients are the main functions of magnesium (Pasternak, Kocot et al. 2010). In addition, magnesium is responsible for neurotransmission by regulating and transferring the ions, such as potassium and calcium, through pumps and channels (Romani 2007). There is also an association between magnesium and the pathophysiology of some neurologic diseases such as migraine, AD, Parkinson's disease and ADHD (de Baaij, Hoenderop et al. 2015; Grober, Schmidt et al. 2015). Magnesium intake may also be related to depression according to a study with 20 years of follow-up (Yary, Lehto et al. 2016). However, an inverse relationship between magnesium intake and depression symptoms was reported in adults (Jacka, Overland et al. 2009). To cope with depression, magnesium preparations can be added to the pharmacological armamentarium (Serefko, Szopa et al. 2013). A study conducted by Rajizadeh et al. (Rajizadeh, Mozaffari-Khosravi et al. 2017) suggested that an intake of 500 mg/day magnesium oxide tablet by patients who suffered from magnesium inadequacy for a period of eight weeks reduced depression symptoms and increased the magnesium levels.

There are many assumptions about the dynamics of the effects of magnesium on brain and mental functions. Significant findings suggest a relationship between magnesium and development of depression. Magnesium is responsible for the modulation of the operation of NMDA and GABA receptors. It also plays a substantial role in the reduction of hippocampal kindling and in synthesizing of the adrenocorticotropic hormone and interacting with the limbic-hypothalamus-pituitary-adrenal (HPA) axis, which often fail to operate regularly in depression patients (Murck 2002). In addition to its crucial role in BDNF production in the brain, magnesium also acts as an inhibitor in regulating the NMDA channels. BDNF, a neurotropic factor widely expressed in the CNS, has a significant role in the brain evolution, survival abilities and neuron maintenance (Rajizadeh, Mozaffari-Khosravi et al. 2017).

13.3   NUTRITION PROBLEMS WHICH MAY BE OBSERVED IN PATIENTS WITH MENTAL ILLNESSES

13.3.1   SCHIZOPHRENIA

Schizophrenia is a severe mental disorder with symptoms such as psychosis and is often accompanied by paranoia and delusions. Delusions, hallucinations or disorganized speech must be observed in order to diagnose this condition. For a person to be diagnosed with schizoaffective disorder, all of the criteria for schizophrenia and for an episode of bipolar disorder or depression must be observed, except for impaired function (Parker and Brotchie 2011).

Impaired cognitive functions and psychotic behavior patterns, which are the common symptoms of schizophrenia, usually withhold patients from having balanced dietary habits. Also, these patients have a hard time selecting healthy food. There is good evidence from studies that suggest that these patients tend to have diets that are rich in energy and contain saturated fat, sugar and cholesterol, and low in fruits and vegetables, fiber, vitamin C and beta carotene when compared to the diets of healthy people (Brown, Birtwistle et al. 1999; McCreadie 2003).

In a study by Arroll et al. (Arroll, Wilder et al. 2014), a link was established between B vitamins deficiency and schizophrenia. Homocysteine levels can be reduced by folate supplementation in schizophrenia patients. In a randomized, double-blind, placebo-controlled, crossover experiment, patients suffering from elevated homocysteine levels were given oral folic acid, B_{12} and pyridoxine for three months, and a placebo afterwards. During the supplement phase of the study, homocysteine levels were reduced as expected and were observed to be associated with clinical improvements in symptoms and neurocognitive execution (Levine, Stahl et al. 2006). Hence, the results are encouraging for vitamin B as an adjunct treatment for those with high homocysteine levels and also for people with a genetic disorder that causes abnormal folate metabolism.

There is also a known correlation between low plasma vitamin D levels and increased risk of schizophrenia. Vitamin D is a good candidate as a therapeutic agent in reducing the symptoms of schizophrenia. A large-scale study from Finland with the participation of 9,114 people showed that vitamin D supplementation in the first year of a newborn was associated with a lower risk of developing schizophrenia by the age of 31 and that also higher doses yielded better results than lower doses (McGrath, Saari et al. 2004).

A better prognosis can be reached with the use of fish oil, with the EPA fraction being more beneficial than the DHA (Marano, Traversi et al. 2013). In a study involving young adults with reported sub-threshold psychotic states, 700 mg of EPA, 480 mg of DHA, 220 mg of other ω-3 PUFAs plus 7.6 mg of mixed tocopherol (Vitamin E) given as a dietary supplement for 12 weeks, as a part of an ω-3 PUFAs supplementation program, reduced the possibility of developing symptoms of a psychotic disorder versus those who received placebo medication (Amminger, Schafer et al. 2010).

It is a known fact that caffeine consumption and a smoking habit are much more frequent among schizophrenia patients than the rest of the population (Strassnig, Brar et al. 2006; de Leon and Diaz 2012). Metabolic side effects such as various degrees of weight gain, dyslipidemia and susceptibility to Type II diabetes are correlated with antipsychotic treatment. In addition, various risk factors such as metabolic syndrome, cardiovascular risk and a shorter period of expected lifetime is pertinent to schizophrenia patients (Newcomer 2005; Vidovic, Dordevic et al. 2013).

Kim et al. (Kim and Wessling-Resnick 2014) proposed that alpha-lipoic acid intake might prove useful for the weight gain issue in schizophrenia patients treated with antipsychotic therapy, by modulating adenosine monophosphate activated protein kinase activity in the hypothalamus and peripheral tissues as this enzyme is involved in cellular energy homeostasis.

One can also suggest that schizophrenia patients often fail to qualify the physical activity recommendation of 150 min/week moderate and vigorous physical activity (Vancampfort, De Hert et al. 2012). Even if the physical activity fails to help with the weight gain problem, it can help improve cardiorespiratory fitness (Vancampfort, Rosenbaum et al. 2015).

It should also be noted that celiac disease is two times more common in schizophrenia, thus highly suggesting a test for celiac disease and, if confirmed, a gluten-free diet should be practiced (Cascella, Kryszak et al. 2011).

13.3.2 AUTISM SPECTRUM DISORDERS

Autism spectrum disorders (ASD) can be diagnosed by the presence of characteristic qualitative behavioral abnormalities in communication, in mutual social intercourse with repetitive patterns, limited and stereotyped interests and actions. These persistent symptoms often appear in early childhood and evolve into deterioration in functioning across various settings (Yates and Le Couteur 2016).

Autism spectrum disorders may be significantly affected by nutritional deficiencies. It is common to observe food sensitivity and selective eating behaviors among children suffering from ASD. Children with autism are more likely to consume less nutrition due to their strict limitation on the food scale (Feucht, Ogata et al. 2010). Thus, this unbalanced nutrition habit may lead to consuming inadequate amounts of dairy, fiber, calcium, iron and vitamins D and E fortified foods, which unfortunately yields to nutritional deficits, hence resulting in metabolic disorders. Autistic children severely suffer from low levels of many vitamins, minerals, essential fatty acids and amino acids (Santhanam and Kendler 2012). This situation causes some children to be diagnosed with digestive problems due to the limited ability to digest some proteins. Therefore, they are prescribed with low protein, specialized diets. It is also possible to detect amino acids in urine and in blood in these patients (Adams 2007).

Nutritional supplementation is of high importance in autistic children due to their eating habits and nutritional deficiencies. According to a study conducted by Adams et al. (Adams, Audhya et al. 2011), increasing nutrient consumption to treat nutritional and metabolic problems may ease the symptoms and comorbidities associated with autism. Some children may positively respond to supplementary intake of ω-3 fats, especially DHA, ranging from 1 g to 3 g per day (Davison, Ng et al. 2012).

In some ASD cases in children, their gut permeability was increased and the cases failed to produce digestive enzymes related to gluten and casein properly. Lack of digestive enzymes can cause failures in the conversion of gluten and casein into amino acids. In addition, leakage into the bloodstream with a risk of passing the brain–blood barrier can be observed due to increased gut permeability (Mulloy, Lang et al. 2010). The risk in this scenario is the disruption in the operation of the nervous system due to the regulation of signal transduction caused by this leakage. Casein is a protein of animal origin and is mainly consumed from milk and dairy products. On the other hand, gluten is of plant origin and commonly consumed from wheat, oats, barley and rye. It is a highly anticipated thought that eliminating gluten and casein from the diet will reduce the symptoms of ASD significantly (Kaluzna-Czaplinska, Michalska et al. 2010). According to another theory, yet less agreed on due to limited evidence, the Body Ecology Diet suggests sustaining and preservation of the inner ecology of the body by rejecting food products which could have disturbed the immune system (Davison, Ng et al. 2012).

The treatment of autism does not fully benefit from a dietitian as they are not largely utilized for this cause. There are two reasons for this inadequacy: lack of awareness of the diet's role and insufficient availability of specialized nutrition services. Nutritional interventions have been reported to crucially alter the situation in a good way for some ASD patients (Rossignol 2009). Researchers have agreed that one should improve the nutritional intake of an autistic child in order to improve their overall health, behavior and brain functions. Similar beliefs are also common to the parents of autistic children as they observe progress with some symptoms in their children, such as the ability to concentrate and make eye contact. As a result, individualized diet and supplementation for autistic children is becoming a hot topic in the literature (Kałużna-Czaplińska and Jóźwik-Pruska 2016). Hence, professional dietary counseling from a registered dietitian can be considered as a part of the treatment.

13.3.3 ANXIETY

Anxiety can be defined as an amplified awareness of one's surroundings and of potential danger. It can be interpreted as an evolutionary tool in terms of dealing with threats (Teitelbaum, Weiss et al. 2017). Baxter et al. (Baxter, Scott et al. 2013) reported that 28.3% of the general population have this condition.

In a study conducted by Lakhan and Vieira (Lakhan and Vieira 2010), nutrition was shown to be an important aspect of anxiety. According to the results of the study, magnesium-containing foods such as beans, nuts, wholegrain foods and green leafy vegetables play a crucial role in a healthy diet. Deficiencies of vitamins D and B and magnesium are very common in anxious people. There is a vicious cycle between magnesium deficiency and anxiety; magnesium deficiency can cause anxiety and anxiety can cause magnesium deficiency (Armstrong, Meenagh et al. 2007). Also, caffeine was observed to play a role in development of anxiety symptoms (Childs, Hohoff et al. 2008).

Body weight can be altered by the increasing or decreasing appetite caused by depressive disorders. A strong correlation has been found between anxiety and obesity according to a recent study (Strine, Mokdad et al. 2008). Obesity is translated into physiological levels like activation of the sympathetic nervous system and HPA axis, by anxiety and stress (Nieuwenhuizen and Rutters 2008). Fat and carbohydrate-rich foods lower the anxiety level through the feedback to the HPA axis, according to the study conducted by Dallman et al. (Dallman, Akana et al. 2003; Dallman, Pecoraro et al. 2003). Chronic activation of those pathways leads to higher levels of cortisol, and this leads to changes with the appetite hormones and weight gain (Vicennati, Pasqui et al. 2009).

There is not enough evidence that suggest an association between anxiety and PUFAs, but this may be expected since comorbidity of anxiety in depressive disorders is high (Ross 2009). According to a recent study, depressed patients with comorbid anxiety have lower ω-3 and PUFA levels than depressed patients without comorbid anxiety (Liu, Galfalvy et al. 2013). There is not much evidence on other anxiety disorders, yet in social phobia, ω-3 and PUFA levels are lower (Green, Hermesh et al. 2006).

Another study suggested a linear relationship between ω-3 and PUFA intake and anxiety, with lower DHA intake correlating to an increased anxiety risk (Jacka, Pasco et al. 2013). Various types of anxiety have been reported to be treated with the help of vitamin B doses above RDA recommendations, regardless of the existence of deficiency (Gaby 2011).

If the anxiety attacks are mostly seen in the late morning or afternoon (being hungry for several hours) or half an hour to one hour after a meal or drink with high sugar, reactive hypoglycemia can be considered as a triggering factor for anxiety (Gaby 2011).

13.3.4 Dementia and Alzheimer's Disease

Dementia is a devastating disease for patients and for their caregivers and families, and is one of the biggest reasons of disability and dependency of the elderly. Caregivers, families and society are suffering from lack of awareness and understanding of dementia, which subsequently prevents the treatment, thus making the families suffer in physical, psychological and economic ways. Dementia should be considered a major problem of public health worldwide and should not be neglected or ignored (WHO 2012).

The most common form of dementia is Alzheimer's disease. 26.6 million people were diagnosed with AD in 2006; the projection of the number by the year 2050 is four times higher, exceeding 100 million, or in other words, 1 in 85 people in the world (Rocca, Petersen et al. 2011).

Dementia in the early stages might affect the behaviors of the patients in shopping, storing and preparing food, hence leading to dysfunction in proper eating patterns. This condition may also ruin the dietary habits of patients and result in a reduced variety of and unbalanced nutrient intake. Through its progression period, the patients may lose their contact with the concept of food, and/or tools to eat them and be left baffled as eating skills are lost (Volkert, Chourdakis et al. 2015). Various eating and swallowing problems may emerge. The nutritional state is highly dependent on eating and swallowing problems and the behavioral and psychological symptoms of dementia. More than 80% of the AD patients suffer from these problems according to Kai et al. (Kai, Hashimoto et al. 2015). In frontotemporal lobar degeneration cases, eating behaviors such as appetite, food preferences, habits and other oral behaviors have been commonly reported to be altered (Ikeda, Brown et al. 2002).

Malnutrition due to eating and swallowing problems may pose as a main problem. A study by Camina Martin et al. (Camina Martin, Barrera Ortega et al. 2012) suggested that dementia can cause malnutrition and the malnutrition level gets worse as dementia progresses. Furthermore, a study by Droogsma et al. (Droogsma, van Asselt et al. 2013) showed that about 14% of community-dwelling older people are facing the risk of malnutrition.

The clinical course of dementia can be worsened by weight loss, which may emerge as a clinical symptom of dementia, and this can lead to more significant functional impairment and dependence, and increases the risk of morbidity, hospitalization, institutionalization and mortality. Body weight should be closely monitored to prevent weight loss, and if weight loss is found, it should be treated immediately (Prince, Albanese et al. 2014).

Other common deficiencies in dementia and AD are micronutrient and macronutrient (protein and energy) deficiencies. Dementia patients mostly use micronutrient supplements, yet there is no convincing evidence to prove that the patients are getting help regarding the condition. According to the latest studies, it is sensible to recommend to patients not to use these supplements for the treatment of AD and dementia. However, B_{12} or folate deficiency, if diagnosed, should be treated (Prince, Albanese et al. 2014).

To administer the most beneficial nutrition treatment, the clinical and nutritional history of the patients should be examined, a physical examination should be made to detect nutrition problems, anthropometric measurements should be conducted and body composition should be evaluated. Each of these parameters will help evaluate the patient's nutrition status properly. Nutrition treatment should be tailored according to the malnutrition status of the patient (Akbulut 2015).

Patients suffering from dementia should receive caregiving service for feeding in order to maintain food hygiene and sufficient nutrition, as the patients may forget to prepare food or fail to identify food. Moreover, the caregiver can monitor the patient during the feeding process, which significantly reduces the risk of aspiration. Another precaution to avoid aspiration is to use viscous fluids or fluid thickeners during feeding (Akbulut 2015).

Meals should be made as entertaining as possible for dementia patients since they might feel anxious or reject food during the feeding process. Feeding the patients with their favorite foods and having the loved ones join their meals are good ways to help the patient with proper nutrition (Akbulut 2015).

High-calorie diets are proven to help dementia patients with their weight gain issues and eating disorders. Treatments like assisted feeding, appetite stimulants and modified foods may be used individually or combined to help with weight gain (Hanson, Ersek et al. 2011).

Patients with mental health conditions, as well as their caregivers, should be supplied with nutritional interventions provided by the registered dietitians, which helps reduce the nutritional side effects of psychiatric drugs, improve cognition, ameliorate self-management of coinciding and comorbid problems, improve food security and enhance the overall occupational, social and psychological performance (Davison, Ng et al. 2012).

ABBREVIATIONS

5-HT:	5-hydroxytryptamine
AD:	Alzheimer's Disease
ALA:	α-linolenic Acid
ADHD:	Attention Deficit Hyperactivity Disorder
ASD:	Autism Spectrum Disorders
BDNF:	Brain-Derived Neurotrophic Factor
CNS:	Central Nervous System

DASH: Dietary Approaches to Stop Hypertension
DHA: Docosahexaenoic Acid
DOPA: Dihydroxyphenylalanine
DSM: The Diagnostic and Statistical Manual of Mental Disorders
EPA: Eicosapentaenoic Acid
FDA: Federal Drug Administration
GABA: Gamma-Aminobutyric Acid
HPA: Hypothalamus-Pituitary-Adrenal Axis
MUFA: Monounsaturated Fatty Acids
NADH: Nicotinamide Adenine Dinucleotide Hydrate
NMDA: N-methyl-D-Aspartic Acid
PUFA: Polyunsaturated Fatty Acids
ROS: Reactive Oxygen Species
SFA: Saturated Fatty Acids
TDP: Thiamine Diphosphate
VDR: Vitamin D Receptors
WHO: World Health Organization

REFERENCES

Adams, J. B. (2007). *Summary of Biomedical Treatments for Autism*, San Diego (CA), Autism Research Institute.

Adams, J. B., T. Audhya, et al. (2011). "Nutritional and metabolic status of children with autism vs. neurotypical children, and the association with autism severity." *Nutr Metab (Lond)* 8(1): 34.

Akbaraly, T. N., S. Sabia, et al. (2013). "Adherence to healthy dietary guidelines and future depressive symptoms: evidence for sex differentials in the Whitehall II study." *Am J Clin Nutr* 97(2): 419–427.

Akbulut, G. (2015). "Psikiyatrik ve mental hastalıklarda tıbbi beslenme tedavisi (Medical nutrition therapy in psychiatric and mental disorders)." *Ankara Nobel Tıp Kitapevleri* (in Turkish)

Al-Amer, R. M., M. M. Sobeh, et al. (2011). "Depression among adults with diabetes in Jordan: risk factors and relationship to blood sugar control." *J Diabetes Complicat* 25(4): 247–252.

Amminger, G. P., M. R. Schafer, et al. (2010). "Long-chain omega-3 fatty acids for indicated prevention of psychotic disorders: a randomized, placebo-controlled trial." *Arch Gen Psychiatry* 67(2): 146–154.

Anglin, R. E., Z. Samaan, et al. (2013). "Vitamin D deficiency and depression in adults: systematic review and meta-analysis." *Br J Psychiatry* 202: 100–107.

Armstrong, D. J., G. K. Meenagh, et al. (2007). "Vitamin D deficiency is associated with anxiety and depression in fibromyalgia." *Clin Rheumatol* 26(4): 551–554.

Arroll, M. A., L. Wilder, et al. (2014). "Nutritional interventions for the adjunctive treatment of schizophrenia: a brief review." *Nutr J* 13: 91.

Bae, J. H. and G. Kim (2018). "Systematic review and meta-analysis of omega-3-fatty acids in elderly patients with depression." *Nutr Res* 50: 1–9.

Bandmann, O., K. H. Weiss, et al. (2015). "Wilson's disease and other neurological copper disorders." *Lancet Neurol* 14(1): 103–113.

Barnard, N. D., A. E. Bunner, et al. (2014). "Saturated and trans fats and dementia: a systematic review." *Neurobiol Aging* 35(Suppl 2): S65–S73.

Batchelor, R. R., G. Kwandou, et al. (2017). "(–)-Riboflavin (vitamin B2) and flavin mononucleotide as visible light photo initiators in the thiol–ene polymerisation of PEG-based hydrogels." *Polym Chem* 8(6): 980–984.

Baxter, A. J., K. M. Scott, et al. (2013). "Global prevalence of anxiety disorders: a systematic review and meta-regression." *Psychol Med* 43(5): 897–910.

Beard, J. L., J. R. Connor, et al. (1993). "Iron in the brain." *Nutr Rev* 51(6): 157–170.

Bellikci Koyu, E. and Z. Buyuktuncer (2015). "Depresyon ve D Vitamini (Depression and vitamin D)." *J Nutr Diet* 43(1): 160–165.

Benton, D. and R. Cook (1991). "The impact of selenium supplementation on mood." *Biol Psychiatry* 29(11): 1092–1098.

Bornstein, R. A., G. B. Baker, et al. (1990). "Plasma amino acids in attention deficit disorder." *Psychiatry Res* 33(3): 301–306.

Brown, S., J. Birtwistle, et al. (1999). "The unhealthy lifestyle of people with schizophrenia." *Psychol Med* 29(3): 697–701.

Buretic-Tomljanovic, A., J. Giacometti, et al. (2008). "Phospholipid membrane abnormalities and reduced niacin skin flush response in schizophrenia." *Psychiatr Danub* 20(3): 372–383.

Camina Martin, M. A., S. Barrera Ortega, et al. (2012). "[Presence of malnutrition and risk of malnutrition in institutionalized elderly with dementia according to the type and deterioration stage]." *Nutr Hosp* 27(2): 434–440.

Cardoso, B. R., T. P. Ong, et al. (2010). "Nutritional status of selenium in Alzheimer's disease patients." *Br J Nutr* 103(6): 803–806.

Cascella, N. G., D. Kryszak, et al. (2011). "Prevalence of celiac disease and gluten sensitivity in the United States clinical antipsychotic trials of intervention effectiveness study population." *Schizophr Bull* 37(1): 94–100.

Cheah, J. H., S. F. Kim, et al. (2006). "NMDA receptor-nitric oxide transmission mediates neuronal iron homeostasis via the GTPase Dexras1." *Neuron* 51(4): 431–440.

Chen, M. H., T. P. Su, et al. (2013). "Association between psychiatric disorders and iron deficiency anemia among children and adolescents: a nationwide population-based study." *BMC Psychiatry* 13: 161.

Childs, E., C. Hohoff, et al. (2008). "Association between ADORA2A and DRD2 polymorphisms and caffeine-induced anxiety." *Neuropsychopharmacology* 33(12): 2791–2800.

Conner, T. S., A. C. Richardson, et al. (2015). "Optimal serum selenium concentrations are associated with lower depressive symptoms and negative mood among young adults." *J Nutr* 145(1): 59–65.

Crayton, J. W. and W. J. Walsh (2007). "Elevated serum copper levels in women with a history of post-partum depression." *J Trace Elem Med Biol* 21(1): 17–21.

Dallman, M. F., S. F. Akana, et al. (2003). "A spoonful of sugar: feedback signals of energy stores and corticosterone regulate responses to chronic stress." *Physiol Behav* 79(1): 3–12.

Dallman, M. F., N. Pecoraro, et al. (2003). "Chronic stress and obesity: a new view of 'comfort food'." *Proc Natl Acad Sci USA* 100(20): 11696–11701.

Davison, K. M., E. Ng, et al. (2012). *Promoting Mental Health Through Healthy Eating and Nutritional Care*, Toronto, Dietitians of Canada.

de Baaij, J. H., J. G. Hoenderop, et al. (2015). "Magnesium in man: implications for health and disease." *Physiol Rev* 95(1): 1–46.

de Leon, J. and F. J. Diaz (2012). "Genetics of schizophrenia and smoking: an approach to studying their comorbidity based on epidemiological findings." *Hum Genet* 131(6): 877–901.

de Souza, R. J., A. Mente, et al. (2015). "Intake of saturated and trans unsaturated fatty acids and risk of all cause mortality, cardiovascular disease, and type 2 diabetes: systematic review and meta-analysis of observational studies." *BMJ* 351: h3978.

Droogsma, E., D. Z. van Asselt, et al. (2013). "Nutritional status of community-dwelling elderly with newly diagnosed Alzheimer's disease: prevalence of malnutrition and the relation of various factors to nutritional status." *J Nutr Health Aging* 17(7): 606–610.

Estherlydia, D. and S. John (2009). "Soft drink consumption and mental health outcomes among boys and girls in the age group of 17–23 years." *Indian J Nutr Diet* 46(7): 278–289.

Fattal, I., N. Friedmann, et al. (2011). "The crucial role of thiamine in the development of syntax and lexical retrieval: a study of infantile thiamine deficiency." *Brain* 134(Pt 6): 1720–1739.

Feucht, S., B. Ogata, et al. (2010). "Nutrition concerns of children with autism spectrum disorders." *Nutr Focus* 25(4): 1–13.

Ford, P. A., K. Jaceldo-Siegl, et al. (2016). "Trans fatty acid intake is related to emotional affect in the Adventist Health Study-2." *Nutr Res* 36(6): 509–517.

Fulkerson, J. A., N. E. Sherwood, et al. (2004). "Depressive symptoms and adolescent eating and health behaviors: a multifaceted view in a population-based sample." *Prev Med* 38(6): 865–875.

Gaby, A. R. (2011). *Nutritional Medicine*, Concord, New Hampshire, Fritz Perlberg Publishing.

Ginter, E. and V. Simko (2016). "New data on harmful effects of trans-fatty acids." *Bratisl Lek Listy* 117(5): 251–253.

Giudetti, A. M. and R. Cagnazzo (2012). "Beneficial effects of n-3 PUFA on chronic airway inflammatory diseases." *Prostaglandins Other Lipid Mediat* 99(3–4): 57–67.

Green, P., H. Hermesh, et al. (2006). "Red cell membrane omega-3 fatty acids are decreased in nondepressed patients with social anxiety disorder." *Eur Neuropsychopharmacol* 16(2): 107–113.

Grober, U., K. Kisters, et al. (2013). "Neuroenhancement with vitamin B12-underestimated neurological significance." *Nutrients* 5(12): 5031–5045.

Grober, U., J. Schmidt, et al. (2015). "Magnesium in prevention and therapy." *Nutrients* 7(9): 8199–8226.

Gronli, O., J. M. Kvamme, et al. (2013). "Zinc deficiency is common in several psychiatric disorders." *PLoS One* 8(12): e82793.

Grosso, G., A. Pajak, et al. (2014). "Role of omega-3 fatty acids in the treatment of depressive disorders: a comprehensive meta-analysis of randomized clinical trials." *PLoS One* 9(5): e96905.

Hanson, L. C., M. Ersek, et al. (2011). "Oral feeding options for people with dementia: a systematic review." *J Am Geriatr Soc* 59(3): 463–472.

Heninger, G. R., P. L. Delgado, et al. (1996). "The revised monoamine theory of depression: a modulatory role for monoamines, based on new findings from monoamine depletion experiments in humans." *Pharmacopsychiatry* 29(1): 2–11.

Herbison, C. E., S. Hickling, et al. (2012). "Low intake of B-vitamins is associated with poor adolescent mental health and behaviour." *Prev Med* 55(6): 634–638.

Hoerr, J., J. Fogel, et al. (2017). "Ecological correlations of dietary food intake and mental health disorders." *J Epidemiol Glob Health* 7(1): 81–89.

Hoogeveen, E. K., J. M. Geleijnse, et al. (2015). "No effect of n-3 fatty acids supplementation on NT-proBNP after myocardial infarction: the Alpha Omega Trial." *Eur J Prev Cardiol* 22(5): 648–655.

Howard, A. L., M. Robinson, et al. (2011). "ADHD is associated with a 'Western' dietary pattern in adolescents." *J Atten Disord* 15(5): 403–411.

Ikeda, M., J. Brown, et al. (2002). "Changes in appetite, food preference, and eating habits in frontotemporal dementia and Alzheimer's disease." *J Neurol Neurosurg Psychiatry* 73(4): 371–376.

Jacka, F. N., S. Overland, et al. (2009). "Association between magnesium intake and depression and anxiety in community-dwelling adults: the Hordaland Health Study." *Aust N Z J Psychiatry* 43(1): 45–52.

Jacka, F. N., J. A. Pasco, et al. (2013). "Dietary intake of fish and PUFA, and clinical depressive and anxiety disorders in women." *Br J Nutr* 109(11): 2059–2066.

Jenkins, T. A., J. C. Nguyen, et al. (2016). "Influence of tryptophan and serotonin on mood and cognition with a possible role of the gut-brain axis." *Nutrients* 8(1).

Ju, S. Y., Y. J. Lee, et al. (2013). "Serum 25-hydroxyvitamin D levels and the risk of depression: a systematic review and meta-analysis." *J Nutr Health Aging* 17(5): 447–455.

Kai, K., M. Hashimoto, et al. (2015). "Relationship between eating disturbance and dementia severity in patients with Alzheimer's disease." *PLoS One* 10(8): e0133666.

Kałużna-Czaplińska, J. and J. Jóźwik-Pruska (2016). "Nutritional strategies and personalized diet in autism-choice or necessity?" *Trends Food Sci Technol* 49: 45–50.

Kaluzna-Czaplinska, J., M. Michalska, et al. (2010). "Determination of tryptophan in urine of autistic and healthy children by gas chromatography/mass spectrometry." *Med Sci Monit* 16(10): CR488–CR492.

Kassebaum, N. J., R. Jasrasaria, et al. (2014). "A systematic analysis of global anemia burden from 1990 to 2010." *Blood* 123(5): 615–624.

Khayyatzadeh, S. S., M. Mehramiz, et al. (2018). "Adherence to a Dash-style diet in relation to depression and aggression in adolescent girls." *Psychiatry Res* 259: 104–109.

Khedr, E., S. A. Hamed, et al. (2008). "Iron states and cognitive abilities in young adults: neuropsychological and neurophysiological assessment." *Eur Arch Psychiatry Clin Neurosci* 258(8): 489–496.

Kiecolt-Glaser, J. K., M. A. Belury, et al. (2007). "Depressive symptoms, omega-6:omega-3 fatty acids, and inflammation in older adults." *Psychosom Med* 69(3): 217–224.

Kim, J. and M. Wessling-Resnick (2014). "Iron and mechanisms of emotional behavior." *J Nutr Biochem* 25(11): 1101–1107.

Konofal, E., M. Lecendreux, et al. (2008). "Effects of iron supplementation on attention deficit hyperactivity disorder in children." *Pediatr Neurol* 38(1): 20–26.

La Rovere, M. T. and J. H. Christensen (2015). "The autonomic nervous system and cardiovascular disease: role of n-3 PUFAs." *Vascul Pharmacol* 71: 1–10.

Lakhan, S. E. and K. F. Vieira (2010). "Nutritional and herbal supplements for anxiety and anxiety-related disorders: systematic review." *Nutr J* 9: 42.

Lam, P. K., D. Kritz-Silverstein, et al. (2008). "Plasma trace elements and cognitive function in older men and women: the Rancho Bernardo study." *J Nutr Health Aging* 12(1): 22–27.

Lesage, A., H. Vasiliadis, et al. (2006). *Prevelance of Mental Illnesses and Related Service Utilization in Canada: An Analysis of the Canadian Community Health Survey [Internet]*, Mississauga (ON), Canadian Collaborative Mental Health Initiative; cited 2012 July 30.

Levine, J., Z. Stahl, et al. (2006). "Homocysteine-reducing strategies improve symptoms in chronic schizophrenic patients with hyperhomocysteinemia." *Biol Psychiatry* 60(3): 265–269.

Li, G., L. Mbuagbaw, et al. (2014). "Efficacy of vitamin D supplementation in depression in adults: a systematic review." *J Clin Endocrinol Metab* 99(3): 757–767.

Li, Z., W. Wang, et al. (2018). "Association of total zinc, iron, copper and selenium intakes with depression in the US adults." *J Affect Disord* 228: 68–74.

Lien, L., N. Lien, et al. (2006). "Consumption of soft drinks and hyperactivity, mental distress, and conduct problems among adolescents in Oslo, Norway." *Am J Public Health* 96(10): 1815–1820.

Lin, S. H., L. T. Lee, et al. (2014). "Serotonin and mental disorders: a concise review on molecular neuroimaging evidence." *Clin Psychopharmacol Neurosci* 12(3): 196–202.

Liu, J. J., H. C. Galfalvy, et al. (2013). "Omega-3 polyunsaturated fatty acid (PUFA) status in major depressive disorder with comorbid anxiety disorders." *J Clin Psychiatry* 74(7): 732–738.

Lu'o'ng, K. V. Q. and L. T. H. Nguyễn (2011). "Role of thiamine in Alzheimer's disease." *Am J Alzheimer's Dis Other Dement®* 26(8): 588–598.

Luchsinger, J. A., M. X. Tang, et al. (2007). "Relation of higher folate intake to lower risk of Alzheimer disease in the elderly." *Arch Neurol* 64(1): 86–92.

Maes, M., P. Galecki, et al. (2011). "A review on the oxidative and nitrosative stress (O&NS) pathways in major depression and their possible contribution to the (neuro)degenerative processes in that illness." *Prog Neuropsychopharmacol Biol Psychiatry* 35(3): 676–692.

Mahadik, S. P., D. Evans, et al. (2001). "Oxidative stress and role of antioxidant and omega-3 essential fatty acid supplementation in schizophrenia." *Prog Neuropsychopharmacol Biol Psychiatry* 25(3): 463–493.

Marano, G., G. Traversi, et al. (2013). "Omega-3 fatty acids and schizophrenia: evidences and recommendations." *Clin Ter* 164(6): e529–e537.

Markley, H. G. (2012). "CoEnzyme Q10 and riboflavin: the mitochondrial connection." *Headache* 52(Suppl 2): 81–87.

McCreadie, R. G. (2003). "Diet, smoking and cardiovascular risk in people with schizophrenia: descriptive study." *Br J Psychiatry* 183: 534–539.

McGrath, J., K. Saari, et al. (2004). "Vitamin D supplementation during the first year of life and risk of schizophrenia: a Finnish birth cohort study." *Schizophr Res* 67(2–3): 237–245.

Mizunoya, W., K. Ohnuki, et al. (2013). "Effect of dietary fat type on anxiety-like and depression-like behavior in mice." *Springerplus* 2(1): 165.

Młyniec, K., M. Gaweł, et al. (2015). "Essential elements in depression and anxiety. Part II." *Pharmacol Rep* 67(2): 187–194.

Mokhber, N., M. Namjoo, et al. (2011). "Effect of supplementation with selenium on postpartum depression: a randomized double-blind placebo-controlled trial." *J Matern Fetal Neonatal Med* 24(1): 104–108.

Momčilović, B., J. Prejac, et al. (2010). "An essay on human and elements, multielement profiles, and depression." *Transl Neurosci* 1(4): 322–334.

Mulloy, A., R. Lang, et al. (2010). "Gluten-free and casein-free diets in the treatment of autism spectrum disorders: a systematic review." *Res Autism Spectr Disord* 4(3): 328–339.

Murakami, K., Y. Miyake, et al. (2010). "Dietary folate, riboflavin, vitamin B-6, and vitamin B-12 and depressive symptoms in early adolescence: the Ryukyus Child Health Study." *Psychosom Med* 72(8): 763–768.

Murck, H. (2002). "Magnesium and affective disorders." *Nutr Neurosci* 5(6): 375–389.

Nadalin, S., A. Buretic-Tomljanovic, et al. (2010). "Niacin skin flush test: a research tool for studying schizophrenia." *Psychiatr Danub* 22(1): 14–27.

Newcomer, J. W. (2005). "Second-generation (atypical) antipsychotics and metabolic effects: a comprehensive literature review." *CNS Drugs* 19(Suppl 1): 1–93.

Nieuwenhuizen, A. G. and F. Rutters (2008). "The hypothalamic-pituitary-adrenal-axis in the regulation of energy balance." *Physiol Behav* 94(2): 169–177.

Nowak, G., M. Siwek, et al. (2003). "Effect of zinc supplementation on antidepressant therapy in unipolar depression: a preliminary placebo-controlled study." *Pol J Pharmacol* 55(6): 1143–1147.

Oldham, M. A. and A. Ivkovic (2012). "Pellagrous encephalopathy presenting as alcohol withdrawal delirium: a case series and literature review." *Addict Sci Clin Pract* 7: 12.

Osredkar, J. and N. Sustar (2011). "Copper and zinc, biological role and significance of copper/zinc imbalance." *J Clin Toxicol* S 3(001).

Owen, C., A.-M. Rees, et al. (2008). "The role of fatty acids in the development and treatment of mood disorders." *Curr Opin Psychiatry* 21(1): 19–24.

Pan, X., C. Zhang, et al. (2011). "Soft drink and sweet food consumption and suicidal behaviours among Chinese adolescents." *Acta Paediatr* 100(11): e215–e222.

Park, J. Y., J. S. You, et al. (2010). "Dietary taurine intake, nutrients intake, dietary habits and life stress by depression in Korean female college students: a case-control study." *J Biomed Sci* 17(Suppl 1): S40.

Parker, G. and H. Brotchie (2011). "Mood effects of the amino acids tryptophan and tyrosine: 'Food for Thought' III." *Acta Psychiatr Scand* 124(6): 417–426.

Parletta, N., C. M. Milte, et al. (2013). "Nutritional modulation of cognitive function and mental health." *J Nutr Biochem* 24(5): 725–743.

Pasco, J. A., F. N. Jacka, et al. (2012). "Dietary selenium and major depression: a nested case-control study." *Complement Ther Med* 20(3): 119–123.

Pasternak, K., J. Kocot, et al. (2010). "Biochemistry of magnesium." *J Elementol* 15(3): 601–616.

Payne, M. E., S. E. Steck, et al. (2012). "Fruit, vegetable, and antioxidant intakes are lower in older adults with depression." *J Acad Nutr Diet* 112(12): 2022–2027.

Prakash, R., S. Gandotra, et al. (2008). "Rapid resolution of delusional parasitosis in pellagra with niacin augmentation therapy." *Gen Hosp Psychiatry* 30(6): 581–584.

Prince, M., E. Albanese, et al. (2014). *Nutrition and Dementia*, A. s. D. International, London, Alzheimer's Disease International.

Rajizadeh, A., H. Mozaffari-Khosravi, et al. (2017). "Effect of magnesium supplementation on depression status in depressed patients with magnesium deficiency: a randomized, double-blind, placebo-controlled trial." *Nutrition* 35: 56–60.

Rao, T. S., M. R. Asha, et al. (2008). "Understanding nutrition, depression and mental illnesses." *Indian J Psychiatry* 50(2): 77–82.

Rayman, M., A. Thompson, et al. (2006). "Impact of selenium on mood and quality of life: a randomized, controlled trial." *Biol Psychiatry* 59(2): 147–154.

Riemer, S., M. Maes, et al. (2010). "Lowered ω-3 PUFAs are related to major depression, but not to somatization syndrome." *J Affect Disord* 123(1–3): 173–180.

Rocca, W. A., R. C. Petersen, et al. (2011). "Trends in the incidence and prevalence of Alzheimer's disease, dementia, and cognitive impairment in the United States." *Alzheimers Dement* 7(1): 80–93.

Roman, M., P. Jitaru, et al. (2014). "Selenium biochemistry and its role for human health." *Metallomics* 6(1): 25–54.

Romani, A. (2007). "Regulation of magnesium homeostasis and transport in mammalian cells." *Arch Biochem Biophys* 458(1): 90–102.

Rondanelli, M., A. Giacosa, et al. (2010). "Effect of omega-3 fatty acids supplementation on depressive symptoms and on health-related quality of life in the treatment of elderly women with depression: a double-blind, placebo-controlled, randomized clinical trial." *J Am Coll Nutr* 29(1): 55–64.

Ross, B. M. (2009). "Omega-3 polyunsaturated fatty acids and anxiety disorders." *Prostaglandins Leukot Essent Fatty Acids* 81(5–6): 309–312.

Rossignol, D. A. (2009). "Novel and emerging treatments for autism spectrum disorders: a systematic review." *Ann Clin Psychiatry* 21(4): 213–236.

Russo, A. J. (2011). "Analysis of plasma zinc and copper concentration, and perceived symptoms, in individuals with depression, post zinc and anti-oxidant therapy." *Nutr Metab Insights* 4: 19–27.

Sanchez-Villegas, A., L. Verberne, et al. (2011). "Dietary fat intake and the risk of depression: the SUN Project." *PLoS One* 6(1): e16268.

Santhanam, B. and B. Kendler (2012). "Nutritional factors in autism: an overview of nutritional factors in the etiology and management of autism." *Integr Med* 11(1): 46.

Sartorius, T., C. Ketterer, et al. (2012). "Monounsaturated fatty acids prevent the aversive effects of obesity on locomotion, brain activity, and sleep behavior." *Diabetes* 61(7): 1669–1679.

Serefko, A., A. Szopa, et al. (2013). "Magnesium in depression." *Pharmacol Rep* 65(3): 547–554.

Sever, Y., A. Ashkenazi, et al. (1997). "Iron treatment in children with attention deficit hyperactivity disorder." *Neuropsychobiology* 35(4): 178–180.

Shaffer, J. A., D. Edmondson, et al. (2014). "Vitamin D supplementation for depressive symptoms: a systematic review and meta-analysis of randomized controlled trials." *Psychosom Med* 76(3): 190–196.

Shi, Z., A. W. Taylor, et al. (2010). "Soft drink consumption and mental health problems among adults in Australia." *Public Health Nutr* 13(7): 1073–1079.

Skarupski, K. A., C. Tangney, et al. (2010). "Longitudinal association of vitamin B-6, folate, and vitamin B-12 with depressive symptoms among older adults over time." *Am J Clin Nutr* 92(2): 330–335.

Spedding, S. (2014). "Vitamin D and depression: a systematic review and meta-analysis comparing studies with and without biological flaws." *Nutrients* 6(4): 1501–1518.

Strassnig, M., J. S. Brar, et al. (2006). "Increased caffeine and nicotine consumption in community-dwelling patients with schizophrenia." *Schizophr Res* 86(1–3): 269–275.

Strine, T. W., A. H. Mokdad, et al. (2008). "Depression and anxiety in the United States: findings from the 2006 Behavioral Risk Factor Surveillance System." *Psychiatr Serv* 59(12): 1383–1390.

Su, K. P. (2008). "Mind-body interface: the role of n-3 fatty acids in psychoneuroimmunology, somatic presentation, and medical illness comorbidity of depression." *Asia Pac J Clin Nutr* 17(Suppl 1): 151–157.

Swardfager, W., N. Herrmann, et al. (2013). "Potential roles of zinc in the pathophysiology and treatment of major depressive disorder." *Neurosci Biobehav Rev* 37(5): 911–929.

Szewczyk, B., M. Kubera, et al. (2011). "The role of zinc in neurodegenerative inflammatory pathways in depression." *Prog Neuropsychopharmacol Biol Psychiatry* 35(3): 693–701.

Takeda, A., A. Minami, et al. (2001). "Zinc homeostasis in the brain of adult rats fed zinc-deficient diet." *J Neurosci Res* 63(5): 447–452.

Takeda, A. and H. Tamano (2009). "Insight into zinc signaling from dietary zinc deficiency." *Brain Res Rev* 62(1): 33–44.

Teitelbaum, J., A. Weiss, et al. (2017). "MNT in psychiatric and cognitive disorders." In: L. K. Mahan and J. L. Raymond (Eds) *Krause's Food the Nutrition Care Process*, Elsevier: 839–866.

Toth, K. (2011). "Zinc in neurotransmission." *Annu Rev Nutr* 31: 139–153.

Tseng, P. T., Y. S. Cheng, et al. (2018). "Peripheral iron levels in children with attention-deficit hyperactivity disorder: a systematic review and meta-analysis." *Sci Rep* 8(1): 788.

Turnlund, J. R., W. R. Keyes, et al. (2005). "Long-term high copper intake: effects on copper absorption, retention, and homeostasis in men." *Am J Clin Nutr* 81(4): 822–828.

Ullah, H., G. Liu, et al. (2018). "Developmental selenium exposure and health risk in daily foodstuffs: a systematic review and meta-analysis." *Ecotoxicol Environ Saf* 149: 291–306.

Vancampfort, D., M. De Hert, et al. (2012). "International Organization of Physical Therapy in Mental Health consensus on physical activity within multidisciplinary rehabilitation programmes for minimising cardio-metabolic risk in patients with schizophrenia." *Disabil Rehabil* 34(1): 1–12.

Vancampfort, D., S. Rosenbaum, et al. (2015). "Exercise improves cardiorespiratory fitness in people with schizophrenia: a systematic review and meta-analysis." *Schizophr Res* 169(1–3): 453–457.

Varghese, S., B. Shameena, et al. (2001). "Polyunsaturated fatty acids (PUFA) regulate neurotransmitter contents in rat brain." *Indian J Biochem Biophys* 38(5): 327–330.

Vicennati, V., F. Pasqui, et al. (2009). "Stress-related development of obesity and cortisol in women." *Obesity (Silver Spring)* 17(9): 1678–1683.

Vidovic, B., B. Dordevic, et al. (2013). "Selenium, zinc, and copper plasma levels in patients with schizophrenia: relationship with metabolic risk factors." *Biol Trace Elem Res* 156(1–3): 22–28.

Volkert, D., M. Chourdakis, et al. (2015). "ESPEN guidelines on nutrition in dementia." *Clin Nutr* 34(6): 1052–1073.

Wang, Y., L. Huang, et al. (2017). "Iron status in attention-deficit/hyperactivity disorder: a systematic review and meta-analysis." *PLoS One* 12(1): e0169145.

Westover, A. N. and L. B. Marangell (2002). "A cross-national relationship between sugar consumption and major depression?" *Depress Anxiety* 16(3): 118–120.

WHO (2012). *Dementia: A Public Health Priority*, World Health Organization, Alzheimer's Disease International.

WHO (2014a). *Investing in the World's Health Organization: Taking Steps Towards a Fully-Funded Programme Budget 2014–2015*, World Health Organization.

WHO (2014b). *Mental Health: A State of Well-Being*, World Health Organization.

Wray, D. W., S. K. Nishiyama, et al. (2012). "Acute reversal of endothelial dysfunction in the elderly after antioxidant consumption." *Hypertension* 59(4): 818–824.

Yao, J. K., G. G. Dougherty, Jr., et al. (2016). "Prevalence and specificity of the abnormal niacin response: a potential endophenotype marker in schizophrenia." *Schizophr Bull* 42(2): 369–376.

Yary, T., S. M. Lehto, et al. (2016). "Dietary magnesium intake and the incidence of depression: a 20-year follow-up study." *J Affect Disord* 193: 94–98.

Yates, K. and A. Le Couteur (2016). "Diagnosing autism/autism spectrum disorders." *Paediatr Child Health* 26(12): 513–518.

Yu, S., Y. Feng, et al. (2011). "Diet supplementation with iron augments brain oxidative stress status in a rat model of psychological stress." *Nutrition* 27(10): 1048–1052.

Zaalberg, A., H. Nijman, et al. (2010). "Effects of nutritional supplements on aggression, rule-breaking, and psychopathology among young adult prisoners." *Aggress Behav* 36(2): 117–126.

Zatta, P., D. Drago, et al. (2008). "Accumulation of copper and other metal ions, and metallothionein I/II expression in the bovine brain as a function of aging." *J Chem Neuroanat* 36(1): 1–5.

Zhang, G., H. Ding, et al. (2013). "Thiamine nutritional status and depressive symptoms are inversely associated among older Chinese adults." *J Nutr* 143(1): 53–58.

Index

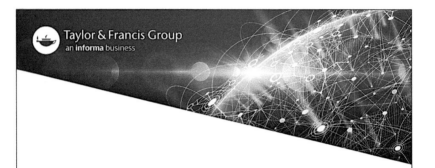